Chronic Pain

Edited by
Saifudin Rashiq
Donald Schopflocher
Paul Taenzer
Egon Jonsson

Chronic Pain

A Health Policy Perspective

Edited by
Saifudin Rashiq, Donald Schopflocher,
Paul Taenzer, and Egon Jonsson

WILEY-VCH Verlag GmbH & Co. KGaA

Volume Editors

Prof. Saifudin Rashiq
University of Alberta
Department of Anesthesiology and
Pain Medicine
Clinical Sciences Bldg. 8-120
Edmonton, Alberta T6G 2G3
Canada

Prof. Donald Schopflocher
University of Alberta
Nursing Research Office
Clinical Science Bldg. 4-103
Edmonton, Alberta T6G 2G3
Canada

Prof. Paul Taenzer
University of Calgary
Faculty of Medicine
2nd Street SW 100-2310
Calgary, Alberta T2S 3C4
Canada

Prof. Egon Jonsson
University of Alberta
Department of Public Health Science
Institute of Health Economics
10405 Jasper Ave
Edmonton, Alberta T5J 3N4
Canada

Series Editor

Prof. Egon Jonsson
University of Alberta
Department of Public Health Science
Institute of Health Economics
10405 Jasper Ave
Edmonton, Alberta T5J 3N4
Canada

All books published by Wiley-VCH are carefully
produced. Nevertheless, authors, editors, and
publisher do not warrant the information
contained in these books, including this book,
to be free of errors. Readers are advised to keep
in mind that statements, data, illustrations,
procedural details or other items may
inadvertently be inaccurate.

Library of Congress Card No.: applied for

British Library Cataloguing-in-Publication Data
A catalogue record for this book is available from
the British Library.

**Bibliographic information published by the
Deutsche Nationalbibliothek**
Die Deutsche Nationalbibliothek lists this
publication in the Deutsche Nationalbibliografie;
detailed bibliographic data are available in the
Internet at http://dnb.d-nb.de

© 2008 WILEY-VCH Verlag GmbH & Co. KGaA,
Weinheim

Printed in the Federal Republic of Germany
Printed on acid-free paper

Cover design   Adam Design, Weinheim
Typesetting   SNP Best-set Typesetter Ltd.,
Hong Kong
Printing   betz-druck GmbH, Darmstadt
Bookbinding   Litges & Dopf GmbH,
Heppenheim

ISBN: 978-3-527-32382-1
ISSN: 1864-9947

Foreword

Chronic pain is a major health care problem in every developed country, but has largely been ignored by those who make health care policy decisions. As the profile of chronic pain and those who suffer from it has risen, therapists and researchers have begun to think of it in a much broader way. No longer is it possible to think of chronic pain as just a medical matter: the costs to society in loss of productivity, wages, taxes, and the burdens of wage replacement, disability payments and health care are enormous. Since the population is aging, there will be more chronic disease and chronic pain among its citizens; costs will increase both on a population basis and as new technology is introduced into medicine. A prudent society should attempt to study this problem and design cost-effective programs both for prevention and for remediation.

Health care for patients with chronic pain is usually grafted onto existing national and regional health care programs without significant efforts to assess the actual needs or design programs that are aimed at the problems faced by those who suffer from it. This book addresses these deficiencies and begins the process of creating programs that will make efficient use of limited resources to reduce the burden of chronic pain on both patients and society.

The editors have assembled an all-star cast to create this volume, and the resulting book is unique among texts on chronic pain. Furthermore, almost all of the contributors are from Canada, providing a unique perspective that is, at the same time, highly relevant to other jurisdictions.

Of course, we cannot lose sight of the individual patient when we plan for an effective system to deal with chronic pain patients. The chapter by Brown focuses on the lived experience of the individual, which is essential in the management of chronic pain. At the other extreme, Buchbinder focuses on the whole population in her account of the mass media project on low back pain that was implemented in Victoria, Australia. Other chapters address issues in special populations such as the injured worker, the young, the elderly and the aboriginal.

There are also reviews of what is known about the history of pain, epidemiology, efficacy of pharmacological treatments, complementary and alternative medicine, non-pharmacological strategies and multidisciplinary pain management. These describe the current practice and set the stage for looking to the future: what should be done. Another chapter reviews successful integrations of pain manage-

Chronic Pain: A Health Policy Perspective
Edited by S. Rashiq, D. Schopflocher, P. Taenzer, and E. Jonsson
Copyright © 2008 WILEY-VCH Verlag GmbH & Co. KGaA, Weinheim
ISBN: 978-3-527-32382-1

ment into healthcare systems in France and the Veterans Administration in the United States. Certainly, we need to implement better education for all types of health care providers at varying stages of their professional development. Using the information we have from epidemiological studies and methods of delivering health care, pain management must be integrated into whatever system is designed to provide for the health of a region's citizens. Finally, there is an ethical issue in the need to develop pain management facilities accessible to all of the people. Unfortunately, the ethical imperative for a health care provider to alleviate pain whenever possible often does not reach the health policy planner or system administrator who is concerned about the bottom line rather than the alleviation of patient suffering.

A just society cannot ignore pain and suffering. It is a human right to have pain and suffering addressed by health care providers. This book is an excellent resource for healthcare policy planners who seek to understand the problem of chronic pain. It provides a starting point for understanding what the problems are for chronic pain patients, what treatments are effective, what solutions can be implemented, and how to do all this efficiently.

Seattle, August 2008 *John Loeser*

Contents

Chronic Pain: A Health Policy Perspective
Edited by S. Rashiq, D. Schopflocher, P. Taenzer, and E. Jonsson
Copyright © 2008 WILEY-VCH Verlag GmbH & Co. KGaA, Weinheim
ISBN: 978-3-527-32382-1

Preface

This book will be successful if it can convince its readers, and especially health care decision makers, of the truth of the following propositions:

- Chronic pain is a large and growing health problem in Canada as it is elsewhere in the world.
- Effective treatments exist that could bring partial or complete relief to many chronic pain sufferers.
- Efficient provision of these treatments means transforming health care systems to better integrate services and to disseminate existing knowledge more widely.
- Continuous improvement of the quality of life of sufferers of chronic pain will still crucially depend upon the generation of new knowledge and the fostering of innovation in health care delivery.

To do this we must provide strong evidence.

The task of marshalling evidence about Health Care has traditionally fallen to a form of health policy analysis called *Health Technology Assessment* (HTA). HTA is: *"the systematic evaluation of properties, effects, and/or impacts of health care technology"*–National Information Center on Health Services Research and Health Care Technology, National Institutes of Health, United States [1]

In the last 25 years, HTA has expanded enormously as an activity and as a discipline. This growth has been predicated upon.

- An unprecedented growth in scientific research in the medical sciences.
- The development of rigorous scientific methods for the synthesis and review of scientific information (such as systematic review and meta-analysis).
- The general belief that new medical technologies are scientific achievements with a potential to improve the health of patients,
- The desire within publicly funded health care systems to effectively allocate limited resources.

HTAs provide information that educates and assists individual health practitioners, health system providers, such as HMOs, hospital and other health authorities, and health policymakers. The strong interest in HTA among health and medical researchers, health practitioners, health administrators and government policy makers makes HTA a fertile ground for discourse on knowledge transfer strategies between the research and policy worlds.

HTA is typically conducted by interdisciplinary groups using explicit and rigorous analytical and scientific frameworks. The discipline has attracted many researchers operating within academic settings from around the globe. Health Technology Assessment International (HTAi), the professional society for HTA professionals is located at the Institute of Health Economics (IHE) in Edmonton Alberta and has over 2000 members. The Editorial Office of *The International Journal of Technology Assessment in Healthcare* is also located at IHE.

Recently, there has been a trend for HTAs to become broader in scope. There have been calls for an expansion of the types of contextual information included in HTAs, including official statistics. Researchers are developing methods to include qualitative information into systematic reviews. Some agencies are expanding the scope of the questions being addressed by HTAs. For example the Swedish Council on Technology Assessment in Health Care (SBU) has published full reviews of disease management practices for complete diseases.

In Alberta, the Health Technologies Decision Process [2] has been a leader in integrating contextual information and qualitative evidence with the more usual quantitative and experimental evidence leading to better informed policy decisions. A model was developed that recognizes that policy makers require a wide range of information in order to make their decisions. The categories under which this information can be organized and presented are captured by the mnemonic STEEPLE:

- Social and system demographics
- Technology effects and effectiveness
- Ethics
- Economics
- Politics
- Legislation
- Environment

In this book, information of this type and scope is organized and presented in four sections. In the first, information and evidence are presented to contextualize the dilemmas (personal, societal, economic, ethical, scientific) surrounding chronic pain. In the second, evidence about treatments (medical and non-medical) is reviewed. The focus is on what is known about treatment effectiveness from the most stringent medical research. The third section describes some additional challenges in the management of chronic pain presented by populations with special characteristics and circumstances. The final section concentrates upon broader issues such as the prevention of chronic pain, the optimal organization of health care service delivery for the management of chronic pain, and processes that might hasten our arrival at that optimal organization. While rigorous evidence of effectiveness is much less plentiful, and the scale of these activities makes them complex and time-consuming to pursue, it is here that the greatest potential to alleviate chronic pain may lie.

This book is aimed initially at an audience of policy and decision makers in health care. As the authors are primarily working in Canada, the final chapter presents some specific recommendations for the future in Canada and in Alberta that we hope may also be useful to the broader health policy community in the Americas, in Europe, and elsewhere. But we believe that the book will also be useful to everyone who is touched either directly or indirectly by chronic pain. It provides a snapshot of what we now know (and don't know) about the best ways to alleviate chronic pain, and the most promising ways to ensure that all can be properly served by the health care systems of the future.

Specific Contents

In Chapter 1, Cary Brown introduces the reader to people in pain and their valiant and often unheard attempts to communicate the nature of their suffering. Too often, these people remain invisible, in doubt about themselves, and uncertain about their future. To overcome this alienation, health care professionals could practise greater compassion by engaging with these lived experiences of pain that are now increasingly available in scholarly writing, and in electronic form on the internet.

In Chapter 2, Harold Merskey shows that the struggle to understand pain is present even in the earliest writings of Western civilization. While it is universally accepted that pain is a personal, private and inherently valid experience, the language of pain has always referred to both physical and emotional qualities. Much speculation has been given to the possibility of a psychological causation of pain. Available data suggest that it is a rare phenomenon. The International Association for the Study of Pain has crafted a definition that is widely accepted today *"Pain is an unpleasant experience associated with actual or potential tissue damage or described in terms of such damage"*.

In Chapter 3, Donald Schopflocher and Christa Harstall report that existing chronic pain prevalence estimates vary widely because they fail to use clear case definitions, and well-validated and reliable data collection tools or to collect other information that might explain this variability. Canadian studies using large-scale population surveys suggest that over 5% of the adult population suffers moderate to severe chronic pain with serious effects on health, disability, distress and the propensity to suffer depression. In Canada, these figures will rise as the population ages.

In Chapter 4, Ceri Phillips and Donald Schopflocher demonstrate that chronic pain leads to dramatic increases in health care utilization and therefore direct costs to the health care system. Yet societal costs, especially those resulting from work absence and disability, dwarf these direct costs, and additional costs to the individual such as those for uninsured treatments, informal care, or intangibles related to loss of quality of life are substantial but difficult to estimate. Improving the management of chronic pain lies in analyses of effectiveness, efficiency, and equity applied to high quality evidence across the whole health care system.

In Chapter 5, Dick Sobsey and Derek Truscott show that ethical issues abound in the management of chronic pain, in part because patient and caregiver are sometimes at odds. Fundamental ethical challenges that pain practitioners wrestle with daily include the undertreatment of chronic pain, maintaining a balance between treatment and overtreatment, and consequences if potentially addictive medications are prescribed.

In Chapter 6, James Henry reminds us that pain is a complex phenomenon that combines information from the nervous system with thoughts, emotions and social context. While laboratory research has taught us a great deal about the way in which the nervous system creates and processes the signals for pain, each discovery raises a whole host of further questions. The development of new and more effective treatments for chronic pain will be accelerated as more is learned, but it is impossible to predict how long this may take, even if current research funding levels are increased.

In Chapter 7, Joel Katz, Colin McCartney, and Saifudin Rashiq point out that every chronic pain was once an acute pain and argue that early intervention is a key to preventing acute pain from becoming chronic. This includes psychological support, since depression, anxiety and catastrophizing make the development of chronic pain more likely after injury or surgery. It also includes a number of pre-surgical precautions, as well as the use of advanced surgical techniques that are less likely to cause severe and chronic pain than traditional operations.

In Chapter 8, Chris Spanswick introduces the reader to the range of patients who are seen in specialized chronic pain clinics. The most common disorders include low back, including sciatica and neck pain or whiplash associated disorder, fibromyalgia, neuropathic pain, including pain associated with diabetes, and pain associated with systemic disease such as arthritis. Often the journey has been a long and frustrating one, and patients are very distressed and frustrated by a long history of unsuccessful treatment which adds to the complexity of care. Factors underlying the referral processes are also introduced.

In Chapter 9, Mary Lynch reviews the evidence for the effective use of drugs for the treatment of chronic pain of many kinds but details the significant limitations that most of these have. Unfortunately, many Canadian physicians do not know how best to treat chronic pain using currently available medicines, and it is difficult or impossible in many jurisdictions to obtain timely consultation with a pain specialist. Canada also lacks a national pain drug strategy leaving several effective, inexpensive pain drugs unavailable and putting drug education and research primarily in the hands of pharmaceutical manufacturers.

In Chapter 10, Alexander Clark reviews the evidence in support of the use of non-pharmacological treatments for chronic pain. Unfortunately, the quality of the research in this area is modest at best and most data are for short-term outcomes only. Treatment strategies with some empirical support include: acupuncture for short-term relief of chronic low back and neck pain; individual and group exercise therapy for a variety of chronic pain syndromes; laser therapy for short-term relief of rheumatoid and osteoarthritis and chronic low back pain; massage therapy for chronic low back pain; relaxation therapy for chronic low back pain; and Transcutaneous Electrical Nerve Stimulation (TENS) for short-term relief of osteoarthritis of the knee and chronic low back pain.

In Chapter 11, Alexander Clark and Christopher Spanswick conclude that there is very little quality research published on the efficacy of injection therapies for chronic pain, and that although adverse events from these treatments are rare they can be catastrophic. On the other hand, the efficacy of spinal cord stimulation is supported by research evidence for a variety of intractable chronic pain conditions but technical and surgical problems are still common.

In Chapter 12, Eldon Tunks reviews the effectiveness of multidisciplinary pain clinics which involve collaboration among professionals with complementary training and skills to address the multifaceted problems associated with chronic pain. These typically involve psychological and rehabilitation models with a wide variety of treatment modalities and incorporate principles of support, learning how to cope, promoting function, and sustaining function in the community. There is evidence that the individual treatment approaches used by multidisciplinary pain clinics are effective as stand-alone interventions and there is strong evidence that multidisciplinary pain clinics are clinically and economically effective.

In Chapter 13, Mark Ware reviews the use of complementary and alternative medicines (CAM) for pain. While several good quality research trials and reviews have been conducted, many of the therapies offered by practitioners and used by patients have not been rigorously evaluated. Yet increasing knowledge about CAM is important because patients may be spending resources and time on CAM; be at risk of drug–herb interactions; be unaware of safety aspects of the CAM therapies being used; be seeking treatment of symptoms not known to the treating physician; have attitudes towards conventional treatments which may influence compliance, and may be using self care practices to foster independence. Existing CAM policy and regulatory frameworks in Canada could facilitate the evaluation of CAM and pilot projects in academic multidisciplinary pain clinics could provide valuable effectiveness assessments.

In Chapter 14, Michael McGillion, Sandra LeFort, and Jennifer Stinson highlight pain self-management education as an accessible, community-based approach to enabling people to better deal with everyday problems that result from chronic pain. These interventions emphasize collaborative partnerships between patients and health care professionals, and the most successful ones bolster individuals' confidence to achieve optimal functioning, and to accept illness-induced limitations. Evidence is clear that these interventions are highly effective, cost-effective adjuncts to the usual care for reducing chronic pain and related disability.

In Chapter 15, a number of additional challenges in the management of chronic pain presented by populations with special characteristics and circumstances are discussed. Charl Els reports that chronic pain is associated with a wide range of emotional responses and psychiatric disorders including depression and that available data indicate that depression is a consequence rather than a cause of chronic pain. However, patients with co-morbid chronic pain and psychiatric disorders are poorly served by the current chronic pain and psychiatric care systems as both are under resourced and have little experience or expertise related to these comorbidities. New models of care that encourage the collaboration of mental health and chronic pain professionals to develop a strong evidence-based approach to managing patients with these co-morbidities are required. A number of new pharmacological therapies and treatment strategies have particular promise in serving this population and can be explored if targeted research funding becomes available.

In Chapter 16, Douglas Gross, Martin Mrazik, and Iain Muir report that ongoing pain is a major reason for work disability. Successful management strategies include continuum of care models that include early assessment, identification of recovery barriers, and targeted interventions for those workers most at risk of prolonged work loss. Unfortunately, coordinated management approaches are very different from the unidisciplinary primary care currently provided in most jurisdictions and would require major funding and policy changes.

In Chapter 17, Katherine Diskin and Nady el-Guebaly report that there is a high prevalence of addictions of various types in the chronic pain population, though this is largely due to lifestyle risks rather than to the prescription of opioid medications. Where it can be done safely, treatment with opioid medications may be appropriate as there are well developed strategies that can be employed to safely manage opioid treatment for patients with co-morbid addictions.

In Chapter 18, Bruce Dick shows that poorly managed chronic pain can have pervasive and very significant consequences on a developing child. Biopsychosocial treatment approaches have been validated as standard clinical practice for managing complex chronic pain in children and adolescents. However, pediatric chronic pain is still often poorly managed. Available evidence suggests that education of health care providers and use of clinical guidelines, public media campaigns, and the development of multidisciplinary pain management programs are all warranted strategies.

In Chapter 19, Thomas Hadjistavropoulos and Greg Marchildon report that pain is poorly assessed and managed in older persons whether they are dwelling in the community or residents of long-term care (LTC) facilities. Pain management teams within LTC facilities should be established and special training in pain assessment and management for seniors in LTC should be facilitated. Moreover, given the aging of Canada's population and the high prevalence of pain problems, specialized geriatric pain assessment and management training should be specifically integrated in the experience of all health professionals.

In Chapter 20, Louise Crane uses her own experience to illustrate that cultural perspectives and biases of both aboriginal patients and their non-aboriginal providers create barriers to effective care. More data are needed about indigenous communities and about the determinants of health and chronic pain in order to plan effective care for these communities. Programs which stress interdisciplinary collaboration between western and traditional healers would provide culturally sensitive management of chronic pain in indigenous communities nationwide.

In Chapter 21, Rachelle Buchbinder, Douglas Gross, Erik Werner and Jill Hayden argue that population-based interventions targeting low back pain disability have the potential for huge societal improvements in health and well being.

Mass media campaigns designed to alter societal views about back pain and promote behavior change have now been performed in several countries with mixed results. There is limited empirical understanding of the characteristics of effective (or ineffective) health campaigns though more intensive and expensive media campaigns may be more effective than low-budget campaigns. However, using a theoretical framework in future campaigns may improve the planning and evaluation of mass media interventions for low back pain.

In Chapter 22, Elizabeth Peter and Judy Watt-Watson report that an integrated, interprofessional, pre-licensure undergraduate pain curriculum can lead to changed knowledge and beliefs and positive student experiences; as can team-based post-licensure workshops and programs. Unfortunately, consistent application of pain policies, standards, and guidelines has been problematic in practice, perhaps because they are difficult to enforce. Accreditation standards may be a meaningful policy mechanism to further increase educational offerings related to pain post-licensure.

In Chapter 23, Paul Taenzer, Donald Schopflocher, Saifudin Rashiq, and Christa Harstall describe such a post-licensure program to deliver short workshops to primary health care providers about the effective treatment of chronic pain. HTA specialists working with clinical health educators used systematic reviews to develop the workshops and associated materials. Independent evaluation reported promising results for increased knowledge and clinical practice change.

In Chapter 24, Patricia Dobkin and Lucy Boothroyd begin by highlighting challenges to health system organization for the optimal chronic pain care, such as timing of treatment, access to services, and continuity of care. Two health care systems, the national health care system in France and the Veteran's Administration, that successfully reorganized pain services are described. Common features included structural elements such as a hierarchy of services, primary care services for timely diagnosis and treatment, a gradation of specialized services including comprehensive multidisciplinary care, and education programs. Common features also included process features such as an interdisciplinary approach, care pathways and discharge protocols to facilitate continuity of care, and educational processes including self-care.

In Chapter 25, Pierre Bouchard describes a planning process used by the Québec Ministry of Health and Social Services to develop a provincial strategy for managing chronic pain. The key success factors included the use of tested and recognized processes that had previously guided the development of similar continuums of services, joint ministerial leadership fostering the emergence of a productive synergy among clinicians, researchers, managers and users of the anticipated services, the presence of credible practitioners on the advisory committee who actively strove to move the issue forward and championed the cause, the use of evidence-based data from the scientific literature and the involvement of the Quebec HTA agency, access to the government's administrative apparatus, and the capacity of the engaged patient organization to advocate at the political level.

In the Conclusion, Paul Taenzer, Donald Schopflocher, and Saifudin Rashiq return to the experiences of chronic pain sufferers, this time as successful patients of a multidisciplinary pain clinic to show what successful treatment can ultimately provide. Finally, specific recommendations to health policy makers in Alberta and Canada are provided in the hope of initiating a positive change in the management of chronic pain.

Edmonton and Calgary, August 2008

Saifudin Rashiq
Donald Schopflocher
Paul Taenzer
Egon Jonsson

References

1 National Information Center on Health Services Research and Health Care Technology, National Institutes of Health, United States (2008). Available at: http://www.nlm.nih.gov/nichsr/hta101/ta101014.html, accessed on January 31, 2008.

2 Brehaut, J.D. and Juzwishin, D. (2005) *Bridging the Gap: the Use of Research Evidence in Policy Development,* AHFMR.

List of Contributors

Lucy J. Boothroyd
AETMIS
2021 Union Avenue
Montreal, Quebec H2W 1S6
Canada

Pierre Bouchard
Chargé de Projet
5 Chemin St. Catherine
Baie-Saint-Paul, Quebec G3Z 2E6
Canada

Cary Brown
University of Alberta
Department of Occupational
Therapy
Edmonton, Alberta T6G 2G4
Canada

Rachelle Buchbinder
Cabrini Hospital and Monash
University
School of Public Health and
Preventive Medicine
183 Wattletree Road
Malvern, Victoria 3144
Australia

Alexander J. Clark
University of Calgary
Calgary Health Region Chronic
Pain Centre
160-2210 2nd Street SW
Calgary, Alberta T2S 3K3
Canada

John Clark
University of Calgary
Calgary Health Region Chronic
Pain Centre
160-2210 2nd Street SW
Calgary, Alberta T2S 3K3
Canada

Louise Crane
3637 39th Street NE
Calgary, Alberta T1Y 5H4
Canada

Bruce Dick
University of Alberta
Department of Anesthesiology and
Pain Medicine
8-120 CSB
Edmonton, Alberta T6G 2B7
Canada

Chronic Pain: A Health Policy Perspective
Edited by S. Rashiq, D. Schopflocher, P. Taenzer, and E. Jonsson
Copyright © 2008 WILEY-VCH Verlag GmbH & Co. KGaA, Weinheim
ISBN: 978-3-527-32382-1

Katherine Diskin
University of Calgary
Addiction Division
Foothills Medical Centre
1403-29 Street NW
Calgary, Alberta T2N 2T9
Canada

Patricia L. Dobkin
McGill University
Department of Medicine
McGill Programs in Whole
Person Care
546 Pine Avenue West
Montreal, Quebec H2W 1S6
Canada

Charl Els
University of Alberta
Department of Psychiatry
Mackenzie Health Science Centre
Edmonton, Alberta T5G 2G4
Canada

Douglas P. Gross
University of Alberta
Faculty of Rehabilitation
Medicine
Department of Physical Therapy
Corbett Hall
Edmonton, Alberta T6G 2G4
Canada

Nady el-Guebaly
University of Galgary
Addiction Division
Foothills Medical Centre
1403–29 Street NW
Calgary, Alberta T2N 2T9
Canada

Thomas Hadjistavropoulos
University of Regina
Department of Psychology
Regina, Saskatchewan S4S 0A2
Canada

Christa Harstall
University of Alberta
Institute of Health Economics
10405 Jasper Avenue
Edmonton, Alberta T5J 3N4
Canada

Jill A. Hayden
University of Toronto
Department of Health Policy,
Management & Evaluation
399 Bathurst Street
Toronto, Ontario M5T 2S8
Canada

James L. Henry
McMaster University
Michael G. DeGroote Institute
for Pain Research and Care
Health Sciences Centre
1200 Main Street West
Hamilton, Ontario L8N 3Z5
Canada

Egon Jonsson
University of Toronto
Department of Health Policy,
Management & Evaluation
399 Bathurst Street
Toronto, Ontario M5T 2S8
Canada

Joel Katz
York University
Faculty of Health
Department of Psychology
4700 Keele Street
Toronto, Ontario M3J 1P3
Canada

Sandra LeFort
Memorial University of
Newfoundland
School of Nursing
300 Prince Philip Drive
St. John's, Newfoundland A1B 3V6
Canada

Mary E. Lynch
Dalhousie University
Queen Elizabeth II Health
Sciences Centre
Fourth Floor Dickson Center
5820 University Avenue
Halifax, Nova Scotia B3H 1V7
Canada

Gregory Marchildon
University of Regina
Johnson-Shoyama Graduate
School of Public Policy
Regina, Saskatchewan S4S 0A2
Canada

Colin J. L. McCartney
University of Toronto
Department of Anesthesia
2075 Bayview Avenue
Toronto, Ontario M4N 3M5
Canada

Michael Hugh McGillion
McMaster University
FUTURE Program for
Cardiovascular Nurse Scientists
Canada

Harold Merskey
71 Logan Avenue
London, Ontario N5Y 2P9
Canada

Martin Mrazik
University of Alberta
Faculty of Education
Department of Educational Psychology
6-135 Education North
Edmonton, Alberta T6G 2G5
Canada

Iain Muir
University of Alberta
Faculty of Rehabilitation Medicine
Department of Physical Therapy
3-12 Corbett Hall
Edmonton, Alberta T6G 2G4
Canada

Elizabeth Peter
University of Toronto
Lawrence Bloomberg Faculty
of Nursing
Joint Centre for Bioethics
215-155 College Street
Toronto, Ontario M5T 1P8
Canada

Ceri J. Phillips
Swansea University
Institute for Health Research
School of Health Sciences
Suigleton Park
Swansea SA2 8PP
United Kingdom

Saifudin Rashiq
University of Alberta
Department of Anesthesiology and
Pain Medicine
Clinical Science Building 8-120
Edmonton, Alberta T6G 2G3
Canada

Donald Schopflocher
University of Alberta
Nursing Research Office
Clinical Science Building 4-103
Edmonton, Alberta T6G 2G3
Canada

Dick Sobsey
University of Alberta
Department of Educational
Psychology
6-102 Education North
Edmonton, Alberta T6G 2G5
Canada

Christopher Spanswick
University of Calgary
Chronic Pain Centre
Calgary Health Region
Department of Anaesthesia
160, 2210 2nd Street SW
Calgary, Alberta T2S 3C3
Canada

Jennifer Stinson
University of Toronto
Lawrence Bloomberg Faculty
of Nursing
155 College Street
Toronto, Ontario M5T 1P8
Canada

Paul Taenzer
University of Calgary
Faculty of Medicine
2nd Street SW 100-2310
Calgary, Alberta T2S 3C4
Canada

Derek Truscott
University of Alberta
Department of Educational Psychology
6-102 Education North
Edmonton, Alberta T6G 2G5
Canada

Eldon Tunks
HHSC–Chedoke Rehabilitation
Holbrook Building D-165
Sanatorium Road
Hamilton, Ontario L8N 3Z5
Canada

Mark A. Ware
Montreal General Hospital
Pain Clinic
1650 Cedar Avenue
Montreal, Quebec H3G 1A4
Canada

Judy Watt-Watson
University of Toronto
Lawrence Bloomberg Faculty
of Nursing
155 College Street
Toronto, Ontario M5T 1P8
Canada

Erik L. Werner
University of Bergen
Department of Public Health
and Primary Health Care
5020 Bergen
Norway

Part One
Context

1
The Lived Experience of Chronic Pain:
Evidence of People's Voices

Cary Brown

1.1
Introduction

Box 1: Key points

- People in pain are all around us – they speak through literature, in scholarly writing, through visual art and, increasingly, in electronic form on the internet.

- Too often the individual in pain remains unacknowledged, isolated in a downward spiral of suffering.

- Common themes of experiential accounts emphasize doubt, invisibility, and unpredictability.

- Confronting lived experiences of pain and these themes directly in health education may promote compassion in health care providers that can affirm the individual and increasingly overcome alienation and its negative effects.

Modern health care responds to an ever-expanding set of scientific evidence about treatments, drugs and program delivery models and yet the number of people who live in chronic pain continues to grow. This is an alarming puzzle.

Perhaps we are using the wrong tools to address the problem of pain [1, 2]. Perhaps we have just not heard the voices of people in pain, or understood and appropriately responded to them. In her classical work, *The Body in Pain*, Elaine Scarry [3] tells us that

> "*to have pain is to have certainty; to hear about pain is to have doubt' and that 'the doubt of other persons . . . amplifies the suffering of those already in pain*"

Chronic Pain: A Health Policy Perspective
Edited by S. Rashiq, D. Schopflocher, P. Taenzer, and E. Jonsson
Copyright © 2008 WILEY-VCH Verlag GmbH & Co. KGaA, Weinheim
ISBN: 978-3-527-32382-1

Perhaps it is just difficult for the person in pain to speak. Philosopher David Morris [2] said that

> ". . . *pain not only hurts but more often then not frustrates, baffles, and resists us.*"

In *On Being Ill*, Virginia Woolf [4] commented:

> "*the merest schoolgirl when she falls in love has Shakespeare and Keats to speak for her, let a sufferer try to describe the pain in his head and language at once runs dry*"

Nor is this just an academic or literary message (Box 2).

Box 2: Brenda's story

"*I have been known to . . . Storm out [of the office], be . . . what's the word, a difficult patient because I'm frustrated. You know, if you feel you're not being listened to, it's frustrating, really frustrating.*"

Credits: DIPEx© – http://www.dipex.org/DesktopDefault.aspx

'Brenda' (a pseudonym) has made her story available through an online repository of patient narratives [5]. Her's is one of dozens of stories about chronic pain and other health conditions. To provide the pain sufferer's own words, to convey the lived experience of pain as a reality in daily life, I will weave Brenda's story through this chapter.

The first section tells how to better access accounts of the experience of living with chronic pain and offers samples which identify some persistent themes. The second section explores the legitimacy of evidence offered through lived experience and the techniques that researchers in the social sciences use to understand meanings, values and beliefs. The third section identifies barriers and discusses the consequences of not hearing the voices of people with pain. To conclude I note some promising initiatives that can serve as models for policy development.

1.2
The Perspective of People with Pain

Box 3: Brenda's story

Brenda is 52 years old and worked for a number of years as a medical receptionist prior to being involved in a car accident where she was badly injured and was unable to return to work. She experiences back pain and has had multiple interventions including: Facet joint injections, physiotherapy and TENS. Her current medication is: dihydrocodeine, co-proxomol, buprenorphine (Temgesic) for flare-ups, and diazepam for muscle spasms. Brenda lives with her dog who she credits with keeping her motivated and active.

Credits: DIPEx© – http://www.dipex.org/DesktopDefault.aspx

1.2.1
Unfiltered Voices

There are many places where we can learn about the chaos of lives in pain without the filter of a third-party. Autobiographies are an example. Oliver Sacks, a physician, experienced prolonged pain as a result of a climbing accident that left him bed-ridden for many months [6]. He emphasized the isolation of people suffering from migraine pain

> *"If migraine patients have a common and legitimate second complaint besides their migraines, it is that they have not been listened to by physicians. Looked at, investigated, drugged, charged, but not listened to."*

Arthur Frank wrote about his experience of living with cancer [7]

> *"In writing about the incoherence of pain, one risks becoming incoherent all over again. Language easily goes wrong. I could write that at night in pain I came to know illness face to face.*

> *But this metaphor distorts that experience . . . at night I faced only myself"*

Lucy Grealy was diagnosed at age nine with Ewing's sarcoma and wrote about the physical and psychic pain that marked her life from that point forward as she endured over 30 surgical procedures to reconstruct her jaw [8]

> "This presented a curious reversal of fear for me, because I already understood that with other types of pain the fear of not knowing about it usually brought about more suffering than the thing itself. This was different, this was dread. It wasn't some unknown black thing hovering and threatening in the shadows; it had already revealed itself to me and, knowing that I knew I couldn't escape, took its time stalking me. This was everything I ever needed to know about fate."

Not being listened to, 'language gone wrong', fear of the known – these are all common themes that emerge from the stories of people with pain. Because 'Brenda' used to work for a doctor and understands the time pressure doctors are under, she works to overcome these problems . . .

Box 4: Brenda's story

"So if I go to a GP and I'm not being understood or I'm not being . . . thinking "Okay here comes another one" you know, or whatever, whatever they don't know about chronic pain, I'll try and be different and say "Well actually it . . ." you know, and go overboard to try and make them understand a little more of what my situation is like".

Credits: DIPEx© – http://www.dipex.org/DesktopDefault.aspx

1.2.2
Literary Routes to the Lived Experience of Pain

Literature is a powerful way to convey the lived experience of pain. Victor Hugo [9] pointed out the idiosyncratic nature of pain

> *"Pain is as diverse as man. One suffers as one can."*

At the end of the nineteenth century, Leo Tolstoy [10] wrote

> *"The gnawing, unmitigated, agonizing pain, never ceasing for*
> *an instant, the consciousness of life inexorably waning but not*
> *yet extinguished, the approach of that ever dreaded and hateful*
> *Death which was the only reality, and always the same falsity."*

This stark image of pain from *The Death of Ivan Ilyich*, written in 1886, is strikingly congruent with what Lucy Grealy says about pain and dread despite being 100 years and a universe of medical technology apart.

Contemporary non-fiction also makes people's description of the pain experience available. Marni Jackson's recent journalistic work *Pain: The Fifth Vital Sign* [11] is intended to make diverse information about the complexity of pain accessible to a wide audience. Jackson frequently uses her own and others' voices to illustrate what it is like to live with pain.

1.2.3
Lived Experience of Pain in Scholarly Publications

Professional journals (e.g. *BMC Medical Ethics* at www.biomedcentral.com/ bmcmedethics) are increasingly available in open-access, on-line formats. Because the financial burden of subscribing to these journals is removed, everyone (clinicians, policy makers and the general public) can examine emerging issues and share in understanding the pain experiences these forums provide. Some health-care journals regularly contain first-person accounts from people with a range of life-altering medical conditions (like chronic pain) to illustrate points within a medically focused discussion. For example, in a *British Medical Journal* (BMJ) article, Chloe talks about her doctor's disbelief about her arthritis [12]

> "I made an appointment with my (now ex) GP, who when I
> informed her that I had been diagnosed as having RA said
> "Nonsense you probably just have flu." I finally persuaded her to
> order blood tests, which reconfirmed the diagnosis. However she
> flatly refused to allow me to keep the rheumatology appointment
> and insisted that I attend a hospital with an over one year
> waiting list, even for "urgent" first appointments."

Other healthcare journals include qualitative research reports that depend on quotations from study participants to illustrate the researcher's conclusions. For example, in the journal *Work*, a young student musician's words illustrate his reaction to his physician [13]

> ". . . *figured if I had pain and it was something that's going to affect my career, not only in school, but my performing career, that it should be taken a little more seriously* [by the physician I went to].

1.2.4
The Visual Arts

People hear differently. What is clear to one, may not be clear to another. This is well illustrated by the simple children's game of 'whispers' where a message is started by one child and whispered around the circle until it returns to the origina- tor. What started off as 'see the dog run away fast' may come back as a nonsensical statement, about 'cooking wild duck on an outdoor potato hat'. Hearing about other people's pain may be rather like that; the message is garbled because people hear in different ways. This makes it important that as many alternative media as possible are used to convey the message. People who don't read Tolstoy or Victor Hugo may find pain messages accessible through a range of electronic media available on personal computers. For example, a 'Google' search of the internet locates many 'pain art' projects.

Of particular interest are two projects that show how we can directly access people's pain expression through visual art. Debra Padfield's *Perceptions of Pain* (http://ije.oxfordjournals.org/cgi/content/full/32/5/704) started when people from a pain clinic worked with an artist to create images of their pain. The result- ing exhibition was published as a book with the assistance of *Novartis Pharma AG* and is now owned by *Napp Educational Foundation*, in Cambridge, UK. Selected images from the project (see Figure 1.1) are available through a research project for healthcare providers in primary care to use with patients attending doctors' appointments. Perhaps when words fail, as they often do for people with pain, images can be used to give pain a presence accessible to others. The images in *Perceptions of Pain* were created in an effort to express the multilayered experience of living with chronic pain. As one participant said [14]

> ". . . *when I first saw the images that Deborah and I produced together I felt a shiver of recognition mixed with feelings of anger and sadness. But for the first time I was able to point at something and say 'that's my pain'."*

The *PAIN Exhibit* (www.painexhibit.com) is an online collection of works created to express pain experience. The artists are international and the work is often accompanied by commentary to increase the likelihood that the viewer will accu-

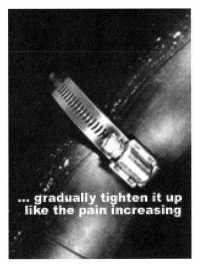

Figure 1.1 Pain expression through visual art: *Jubilee clip* from Debra Padfield's *Perceptions of Pain* (with kind permission from Dewi Lewis Publishing).

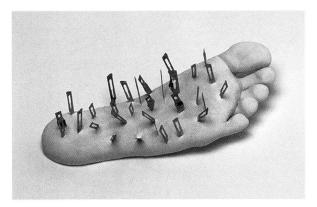

Figure 1.2 Pain expression through visual art: *CPII* by Mark R. Collen (with kind permission from the artist).

rately decode the experiential evidence. Mark R. Collen began the collection in 2001 as a response to the years of under-treatment he endured (see Figures 1.2 and 1.3). Collen was motivated to exhibit his and over 500 other works of pain-related art because they are "far more effective at communicating the pain experience than words". (www.painexhibit.com/aboutexhibit.html). The exhibit is categorized into sub-themes (for example 'normalcy', 'suffering' and 'hope') so that viewers can more readily access works related to the evidence they are seeking. The *PAIN Exhibit* has been used widely in North American and Europe for teach-

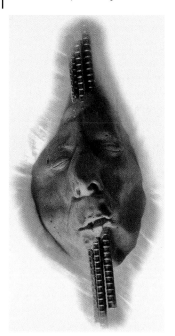

Figure 1.3 Pain expression through visual art: *CPIII-Trapped in Hell* by Mark R. Collen (with kind permission from the artist).

ing healthcare providers. To date, the work has received attention in both the popular press and academic journals.

1.2.5
Pain Voices in Electronic Media

Electronic media allow people with pain and their significant others to share experiences and information across continents and time-zones via the internet. Brenda is an example. Brenda's story and more than a hundred others are freely accessible from the DIPEx website (www.DIPEx.org). This very rich resource contains audio-video clips of people talking about living with a chronic condition. The viewer can select by type of health condition and/or by type of life issue (family, employment, sexuality etc.) and then watch short clips. One can see the guarded postures, the grimaces and the smiles; all the non-verbal communication of ordinary people talking about dealing with extraordinary challenges. These clips are all the more powerful for pointing to what is not talked about. Here are Brenda's experiences with facet joint injections.

Box 5: Brenda's story

"But I spoke to a woman, a nurse at the pain clinic and she said that I was on the right track. Some people, unfortunately, don't exercise, don't watch their diet and expect the injections to do everything for them without, without you know adding to it, you know being their own doctor as it were and that's unfortunate."

Credits: DIPEx© – http://www.dipex.org/DesktopDefault.aspx

Being able to see Brenda as she talks about the experience tells us even more. Brenda becomes a real person, sincere, with a sense of humor, and a conviction that she has an important story to tell. These types of resources – graphic and compelling – communicate a richness that is inaccessible through textbooks, lectures or simulated patient exercises. They can be invaluable teaching tools.

1.3
Quality

How do we know that patients' voices and art work can be considered as good quality evidence about the experience of pain? Sackett, a foundational thinker in evidence-based medicine, does suggest that [15]

> *"Evidence based medicine is not restricted to randomized trials*
> *and meta-analyses. It involves tracking down the best external*
> *evidence with which to answer our clinical questions . . ."*

We have located exemplars of the patients' voices. But how do we know that these voices are '*best evidence*'? Perhaps the patient is speaking of a highly individual experience, or has some secondary or hidden agenda or bias behind the recounting of the story. Even evidence from experimental studies that appear to be carefully designed and rigorously controlled are sometimes found to have serious flaws. What are the indicators of quality in the type of experiential evidence discussed here?

First, we look for *transparency* [16]. Is the source credible? Have contributors explicitly stated their perspective and background? For example, the patient narratives used to illustrate points in the previously cited BMJ article [12] are very transparent. We know where they come from and a context has been provided. Mark Collen's *PAIN Exhibit* provides the background of the project and tells us exactly why the project was started. We know that this evidence is presented from the perspective of someone who has experienced on-going pain. Collen has made the biases explicit. Similarly, Deborah Padfield provides information about herself and the commercial interests (Novartis Pharma in this case) that sponsored the work. Arthur Frank, Lucy Grealy and Oliver Sacks' publishers all provide bibliographic information to help the reader understand the context and bias of the authors. Given that there is no 'bias-free' evidence; an indicator of quality is how transparent the author is about their bias.

A second criteria of quality is the *credibility* of the sources presenting the evidence [16]. Journals like *BMJ* and *Work* that bring us the patients' voices are credible in that they are peer-reviewed and are regarded as being 'high-quality' by their readers. Narrative evidence presented in books from large publishing houses like Harpers, who published Grealy's work, have more credibility then those that are self-published. Morris [2], Scarry [3], Good [17] and Frank [6], all present their academic credentials as demonstration of their credibility. In the case of Brenda's story, the DIPEx site [4] is supported by the National Health Services (NHS) in the United Kingdom and is a registered charity. DIPEx makes available information about its structure and board on the website. DIPEx was also voted by two leading national newspapers in the UK as being among the top health websites providing evidence that the resource is well respected by credible reviewers.

Trustworthiness of experiential evidence is also related to what qualitative researchers call *triangulation* [16]. This means that consistent themes emerge from across the range of resources being reviewed. Where there is more consistency, there is stronger evidence that what the author is articulating is a theme that is universal, not isolated or idiosyncratic. When we look at the evidence presented through the voices of people with pain there are definite consistencies. *Disbelief, feeling 'voiceless' and unheard,* and *the desire to have others legitimize pain* are all themes that emerge consistently from stories about living with chronic pain. These themes are evidence that despite all of the advances in medical technology and the availability of apparently effective interventions, people with pain are still suffering, in part from our neglect and refusal to acknowledge the genuineness of their suffering.

1.4
Barriers to Accessing the Evidence

That healthcare providers and policy makers do not hear the patients' voice is not a problem unique to chronic pain. However, not being heard is even more likely for people with chronic pain because, unlike a fracture, or arthritis, or most other health problems, chronic pain is invisible.

Why don't healthcare professionals and policy makers listen? Historically, the answer is largely embedded in Descarte's idea that the body and the mind were two separate parts of what make us human. When this idea was presented in the mid-seventeenth century, it revolutionized thinking about the body. Many important advances in anatomy and physiology are a result of this perspective. From that point forward 'scientific' thinking about pain focused on understanding the biological/organic components of pain. Over the next 400 years critical advances in our understanding of the biological processes involved in pain and its treatment were achieved. Unfortunately, similar attention was not paid to the psychological processes and we marginalized this type of study as less important and valid. It is only recently that healthcare providers have recognized the importance of viewing health and illness as part of a complex system composed of biological, psychological and social elements, all of which interact and influence illness and disease [18, 19].

Research now suggests that ignoring the psycho-social aspects of chronic pain increases the likelihood that the patient's pain experience will become progressively more complex and resistant to intervention. The interaction between patients who can't express themselves and healthcare providers who do not understand what the person with pain is trying to express results in misunderstanding, negative affective responses and, frequently, mistrust and disbelief [20]. Without effective and accepted pain expression the person with pain cannot communicate his or her experience. Without awareness of the patient's experience a reliable assessment of interdependent bio-psychosocial function cannot be achieved and consequent interventions may be ineffective, wasteful and potentially harmful.

1.5
Consequences of Unvoiced Pain

Box 6: Brenda's story

"I just felt less than human almost. I did, I felt like a cow or something I just felt less than human almost."

Credits: DIPEx© –http://www.dipex.org/DesktopDefault.aspx

There are negative consequences when patients feel they are doubted by their physician and when physicians express distrust of their patients [21].

Physicians, frustrated when treatment isn't working, can begin to blame the patient and the patient can become angry and withdrawn, understanding that the physician doubts their honesty [22, 23]. Existing social stereotypes about people with pain can act as critical barriers to accepting patients' accounts as indicators of legitimate ill health, particularly when the patient does not respond in an anticipated manner [24]. People with pain, and the myriad of stakeholders they interact with may come from different linguistic groups [25]. They may not feel the need/ comfort/ability to share their opinions overtly, may not be allowed to speak for themselves [26], and may be suffering the effects of past miscommunications [27]. Taken together, these barriers to effective communication of the pain experience can be overwhelming.

1.6
Dealing with the Evidence

What strategies will help healthcare providers and policy makers appreciate the role of lived experience in effective programming for chronic pain? We have identified diverse sources for gaining access to the patient's voice. What remains is knowledge translation – *"developing healthcare workers' and policy makers' receptivity to and awareness of lived experience"*. Knowledge translation strategies make explicit the benefits of the evidence available through qualitative inquiry. For example, the University of Alberta's Arts and Humanities in Health and Medicine (AHHM) Program (www.med.ualberta.ca/education/ahhm.cfm) offers healthcare providers a broader socio-cultural perspective on issues of illness and wellbeing. The newly revised *Core Curriculum for Professional Education in Pain* produced by the International Association for the Study of Pain (www.iasp-pain.org), and the expanding range of publications in journals like the *BMJ* that stress the value of listening to the patient's narrative as a critical component of healthcare for people with chronic, complex conditions [28] can also help. A report released by the University of British Columbia, *Where's the patient's voice in health professional education?* [30] is another useful source of background information and strategies.

1.7
Summary

Box 7: Brenda's story

"It just felt awful because I think there are benefit cheats, you know. There are people that are on benefits that could be working and aren't, they see themselves, you know, easy bit of money or something. I don't know, I'm not them and I don't want to talk about them really, but well it just really hurt my feelings and the thing that was in my favour was that, you know, I said to them "Look I want to work" or something'.

Credits: DIPEx© – http://www.dipex.org/DesktopDefault.aspx

Three themes arise from the literature with regularity. The first theme is *doubt*. People with pain experience the doubt of healthcare providers, family, colleagues and policy makers on a regular basis. This doubt exists because people with chronic pain do not respond to biomedical interventions in the expected manner (i.e. 'getting better'!). Tragically, after repeated encounters with social and health providers that seem to mistrust the person's pain account; pain sufferers begin to doubt their own experience, losing confidence in their ability to manage their health and cope with the life-changing events of chronic illness.

The second theme is *invisibility*. People with chronic pain often have no outward medical signs that legitimize their claim and communicate their situation to others. Unlike someone using a wheelchair or with a paralysis, people with pain cannot rely on others' observation as a way to communicate the pain experience. A range of communication tools are that much more important because observable physical signs are absent and cannot 'communicate' for the individual.

The third theme is that people with chronic pain lead an *unpredictable* life. Chronic pain results in significant social, financial, functional and psychological losses. These losses leave people with chronic pain lacking control over many aspects of their lives. Often attempts to regain control (for example assertive demands for assistance) are perceived by others as negative behaviors. Limited predictability and limited agreement about the best route forward are characteristics of complex health conditions like chronic pain [29]. For people living with chronic pain this uncertainty is experienced as a confusing 'yo-yo' of attitudes, advice and undesired reactions.

The literature is rich with examples of the consequences of these core themes; the iatrogenic effects of disbelief and communication breakdown, the impact of social marginalization and role loss, fear and avoidant behaviors related to physical activity because of the unpredictable nature of pain. Embedded within each of these consequences are multiple layers of contextual influences. Chronic pain is truly more than the sum of its parts. Each problem is dynamically intermeshed with many other confounding and interacting factors.

Currently we have many of the tools needed to locate and to listen to the patient's voice. We can access the pain experience through our own patients and through on-line resources like DIPEx, and a range of art projects and literature. We have the IASP core curriculum in pain for healthcare providers; we have access to many on-line, public domain educational resources. We have examples of successful initiatives to build upon.

The core themes of the pain experience (doubt, invisibility and unpredictability) are conceptual, not tangible objects. Biomedical healthcare is highly effective, but not for problems that are relational and existential. Gordon and Dahl [1] suggest that continuing to examine what are actually systems' and interactional problems with clinical tools is akin to trying to break the sound barrier by tinkering with a Model-T Ford car. We already have many of the pieces of the pain experience. Our challenge is to appreciate their therapeutic value. 'Brenda's' voice has been with us throughout this chapter and the final word rightly belongs to her.

Box 8: Brenda's story

"So I'll make an extra effort to, to give them an idea of what it's like to, you know, walk in my shoes."

Credits: DIPEx© – http://www.dipex.org/DesktopDefault.aspx

References

1 Gordon, D.B. and Dahl, J.L. (2004) Quality improvement challenges in pain management, *Pain* **107**, 1–4.

2 Morris, D.B. (1991) *The Culture of Pain*, University of California Press, Berkeley.

3 Scarry, E. (1985) *The Body in Pain: the Making and Unmaking of the World*, Oxford University Press, Oxford.

4 Wolff, V. (1930/2002) On Being Ill, Consortium.

5 DIPEx on line. Personal experiences of health and illness accessed at http://www.dipex.org (August 2007). 'Brenda' [participant CP18]. DIPEx personal experiences of health and illness. Accessed at http://www.DIPEx.org (August 2007).

6 Sacks, O. Bartleby Online Reference Library. Accessed at http://www.bartleyby.com (August 2007).

7 Frank, A.W. (2002) *At the Will of the Body: Reflections on Illness*, Houghton Mifflin Company, New York.

8 Grealy, L. (1994) *Autobiography of a Face*, Houghton Mifflin Company, Boston.

9 Hugo, V. Bartleby Online Reference Library. Accessed at http://www.bartleby.com (August 2007).

10 Tolstoy, L. (1960) *The Death of Ivan Ilyich and Other Stories*, Penguin Books, London.

11 Jackson, M. (2002) *Pain: The Fifth Vital Sign.*, Crown Publishers, New York.

12 Simpson, C., Franks, C., Morrison, C. and Lempp, H. (2005) The patient's journey: rheumatoid arthritis, *British Medical Journal (Clinical Research Edition)* **331**, 887–9.

13 Cortes, M.C., Hollis, C., Amick, B.C. and Katz, J.N. (2002) An invisible disability: Qualitative research on upper extremity disorders in a university community, *Work (Reading, Mass.)* **18**, 315–21.

14 Padfield, D. (2003) *Perceptions of Pain*, Dewi Lewis Publishing, Stockport, UK.

15 Sackett, D.L., Rosenberg, W.M., Gray, J.A., Haynes, R.B. and Richardson, W.S. (1996) Evidence based medicine: what it is and what it isn't, *British Medical Journal (Clinical Research Edition)* **312**, 71–2.

16 Bowling, A. (1997) *Research Methods in Health: Investigating Health and Health Services*, Open University Press, Buckingham.

17 Good, M. (1992) *Pain as Human Experience: An Anthropological Perspective*, University of California Press, Berkeley.

18 Main, C.J., Richards, H.L. and Fortune, D.G. (2000) Why put new wine in old bottles: the need for a biopsychosocial approach to the assessment, treatment, and understanding of unexplained and explained symptoms in medicine, *Journal of Psychosomatic Research* **48**, 511–14.

19 Borrell-Carrio, F., Suchman, A.L. and Epstein, R.M. (2004) The biopsychosocial model 25 years later: principles, practice, and scientific inquiry, *Annals of Family Medicine* **2**, 576–82.

20 Ong, L.M., de Haes, J.C., Hoos, A.M. and Lammes, F.B. (1995) Doctor-patient communication: a review of the literature, *Social Scienceand Medicine* **40**, 903–18.

21 Kouyanou, K., Pither, C.E. and Wessely, S. (1997) Iatrogenic factors and chronic pain, *Psychosomatic Medicine* **59**, 597–604.

22 Dixon-Woods, M. and Critchley, S. (2000) Medical and lay views of irritable bowel syndrome, *Family Practice* **17**, 108–13.

23 Eccleston, C., Williams, A.C.D.C. and Stainton Rogers, W. (1997) Patients' and professionals' understandings of the causes of chronic pain: blame, responsibility and identity protection., *Social Science and Medicine* **45**, 699–709.

24 Sullivan, M. (1999) Between first person and third person accounts of pain in clinical medicine, in *Pain 1999 – An Updated Review: Refresher Course Syllabus* (Max, M., Ed.), IASP Press, Seattle.

25 Roberts, G., Kent, B., Prys, D. and Lewis, R. (2003) Describing chronic pain: towards bilingual practice, *International Journal of Nursing Studies* **40**, 889–902.

26 Carter, B. (2002) Chronic pain in childhood and the medical encounter: professional ventriloquism and hidden voices, *Qualitative Health Research* **12**, 28–41.

27 Hafner, C. (2002) The information we give may be detrimental, in *Topical Issues in Pain* (Gifford, L., Ed.), pp. 101–18, CNS Press, Falmouth.

28 Greenhalgh, T. (1999) Narrative based medicine: narrative based medicine in an evidence based world, *British Medical Journal (Clinical Research Edition)* **318**, 323–5.

29 Brown, C. (2007) The role of paradoxical beliefs in chronic pain: A complex adaptive systems perspective, *Scandinavian Journal of Caring Science* **21**, 207–13.

30 Farrell, C., Towle, A. and Godolphin, W. (2006) *Where is the Patient's Voice in Health Professional Education?* Division of Health Care Communication, University of British Columbia Accessed at http://www. health-disciplines@ubc.ca/DHCC/ (Sept 2007).

2
History and Definition of Pain

Harold Merskey

- Although physicians have struggled to define pain, laypersons have universally understood that pain is a personal, private and inherently valid experience.

- From the earliest writings of Western civilization, the language of pain has always referred to both physical and emotional qualities.

- The International Association for the Study of Pain provided a definition that is widely accepted today "Pain is an unpleasant experience associated with actual or potential tissue damage or described in terms of such damage".

- While much attention and speculation has been given to psychological causation of pain, available data suggest that it is often over-emphasized in clinical practice.

There are several histories of pain but to my knowledge only Rey [1] deals specifically with the history of pain in all its aspects. Other books that stand out include Fülop-Miller [2] and Robinson [3], dealing with the discovery of anesthetics. Philosophers ranging from Aristotle [4] through Descartes [5] to Wittgenstein [6] discuss pain as part of their topic. Meanwhile ordinary folk, including physicians and surgeons, understood the experience of pain all too well, even when they were uncertain of its source. For them pain ranged from a minor, even trivial, experience to an overwhelming disaster that filled consciousness to the exclusion of other material. Whether they thought of it as an extreme torture, a trivial event or something that was somewhere in-between suffering and a minor irritation, these topics attracted less discussion as to how such problems came about, and more on how to respond. Observers who emphasized the external evidence at the cost of denying the ability to acknowledge the internal events ran the obvious risk of rejecting the valid experience of individuals. By and large, along with Bishop Berkeley [7], most people have treated pain, whether slight or searing, as ultimately being private, inaccessible to others, and valid, no matter what external evidence existed.

Chronic Pain: A Health Policy Perspective
Edited by S. Rashiq, D. Schopflocher, P. Taenzer, and E. Jonsson
Copyright © 2008 WILEY-VCH Verlag GmbH & Co. KGaA, Weinheim
ISBN: 978-3-527-32382-1

This did not matter much to anyone until the statement "I have pain" acquired some economic importance or social benefit. When having a physical complaint might allow an escape from compulsory military service, and having pain could allow such an escape or could lead to a pension, it became obligatory to check the bona fides of the speaker or his or her economic benefits. "Not tonight, dear, I have a headache" or "I'm injured, I want compensation" became cases for proof before sympathy or action.

There were those who said pain could arise from psychological causes, for example, von Feuchtersleben [8] and Anstie [9]. If one reviews the history of pain it is not hard to find contradictory opinions about pain and the mind, ranging from "pain changes a person's mind", to the view that "mind can cause pain". In the early 1960s both Graham Spear and I completed medical theses on pain and its psychology while working in the Department of Psychiatry at the University of Sheffield under the direction of Erwin Stengel. A little later we published a book [10] that looked at the psychological causation of pain and emphasized the psychological contribution to the occurrence of pain, and the frequency with which pain, whether of physical or psychological origin, was found among psychiatric patients. Graham had more reservations than I did about its implications, but I increasingly came to feel that too much emphasis had been given to the psychological etiology. Today, without departing from the view that some pain has roots in anxiety or depression – and this is particularly true for headache – I lay much less emphasis on the psychogenesis of pain. While the main cause of some pain may be "mind", physical factors are usually the most important (and perhaps the most often denied).

2.1
Words for Pain

The earliest written literature of Western civilization is either Greek or Hebrew. Rey [1] points out that in the Iliad, Homer's earliest work, five words or word groups can be distinguished which all have a link with pain. Three groups do not necessarily refer to bodily injury or physical harm but to mourning *(Penthos)*, worry, obsession, grief *(Kédos, Algos)*. According to Rey there is also a whole vocabulary based around another word, *Achos*, which expresses a sudden and violent emotion and the confusion of feelings which can lead to despondency.

The word *Péma* is found in contexts associated with *Algos* and both also refer to emotional distress as well as to physical distress. It has a double meaning, on the one hand adversity, scourge, or subjective misfortune, and on the other hand, pain and suffering. Three words may relate to physical change. They are *Oduné, Péma and Algos*. These words may also overlap with emotional states. According to Rey the word *Oduné* seems to explain a sharp, shooting pain, and the equivalent has come into English usage in words such as pleurodynia and allodynia [11]. It is the word most strongly linked to a physical event. Adjectives most often associated with *Oduné* include words meaning sharp, *Oxus*, and *Pikros* which also means pointed or cutting and biting; usually referring back to the instrument that caused the wound or sting, such as a javelin or an arrow. These words in Homer were employed in relation to the effects of fighting and they are characterized by rapidity.

Unlike *Oduné*, the word *Algos* represents a more general type of suffering involving the whole body and, according to Rey, it is because this word is vague and undefined that its derivatives such as cardialgia (heartburn), "cardia" here referring to the upper part of the stomach at the junction with the esophagus, and "cephalalagia" (head pain), serve to give additional information as to the whereabouts of pain. *Algos* is used not only, as its verbal context suggests, to mean enduring or putting up with, or working with pain, but also to indicate a submission to suffering. It seems that although the word *Algos* has been used far more in the modern vocabulary of pain than in Homer's language it is *Oduné* which is the technical term that belongs to the specialized vocabulary of medicine.

Rey summarizes the usages of these words as showing not an antagonism between physical and emotional in their influence upon the intensity of pain but rather two axes on which pain is evaluated, one with respect to time and origin, that is, longlasting, sharp or cutting, and the other referring directly to the instrument which causes it, simultaneously defining the very nature of the sensation. Tragedians such as Sophocles used such words and also words for ailments and illness and spoke of pains in the pleural. The words *Ponos* and *Algéma* appear here together with *Oduné* and indicate strain as well as fatigue. The word *Ponos* has been said to be the source of the Latin word *Poena* but this is not generally accepted. *Poena* has a range of meanings that includes punishment, hardship, suffering and pain and gave rise to the English word "pain" (as physical punishment) and the German "pein".

In the Hippocratic Collection there are stated to be 772 places where the word *Oduné* is used, 14 for *Algos*, and 194 for *Algéma* where a physical change is suggested, with or without an emotional event. In general, the Hippocratic writings might be taken to be our main source of understanding of pain as a physical event, and its emotional implications are simply assumed.

The uses of words in a language often change and any word may acquire different shades of meaning and almost opposite meanings from those for which it was originally employed. However, in general, there is a clear link between words that we use for what we call pain and words used for physical injury or complaint in the Greek authors. Equally, the consequence of being in pain, i.e. emotional distress or sadness, leaks into, or is absorbed into, some of the meanings of those same words. Talking of physical and psychological states in general, we have no trouble in recognizing when words are being used to describe a physical status and when they are being used to describe an emotional status or condition.

Hebrew has been studied less intensively for the usages of pain. I [10] analysed the classic quotation from Jeremiah {Lamentations I: 12–13} "Is it nothing to you, all you that pass by? Behold and see if there is any pain like my pain which is done unto me with which the Lord has afflicted me in the day of his fierce anger. From above He has sent fire into my bones . . . and I am weary and faint all the day". This passage offers pain as a metaphor for suffering, but also contains some acute observations of how grief may give rise to the experience of pain with faintness and physical change. The passage that took this further than the literal meaning is found in Isaiah 53:4. "A man of sorrows acquainted with grief", where the correct translation should probably be "A man of pains acquainted with illness".

In modern Hebrew there is a standard word for pain, *Ke'ev*, and the verb to feel pain forms adjectives and other words related to it, for example, painful, *Makhov*. More complex forms related to it refer primarily to a physical event that is experienced, for example, Genesis 34:25, where Shechem and his associates have been circumcised and they are said to be in pain or sore, *Ko'evim*. Clear usages for the word pain as a physical event can be detected in some seven places in the Hebrew Bible. However, metaphorical uses are much more common although often seeming to refer to a bodily event. In Ecclesiastes 2:23, the Authorized Version translates "For all his days are sorrows and his travail grief" but the word translated here as sorrows is actually the same word as that for pains. The words for pain, or pains, or painful are used repeatedly throughout Proverbs, Job, Psalms, Jeremiah, Ecclesiastes, Ezekiel, and, in particular, Isaiah to reflect metaphorical pain, sometimes in terms which appear to be directly descriptive of bodily experience but which can be translated in accordance with the sense of the writer's words as grief or sorrow or, as in Isaiah 65:14, heartache. Altogether I could count some 24 instances from the standard concordance of the Hebrew Bible [12] where the word for pain is used metaphorically, sometimes with a very obvious origin in the physical event. For example, Jeremiah 51:8, "Babylon is suddenly fallen and destroyed: howl for her: take balm for her pain, if so she may be healed".

The Hebrew Bible spends a lot more time on historical reports and prophecies of good or evil, and on poetry in the Psalms and in many of the prophets, than it does on factual discussions of bodily functions. One may expect metaphor to predominate in the passages that are Psalms, prayers and so forth. It would be a mistake simply to take the numbers of words used for emotion and compare them with the numbers of words used for lesions in the body and assume, therefore, that what we mean by pain in medicine and in daily life is primarily an emotional event. The fallacy is obvious. The common sense conclusion is that the word pain starts with a physical condition and because of the distress that is often associated with it is invaluable to priests, prophets, and poets.

Given that, let us consider how to define pain for medical purposes.

2.2
"Pain" in Medicine

From the intervening period through to the nineteenth century, roughly 2500 years, the ordinary view of pain was that it had physical origins, but sometimes was influenced by mental processes. This was summed up very neatly by Montaigne [13] in the expression "One feels a cut from the surgeon's scalpel more than ten blows of the sword in the heat of battle". We may see later that this particular comparison is like many statements about pain–true but only partially applicable.

2.2.1
1800–2000

It seems that during the last two centuries more debates took place about pain and emotional contributions to the origins of pain than at any other period of European and North American history. In the twentieth century there seems to have been much more discussion about emotional elements in pain than previously. Whether that was because doctors were smarter – or less smart – or because more books were published than in the eighteenth or seventeenth centuries remains open to consideration.

One view is that, up until the advent of anesthesia, pain at its worst was such an overwhelmingly nasty experience that almost anybody might face, that no one could have thought very much about the notion that it was brought upon oneself through one's own mind. Accounts of the pain of operation and vivid discussions of these phenomena have been provided by authors such as Fülop-Miller [2] and Robinson [3] who saw the benefits of anaesthesia, and individuals such as Fanny Burney [14] who became the Marquise d'Arblay, a late eighteenth century and early nineteenth century writer who described her own breast amputation.

During the nineteenth century, a relatively new factor entered into considerations about pain. The first instance of this was probably the development of insurance for workers and others on railways and in different industries. Such insurance spread gradually throughout Western European countries after the advent of railways and increasing concern for the welfare of injured workers. Negative attitudes were seen to emerge about post-traumatic pain and became an issue in compensation claims. The view that patients in railway injuries could sustain "concussion of the spine" with significant pain and few external signs of injury was put forward by Sir John Eric Erichsen [15] and expanded later by him in 1886. Erichsen was opposed, significantly, by physicians who worked for railway companies. Page [16], who was a surgeon to the London and North Western Railway Company, argued that many of the consequences that were dwelt upon by Erichsen were attributable to "nervous shock" and by this he meant that fright was a potent cause of the symptoms. "In men . . . the frequency of hysteria . . . (if such a term may be used in men) is not fully recognized". Most of the arguments on these topics are well set out by Schmiedebach [17] who identified two groups involved in arguments about compensation, in particular doctors who worked principally, if not wholly, for insurance companies, and doctors who were primarily committed to working for their patients, both of course taking pay one way or another from one side or the other. Post-traumatic stress disorder is another condition which bears many resemblances to the sort of problems that we have been discussing with pain. Is it real? How much?

These issues will never be entirely resolved or resolved to the satisfaction of everybody, but a broad consensus can be established and that consensus currently holds that both moderately severe and severe anxiety, and pain of moderate severity or greater, can be understood as resulting from the patient's experiences.

The most troubling issues that have arisen in this respect have to do with the extent to which pain should be compensated, and what views we take of pain when patients come along and seek compensation. There is then a profound tendency in some quarters to attribute pain to the patient's own thoughts and feelings rather than to the injuries that she or he may have sustained. It is with this background, although much less acutely recognized at the time, that definitions of pain were mooted and a definition was developed. My own effort to provide a definition was also powered by a feeling that we ought to be able to say that we know *what* we are talking about or investigating, even if we do not know how it is caused.

2.3
Defining Pain

The attempt to define pain gave rise to substantial discussion in the mid-twentieth century. Beecher [18], the author of notable advances in the understanding of pain, tried to find a satisfactory definition and cited five other distinguished authorities or sought their help in providing one. These included Sir Thomas Lewis [19], Lord Adrian [20], Herbert Gasser [21], George H. Bishop [22], and Wikler [23]. Thomas Lewis wrote "Reflection tells me that I am so far from being able to define pain that the attempt could serve no useful purpose". He was concerned with distinguishing a sensory element and an emotional element which, for example, Hardy *et al.* [24] characterized as follows: "The sensory element was primarily related to physical stimulation while a 'reaction to pain' which was also part of pain would be less constant and would vary independently from the physical trauma".

Beecher, rightly enough, was not happy with that either. He quoted the words of Strong [25] "Whenever we feel a pain, there we have a sensation or idea, distinct from the pain, with reference to which pain is felt, . . . in every actual state of mind we are able to distinguish these two sides, the cognitive and the affective". He too wanted to recognize what might be thought of as a purely sensory factor and what we might call an emotional component. They simply could not work out a satisfactory way to put these two concepts together in one definition. This block was evidently helped by the uncertainty as to how to measure feeling (affect/emotion) and combine it with the notion of sensation. Hall [26], whom Beecher had not consulted, suggested that what we meant by pain should be apparent in each investigation from the description of the experimental conditions and controls, the instructions, the results and the conclusions.

J. H. Bishop [22], at the request of Beecher, made comments which Beecher described as ". . . a valiant effort to define what seems to be patently undefinable: Bishop wrote 'pain is what the subject says hurts . . . it consists . . . of two phenomena (a) pain is a subjective experience, reported as a *sensation* referred specifically to some part of the body and sufficiently unpleasant to be designated as painful by the subject. End definition (a)". Bishop then went on to talk about how the unpleasant sensation would "of course vary with emotional state, anxiety, anticipation of disaster and so on . . . and is almost impossible to deal with quantitatively since it has such a large component of what is referred to as reaction to sensation. I know of people who can throw a sick headache, and so do you, as a protest, and I can't say they don't have one . . ." Bishop went on to talk about pain as a physiological process with a subjective evaluation in addition to perception as a result of stimuli to sensory endings or pathways of two types of fibers and so on.

Others, who wrote at length about the meaning or definition of pain without providing a satisfactory solution, include W.K. Livingstone [27], V.C. Medvei [28], Sir Gordon Holmes [29], S. Kolb [30], and J. L'Hermitte [31]. The range of disciplines is quite large, including a psychiatrist (Kolb), a cardiologist (Lewis), surgeons (Livingstone and L'Hermitte), neurologist (Holmes), endocrinologist (Medvei), addiction specialist (Wikler), and three physiologists of great distinction (Adrian, Gasser and Bishop).

Bishop was not the first nor the last distinguished investigator to fail to tackle what is essentially a simple problem by starting at the wrong place. He did indeed make a valiant attempt and nearly got there when he said "Pain is what a subject says hurts". He failed to achieve his purpose when he tried to tie the physiological basis to the sensation in a definition. A definition may state what we think that something is. More often it tells us how we limit what we believe to be something. All we need say about a definition of pain, in essence, is that it is a subjective sensation that is unpleasant and that is associated with the idea of damage to the body. This is the simplest possible definition.

Devine and I [32] were interested in comparing the descriptions of pain by patients with so-called psychogenic pain and by others who had known physical causes for pain. We showed that there were two common features in the description of pain provided by patients of both types: there was unpleasantness and there was some effect on the body. This was demonstrated both in patients with lesions and in those without lesions. Psychiatric patients more often gave unusual descriptions but physically ill patients also provided some of those. In both groups, simple descriptions of pain predominated and the more severe pain was associated with the more complex descriptions.

It seemed to me that the issue connected with pain was clouded by the failure to appreciate that pain is a psychological concept and not a physical measure. Thus, pain as an experience has to be distinguished from noxious stimulation, the effects of which we can, in part, measure or detect in nerve tissue, both peripherally and centrally. If we keep clear the issue of mechanism from the issue of the way the word pain is used, the solution becomes very simple. All that is needed is to define pain in terms of what we experience as pain and leave it at that. We can go on to study the psychological states of individuals who report pain and the physical state of animals and ourselves when we react to noxious stimuli and not have too much of a problem.

Supposedly "psychogenic" pains included adjectives like throbbing, aching, burning, building up, sore, numb, radiating, bruised, like toothache, stabbing, bursting, cramps, pressing, heavy, pulling, dragging, and 51 out of 150 patients used such terms [32]. More severe and unusual descriptions included words like "as if it clutches", prodding, tightening and heavy, knotting, cutting, like electricity, draining, tantalizing, jumping, crunching, dizzying, striking like a knife, wasting, digging, all "due to blows". Patients with known physical illness produced comparable adjectives, for example, ". . . strong pain a few inches inside my seat – drawing the seat down to the ground – as if I'm going with it' (from a patient with cancer of the rectum just prior to diagnosis). Others had words like ". . . a jumping pain in my chest when I hurry", "It's hurting – not an ache – as though it's eating me", ". . . in arms and legs like electric shock – sort of touches all over", and ". . . burning pain – and then it swells – warming", (unpublished data from the cases in Ref. [32]). In 1964 I wrote an early definition of pain as follows: "An unpleasant experience that we primarily associate with tissue damage or describe in terms of tissue damage or both" [33]. This tells us how we use the word pain. It does not tell us what the mechanisms of pain are and it cannot do so and it was not intended to do so.

It is very important in the definition to get away from the mechanisms of pain because they are not the phenomenon, and trying to link mechanisms with the phenomenon in a definition is a mistake. The phenomenon is what we label and it is only a matter of convenience and labeling with which we are concerned, but that matter has far-reaching consequences.

Spear and I published the definition more widely in 1967 [34] and in the same year the book with Spear *Pain: Psychological and Psychiatric Aspects* appeared and contained the definition [10]. A few people began to notice. In 1974 I was invited by John Bonica to become Chair of the Sub-Committee on Taxonomy, as it was then called, of the International Association for the Study of Pain. My colleagues on the committee wanted to produce definitions of pain terms. Their first interest was in how to use words like threshold, pain tolerance level, hyperalgesia, hyperpathia and so on, but they agreed that it was logical if we were defining pain terms to try and define pain itself. In the course of discussion my original formulation was modified somewhat to the following: *"Pain is an unpleasant experience associated with actual or potential tissue damage or described in terms of such damage".* Notice that the word *potential* reasonably enough came in to indicate those stimuli of which we are aware as signs of potential damage, although the damage has not yet been produced. The fundamentals of the definition remained the same. The Sub-Committee then adopted this definition and it was published by the IASP with other publications of the committee that later became the Task Force on Taxonomy. This definition and others were last published in 1994 by the IASP [35].

Rather than elucidate the point of the definition myself I can quote from a perceptive author, David Morris [36], who says "This brief definition . . . has made it possible for researchers and clinicians working in many different countries, in various languages, and in far-flung disciplines, to possess at least a basic mutual understanding of what they mean (and, equally important, *do not mean*) by the all purpose, ragtag, everyday English word 'pain'".

Morris argues that the IASP definition even allows that tissue damage sometimes simply generates the language we apply to various unpleasant or traumatic sensory and emotional experiences and that pain is not equivalent to nociception, the process by which a signal of tissue injury is transmitted through the nervous system. "The extended annotation [dealing with the definition] begins with the blunt and unequivocal statement that 'pain is always subjective'." He praises the emphasis on the notion that pain is always a subjective psychological state and that, at the same time, the committee stated that pain "most often has a proximate physical cause".

Morris also tackles objections to the definition from those who claim that it is dualist and from others who say that it neglects the ethical dimensions of pain. He rightly says that Descartes has erroneously been identified as the precursor or progenitor of any theory that separates body from mind but adds that an IASP definition of pain implies no such thing. On the contrary it indicates that minds as well as bodies are necessarily involved in the experience of pain, an experience that is multi-dimensional, not the straightforward projection of sensory impulses that Descartes had described. He makes a very pointed argument for not labeling dualisms, that in fact are not the direct legacy of Descartes but flow from nineteenth century positivist science, as Cartesian. He also makes short work of the criticism of the IASP definition, that some have offered, that it ignores ethical concerns implicit in pain because it fails to highlight pain among disempowered or neglected sub-populations such as women, blacks, children and the elderly. He says the IASP definition neither supports nor promotes social injustice and should not be used as alleged to make pain dependent upon "full linguistic competence" ignoring pain in neonates or in animals. As he says "More important, the IASP account treats linguistic competence not as a philosophical prerequisite for pain but as a clinical resource. Its most radical implication lies in valuing the patient's subjective self-report – still too often devalued or dismissed by doctors unable to find an objectively verifiable reason".

I agree with him when he says that "A workable definition of pain need not be – and should not be – a theory of pain".

References

1 Rey, R. (1993) *History of Pain*, Éditions la dé Couverte, Paris.
2 Fülop-Miller, René (1938) *Triumph over Pain*, The Literary Guild of America Inc, New York.
3 Robinson, V. (1946) *Victory over Pain: A History of Anaesthesia*, Henry Schuman, New York.
4 Aristotle, (1931) *De Anima*, (Book II, Chapter 2), (Book III, Chapter 1) (Trans. J.A. Smith, ed. W.D. Ross), Clarendon Press, Oxford.
5 Descartes, R. (1644) *Principles of Philosophy*, cited by Rey in Ref. 1.
6 Wittgenstein, L. (1997) *Philosophical Investigations*, Blackwell, Oxford.
7 Berkeley, G. (2004) *A Treatise Concerning the Principles of Human Knowledge*, Kessinger Publishing.
8 von Feuchtersleben (1845) *Lehrbuch der ärztlichen Seelenkunde Vienna Gerold* (Trans. by H.E. Lloyd (1847) as *The Principles of Medical Psychology*, Sydenham Society, London.
9 Anstie, F.E. (1871) *Neuralgia and the Diseases That Resemble It*, Macmillan, London.

10 Merskey, H. and Spear, F.G. (1967) *Meaning and Definition of Pain*, in *Pain: Psychological and Psychiatric Aspects*, Bailliere, Tindall & Cassell, London, pp. 14–24.

11 Monograph for the Sub-Committee on Taxonomy, International Association for the Study of Pain (1986) *Classification of Chronic Pain: Descriptions of Chronic Pain Syndromes and Definitions of Pain Terms* (ed. H. Merskey) (Pain Supplement 3), Elsevier Science Publishers, Amsterdam.

12 Brown, F., Driver, S.R. and Briggs, C.A. (1906) *A Hebrew and English Lexicon of the Old Testament Based on the Lexicon of William Gesenius*, Clarendon Press, Oxford.

13 (a) Montaigne, M.E. (1580) *Essais* (Book I, Chapter 40), (ed. J.-V. LeClerc) (1865) Garnier Fr res, Paris, pp. 374–5; (b) Montaigne, M.E. (1580) *Essais* (Book I, Chapter 14) (Trans. E.J. Trechmann) (1927) University Press, Oxford. (Note: in some editions Chapters 14 and 40 are transposed.)

14 Burney, F. (1812) A mastectomy, in *Selected Letters and Journals*, (1986) (ed. J. Hemlow), Clarendon Press, Oxford, pp. 126–42.

15 Erichsen, J.E. Sir (1867) *On Concussion of the Spine, Nervous Shock and Other Obscure Injuries to the Nervous System in Their Clinical and Medico-Legal Aspects*, a new and revised edition, William Wood & Co. (1886), New York.

16 Page, H.W. (1883) *Injuries of the Spine and Spinal Cord without Apparent Mechanical Lesion and Nervous Shock in Their Surgical and Medico-Legal Aspects*, J & A Churchill, London.

17 Schmiedebach, H.-P. (1999) Post-traumatic neurosis in nineteenth century Germany: a disease in political, juridical and professional context. *History of Psychiatry*, **10** (37), 27–57.

18 Beecher, H.K. (1959) *Measurement of Subjective Responses. Quantitative Effects of Drugs*, Oxford University Press, New York.

19 Lewis, Sir T. (1942) *Pain*, Macmillan, London.

20 Gasser, H.S. (1956) *Letter to Beecher*, 16 April, cited in Ref. [18].

21 Lord Adrian, E. D. (1956) Personal communication, cited in Ref. [18].

22 Bishop, G. H. (1956) Personal communication, cited in Ref. [18].

23 Wikler, A. (1956) *Letters to Beecher*, 7 August & 6 September, cited in Ref. [18].

24 Hardy, J.D., Wolff, H.G. and Goodell, H. (1952) *Pain Sensations and Reactions*, Williams and Wilkins, Baltimore.

25 Strong, C.A. (1895) The psychology of pain. *Psychological Review*, **2**, 329–47.

26 Hall, K.R.L. (1953) Studies of cutaneous pain: a survey of research since 1940. *British Journal of Psychology*, **44**, 281–94.

27 Livingstone, W.K. (1943) *Pain Mechanisms: A Physiologic Interpretation of Causalgia and Its Related States*, Macmillan, London.

28 Medvei, V.C. (1949) *The Mental and Physical Effects of Pain*, E. and S. Livingstone, Edinburgh.

29 Holmes, G. (1950) Some clinical aspects of pain, in *Pain and Its Problems* (eds H. Ogilvie and W.A.R. Thompson), Eyre & Spottiswoode, London, pp. 18–27.

30 Kolb, S. and Bonner, F.J. (1954) Psychiatric aspects of pain, in *Pain: Its Mechanisms and Neurosurgical Control* (eds J.C. White and W.H. Sweet), Charles C. Thomas, Springfield, IL.

31 L'Hermitte, J. (1957) Les algo-hallucinoses: les hallucinations de la douleur, in *La Douleur et Les Douleurs* (ed. Th. Alajouanine), Masson, Paris.

32 Devine, R. and Merskey, H. (1965) The description of pain in psychiatric and general medical patients. *Journal of Psychosomatic Research*, **9**, 311–16.

33 Merskey, H. (1964) An Investigation of Pain in Psychological Illness. DM Thesis, Oxford.

34 Merskey, H. and Spear, F.G. (1967) The concept of pain. *Journal of Psychosomatic Research*, **11**, 59–67.

35 Merskey, H. and Bogduk, N. (1994) *Classification of Chronic Pain: Descriptions of Chronic Pain Syndromes and Definitions of Pain Terms*, 2nd edn., International Association for the Study of Pain, Seattle, WA.

36 Morris, D.B. (2003) The challenges of pain and suffering, in *Chronic Pain* (eds. T.S. Jensen, P.R. Wilson and A.C.S. Rice), pp. 3–14.

3
The Descriptive Epidemiology of Chronic Pain

Donald Schopflocher and Christa Harstall

- Chronic pain prevalence estimates vary widely (between 10.1 and 55.2%).

- Canadian studies using large-scale population surveys suggest that over 5% of the adult population suffers moderate to severe chronic pain with serious effects on health, disability, distress and the propensity to suffer depression.

- These studies also show an age gradient in Canada that, without improvements in treatment, will lead to a considerably higher prevalence in the future as the population ages.

- Future prevalence studies should conduct concurrent, prospective epidemiological studies using clear-case definitions, and well-validated and reliable data collection tools in order to narrow the range of estimates, and determine the important variables underlying these differing estimates.

3.1
How Prevalent is Chronic Pain?

3.1.1
Systematic Reviews

The systematic review is a widely used method to synthesize the results of multiple studies in the scientific literature. It includes a comprehensive search, selection of studies using criteria established a priori, critical appraisal of the quality of the selected studies, and pooled analysis, either qualitative or quantitative, of the results.

Verhaak and colleagues [1] conducted a systematic review of studies on the epidemiology of CP among adults. The primary aim of the review was to determine which methods were used in the primary studies to determine the prevalence of CP. Studies that exclusively focused on the pediatric and elderly populations were

Chronic Pain: A Health Policy Perspective
Edited by S. Rashiq, D. Schopflocher, P. Taenzer, and E. Jonsson
Copyright © 2008 WILEY-VCH Verlag GmbH & Co. KGaA, Weinheim
ISBN: 978-3-527-32382-1

excluded, as well as those epidemiological studies that addressed acute pain or pain secondary to a defined disease. Fifteen descriptive studies that assessed the prevalence of CP were identified. Thirteen of these studies were general population surveys and the remaining two were primary health care surveys. Data collection methods used in the individual studies included telephone survey (three studies); postal questionnaire (six studies); interview (three studies); and expert assessments (three studies). Data on research methods, definition of CP, prevalence, demographic, and co-morbidity characteristics were summarized for each study. The authors found a wide variability in the estimates of CP prevalence. A "median point prevalence" of 15% (range: 2–40%) was calculated. When the complexity of the definition of CP was considered ("multidimensional" vs. "simple", according to the authors and not clearly defined), the reported median point prevalence values were 13.5% (based on six studies) and 16% (number of studies not stated), respectively. The authors concluded that although the studies used a wide range of CP definitions and yielded widely varying CP estimates, neither the method of data collection nor the definition of CP seemed to systematically affect the prevalence reported. The use of a "median point prevalence" as a pooled measure estimated from the individual studies however is inappropriate. The set of data used to calculate this measure originates from heterogeneous studies with different populations, data collection methods, and definitions of CP. A combined single estimate therefore is not likely to be an accurate reflection of prevalence.

Nickel and Raspe [2] conducted a qualitative systematic review on the epidemiology and use of services in treating CP. Studies on populations receiving treatment for CP were reported separately. Seventeen epidemiological studies were included in the report. Information regarding data collection methods, prevalence estimates, duration of pain, and demographic variables were extracted from individual studies. Data collection methods of the individual studies included: telephone survey (six studies), postal questionnaire (eight studies), and interview (three studies). The review concludes that epidemiology studies are limited by theoretic, methodological, and economic factors and that quantitative comparisons were precluded due to differences in populations, methods of data collection, definition of CP, and reporting of the results. The authors considered that CP was often not clearly defined and the definition was highly variable among the studies. Nonetheless, they reported that the frequency of CP increased with age, with a peak between 45 and 65 years of age. Likewise, higher rates of CP among women were found and an association between social status and frequency of specific types of pain was noted. Given the heterogeneity of the studies, the review did not try to combine results in a quantitative way but reported the results in narrative.

Both systematic reviews point out that there is a wide variation in CP prevalence estimates among primary studies that may be explained by several factors related to the design and the methodology of the individual studies.

To attempt to overcome some of the limitations of these systematic reviews and to update them, Ospina and Harstall [3] conducted an additional systematic review. Studies were analyzed according to the definition for CP, sample size, and response rate. A quality score was calculated for each study [4, 5] When appropriate, prevalence values were combined from several studies into a single estimate (weighted mean prevalence estimates based on sample size).

Standard internationally accepted definitions or criteria for "chronic" or "severe" pain are not available, but several options (based upon quality or intensity of pain) have been proposed. Efforts to determine the prevalence of CP in the general population have been faced with challenges such as variations according to the population sampled, the methods used to collect data, and the criteria to define CP. Differences in the definition of CP are one reason why prevalence estimates differ greatly from one research study to another.

The International Association for the Study of Pain (IASP) provides one of the most frequently referenced definitions of CP that takes into account factors related to duration and "appropriateness" [6]. IASP defines CP as pain that has persisted beyond the normal tissue healing time (usually taken to be three months) without apparent biological value. In the context of chronic widespread pain (CWP) the definition by the American College of Rheumatology (ACR) [7] was also used in the reviewed research literature. CWP is defined when all of the following have been present for at least three months: pain in the left side of the body, pain in the right side of the body, pain above the waist, and pain below the waist. In addition, axial skeletal pain (cervical spine or anterior chest or thoracic spine or low back) must be present.

Thirty two potentially eligible publications were identified and 13 were selected. Studies were excluded if they focused on acute pain, pain by diagnostic categories or by body area involved, or pain secondary to a defined disease or in a particular setting or community. The studies included were published between the years 1991 to 2002. Three studies were conducted in the United Kingdom [8–10], two in Australia [11, 12], and one each in Canada [13], France [14], Israel [15], The Netherlands [16], Scotland [17], Spain [18], and Sweden [19]. A multinational study conducted by the World Health Organization [20] with collaborative centers in Chile, Germany, Brazil, Turkey, France, The Netherlands, England, India, the United States, Italy, China, Greece, Japan, and Nigeria was also included. Eleven of the included studies[8–18] surveyed the general population, and two studies [17, 20] surveyed the population receiving care in primary care settings.

Ten studies [8–11, 13, 15, 17–20] reported separate prevalence estimates for adolescent and adult populations (range: 15–86 years). Two studies [12, 14] provided prevalence data for elderly populations (65 years and over) exclusively. One further study addressed the prevalence of CP in children up to 18 years old [16].

The populations sampled varied from 410 [13] to 17 496 [11] participants of both genders. The number of male participants in those studies that reported raw data by gender ranged from 158 [13] to 2653 [16]. The number of female participants in the studies ranged from 252 [13] to 2770 [16]. Five studies [8, 9, 16, 17, 19] used postal questionnaires. Four studies [10, 11, 13, 18] conducted telephone interviews, and four studies used face-to-face interviews [12, 14, 15, 20] to collect data.

Pain was the main outcome measure in nine studies [8–10, 13–19]. CP data, however, were collected in four studies [11, 12, 14, 20] as part of a broader community survey that assessed several aspects of the general health status of the population. The duration of CP was considered in several ways. Four studies [13, 14, 19, 20] considered six months as a criterion by which to define CP. Among these, one study [19] also considered a three-month criterion within the definition. The remaining nine studies [8–12, 15–18] used three months to define the minimum duration of CP.

When formal criteria were used to define CP, three studies[8, 9, 15] reported that they used the ACR definition of CWP, while seven studies [10–12, 16–19] used the IASP definition of CP (or a close approximation), and three studies [13, 14, 20] used other or unspecified sets of criteria. With regard to other pain descriptors, the studies were heterogeneous. Pain parameters such as location, intensity, frequency, and associated disability were not investigated in all studies. Even when the same definition of CP was used as a basis for case finding, the precise phrasing of questions to assess pain parameters and the sequence in which they were posed were quite different among studies.

3.1.1.1 Studies That Used the IASP Definition of Chronic Pain

Seven studies [10–12, 16–19] were identified that provided a definition of CP equivalent to the IASP definition. Two studies that were conducted only in children [16] or the elderly [12] will be described separately. Based on the information provided by four [10, 17–19] out of the five studies (one study [11] was excluded from the calculations because of the inconsistencies in the reported figures within that study), the weighted mean prevalence of CP was 35.5%. Prevalence estimates ranged from 55.2% [19] to 11.5% [10]. The weighted mean prevalence of CP among male and female populations was 31.0% (range: 54.9–9.1%) and 39.6% (range: 55.5–13.4%), respectively.

Five primary studies [10, 11, 16, 17, 19] provided data on the number of CP sufferers with severe, limiting, or disabling CP. This information was collected in very different ways, and definitions of severity were not directly comparable among the studies. Severity of CP can be defined in terms of disability, interference, or intensity. For example, severity was measured in one study [19] according to a rating scale graded from 1 (weak) to 5 (intense), while in another [17] it was rated from Grade 0 (pain free) to Grade IV (high disability, severely limiting CP). Nonetheless, it may be assumed that a common motive underlies these definitions: the need to identify and characterize a subgroup that may require a greater amount of services within the health care system.

Prevalence of "very frequent and more intense pain" in children from the general population was 8% [16]. Pooling the results of the studies in adults, it can be said that the prevalence of severe CP (however it is defined) in the general population is approximately 11% among adults.

3.1.1.2 Studies That Used the ACR Definition of Chronic Widespread Pain

Three studies reported the prevalence of CWP in the general population [8, 9, 15]. The weighted mean prevalence of CWP was 11.8% (range: 10.1–13%). All the studies provided estimates of prevalence by gender. The weighted mean prevalence of CWP among male and female populations was 7.2% (range: 3–10.5%)

and 14.7% (range: 14.7–14.9%), respectively. Even though the variation of prevalence estimates on CWP is low, these estimates are based on only three studies.

3.1.1.3 Studies in Children and Elderly Populations

One study [16], using the IASP definition, assessed the prevalence of CP in children (up to 18 years old) to be 25%. The distribution of CP by gender was 19.5% for males and 30.4% for females. Two studies [12, 14] described the prevalence of CP in the elderly general population (those aged 65 years and older). One study [12] used the IASP definition and calculated a pooled prevalence for both genders of 50.2%. Prevalence estimates by gender were not reported. The other study [14] calculated a total prevalence of 32.9% and a distribution of CP by gender of 23.7% for males and 40.1% for females.

3.1.2 Recent Studies

An updated search conducted in 2007 identified 21 additional studies that passed an initial screen similar to that used for the Ospina and Harstall [3] review, prompting the suggestion that an update is advisable. Here we present the results for two large scale international surveys to support the contention that country/culture may be a major determiner of CP prevalence.

Breivik and colleagues [21] conducted a two stage telephone survey about chronic pain across 15 European countries and Israel. A total of 46 394 (of 67 733 contacted) participated in the first questionnaire. 19% of adults aged 18 yrs and older (with a mean age of 49.9) experienced moderate to severe CP lasting more than six months, as defined by ratings of 5 or greater on a 10 point numeric rating scale rating the severity of pain. Countries varied quite widely (range: Spain 12% to Norway 30%).

Demyttenaere and colleagues [22] asked questions about chronic neck and back pain as part of a face to face administration of the Composite International Diagnostic Interview (CIDI) used to examine mental health. The survey was carried out in 17 countries in the Americas, Europe, the Middle East and Africa, Asia, and the South Pacific. The total sample size was 85 088 and the weighted average response rate was 70.8%. A subsample of 42 697 completed additional questions including questions about the 12 month presence of, or treatment for, chronic back or neck problems. The mean age of participants by country ranged from 35.2 and 51.4. The overall prevalence was 23.0% and the weighted prevalences ranged widely across country (range: Colombia 9.7% to Ukraine 42.2%).

3.1.3 Conclusions

CP prevalence estimates reported even by the best existing studies vary widely (between 10.1 and 55.2%). The single most important recommendation for the research agenda is to conduct concurrent, prospective epidemiological studies to estimate the CP prevalence using clear-case definitions, and well-validated and reliable data collection tools. Differences exist between studies in

- populations chosen for study (e.g. general population vs. primary care),
- demographic characteristics of participants including age, sex, income, and forms and levels of co-morbidity,
- cultural contexts and regional differences,
- formal criteria used to define CP,
- the mode of data collection,
- whether CP measures are captured as primary or secondary study outcomes,
- type of questions asked including site of pain, frequency and interval of occurrence, duration, intensity, and level of functional limitations or disability,
- order of questions, including for example use of a screening question asking about any pain before intensity is queried.

So far, there is insufficient information available to offer definitive information about the effects that these differences have upon prevalence estimates though there is some support for the findings of a higher prevalence of CP among females (usually from musculoskeletal origin), an age gradient in chronic pain, and wide variation in CP prevalence between countries and cultures.

Regardless, costs associated with CP will be considerable for the health system, the individual, and society. CP and its management must be addressed by health care systems.

3.2
Chronic Pain in Canada and Alberta

The situation has been no different in Canada as study methods and findings have also varied. Millar [23] presents a portrait of chronic pain in Canada based upon responses to the 1994 National Population Health Survey. The National Population Health Survey (NPHS) is a major longitudinal health survey conducted by Statistics Canada. The NPHS is comprehensive in scope, and includes questions relevant to an examination of the prevalence of chronic pain and the characteristics of chronic pain sufferers. Two questions measure pain states. They ask about pain intensity on a three point scale and about degree of activity limitation or prevention on a four category scale. While it is emphasized that questions are being asked about individuals' usual abilities rather than short term states, there is no specific temporal criterion for duration and persistence, so the questions do not strictly meet the IASP definitions of pain. In addition, a wide variety of questions about health status are also asked on the NPHS. Among 16 989 respondents age 15 and over from across Canada, 17% reported some level of persistent pain, including about 11.9% who reported moderate or greater levels. An age gradient was discovered that ranged from 10% in the 15–24 year old group to 35% in the 75+ year old age group. Females on average reported 5% higher rates consistently at all ages.

As the level of pain increased, so too did the levels of self-reported mental distress, sleep disorders, and reported sickness or disability days.

Birse and Lander [13] report on a small random digit dial telephone survey of 410 adults with an average age of 40.8 from Edmonton, Alberta. Of these, 40.4% responded positively to the question "Do you have or have you had since the past six months any pain or discomfort?" Females were 10% more likely to respond positively than males.

Moulin and colleagues [24] surveyed 2012 adults over age 18 from the general population by random digit dial survey. They asked 13 brief questions relating to the origin, duration, and intensity of CP conditions. Of these, 29% reported continuous or intermittent pain lasting for six months or longer. Of these 88% rated the intensity of their pain as 4 or higher on a 10 point scale. Prevalence increased with age from 22% in the 18–34 year old age range to 39% in the group aged 55 and over. Females had a 4% higher prevalence rate than males.

Boulanger and colleagues [25] used the same method and questionnaires three years later to survey 1055 adults over the age of 18 from the general population. This time 25% of the sample reported continuous or intermittent pain lasting for six months or longer. Similar to the earlier survey, 88% rated the intensity of their pain as 4 or higher on a 10 point scale. Again the prevalence increased with age from 17% in the 18–34 year old age range to 33% in the group aged 55 and over. Females had a 5% higher prevalence rate than males. A larger proportion rated their pain as severe (51%) as compared to the earlier survey (32%).

Schopflocher [26] presents detailed information from a large sample of Albertans about the prevalence and descriptive epidemiology of CP derived from the NPHS. In 1996, the Alberta Health and Wellness Ministry commissioned survey responses from an additional cross-sectional sample of individuals in order to examine health status across the province's health regions. The population under study was Albertans aged 12 and over (or aged four to 11 as reported by a parent or proxy). The total sample size was 15 535.

Table 3.1 presents this data as a cross tabulation between the pain severity and the activity limitation by pain questions. The numbers in the cross tabulation table are an estimate of the number of Albertans in each pain category.

As is evident from the table, there is a strong positive association between the two pain questions. This is reflected in the table by the fact that the vast majority of individuals have similar elevations on the two dimensions. As a result, entries in Table 3.1 are presented against four background shades to distinguish four levels of chronic pain that were labeled mild, mild to moderate, moderate, and severe. Overall, 3.95% were consider to have mild CP, 2.58% to have mild to moderate CP, 2.35% to have moderate CP, and 2.28% to have severe CP. The remaining 88.8% report no pain. Note that the inclusion of children as young as four years of age has the effect of lowering the prevalence rates from what they would be in any adult population (as is typical of most studies).

Age–sex relationships were also examined. The prevalence of pain increases markedly with age (from <6% at age 18 to 18% at age 65 in males and 24% in females), and females are more likely to suffer chronic pain than are males at every age over age 18. Additionally, chronic pain levels show a marked gradient by household income quintile with the lowest quintile having rates above 30% and the highest quintile having rates around 12%.

A number of health status variables are also measured by the NPHS and are presented below according to the derived pain classification. The fundamental finding is that all of these variables show a gradation with levels of chronic pain.

Table 3.1 Estimated population by pain categories.

Activity	Severity				Total (%)
	No pain	**Mild**	**Moderate**	**Severe**	
No pain	2 284 477				2 287 447 (88.8)
Doesn't prevent activities		40 248	27 648	2941	70 836 (2.8)
Prevents few activities		33 756	47 181	3460	84 396 (3.3)
Prevents some activities		16 337	52 511	9508	78 356 (3.0)
Prevents most activities		4498	26 952	22 151	53 600 (2.1)
Total	2 287 447	94 838	154 291	38 059	2 571 666
(%)	(88.8)	(3.7)	(6.0)	(1.5)	(100)

Table 3.2 shows that self-reported Health status is markedly decreased in the presence of CP.

Table 3.2 Percentage of each pain group in each self-reported health group.

Self reported health	Pain classification				
	No pain	**Mild**	**Mild to moderate**	**Moderate**	**Severe**
Excellent	32.9	13.3	4.7	4.0	3.0
Very good	38.9	31.2	23.8	11.8	10.1
Good	23.5	37.5	37.9	32.0	19.9
Fair	3.9	16.2	25.6	39.3	31.9
Poor	0.7	1.8	8.0	12.9	35.2

Table 3.3 shows the CP tabulation with the Distress scale which is the sum of six items from the Composite International Diagnostic Interview (CIDI). Scores range from 0 to 24 with higher scores indicating more distress.

Table 3.3 Distress scale score by pain group.

	Pain classification				
	No pain	Mild	Mild to moderate	Moderate	Severe
Distress scale	2.26	3.46	3.74	4.53	6.45

Table 3.4 presents the probability of being diagnosed as a case of clinical depression in an examination by a psychiatrist, as also derived from items from the CIDI.

Table 3.4 Probability of suffering clinical depression by pain group.

	Pain classification				
	No pain	Mild	Mild to moderate	Moderate	Severe
Probability of depression	0.04	0.11	0.14	0.16	0.25

Table 3.5 shows the number of chronic diseases reported (from a list of 16) tabulated against CP levels.

Table 3.5 Percentage of each pain group reporting chronic diseases.

Number of chronic diseases	Pain classification				
	No pain	Mild	Mild to moderate	Moderate	Severe
0	48.1	21.3	10.2	8.3	4.2
1	27.7	27.1	23.4	20.7	17.7
2	13.3	25.2	26.5	20.6	22.2
3	6.2	11.7	14.5	15.4	17.1
4 or more	4.7	14.7	25.4	35.0	38.8

Table 3.6 shows the proportions reporting activity limitations tabulated against CP levels.

Table 3.6 Proportion reporting general activity limitations by pain group.

	Pain classification				
	No pain	Mild	Mild to moderate	Moderate	Severe
Activity limitations	0.10	0.27	0.53	0.64	0.85

Table 3.7 shows the proportions reporting being sedentary tabulated against CP levels.

Table 3.7 Proportion rated inactive (sedentary) by pain group.

	Pain classification				
	No pain	**Mild**	**Mild to moderate**	**Moderate**	**Severe**
Proportion inactive	0.04	0.11	0.14	0.16	0.25

Table 3.8 shows the average number of illness or disability days reported in the previous two weeks tabulated against CP levels.

Table 3.8 Average disability days in the past two weeks by pain group.

	Pain classification				
	No pain	**Mild**	**Mild to moderate**	**Moderate**	**Severe**
Disability days	0.63	1.15	2.35	3.19	6.19

Data from the Canadian Community Health Survey (CCHS) is also available from Statistics Canada. The CCHS is a major cross-sectional health survey first conducted by Statistics Canada in 2001. Over 130 000 Canadians were sampled with the intention of providing health indicators at a regional level for over 135 health regions across Canada. Although the CCHS contains fewer questions than the NPHS, the pain questions described above were included for 2001. The derived chronic pain classification was calculated for the 13 725 subjects (aged 15 and over) to whom the CCHS was administered.

Age–sex specific prevalences for individuals over age 15 were very similar between the 1996 NPHS and 2001 CCHS samples. There appears to be a constant but very slight increase in prevalence for females (about 1%) and a similar increase for males below the age of 50. The age–sex prevalence rates calculated from the 2001 CCHS were applied to population projections previously prepared by Alberta Health and Wellness [27]. If there are no changes in chronic pain incidence or recovery rates, it becomes apparent that there will be a large increase in the number of individuals suffering chronic pain in Alberta. This increase is estimated to be about 70 per cent in the next 25 years. Part will be due to a population increase in Alberta, but the primary impact will come from the aging of the population.

3.2.1
Conclusions

While estimated prevalence levels also vary across different Canadian general population surveys, it is clear that persistent pain is a problem affecting a significant minority of adult Canadians. Within Canada, females report higher levels of persistent pain than do males, and older persons have considerably higher prevalences than younger ones. As the Canadian population continues to age over the next two decades, the total number of CP sufferers will continue to grow. It is also apparent that CP is associated with decreased general health, increased sick and disability days and inactivity, and greater distress and a greater likelihood of suffering depression.

Given that the burdens associated with CP are considerable for the health system, the effective management of CP will become an increasingly urgent priority for health policy makers in Canada and elsewhere.

References

1 Verhaak, P.F.M., Kerssens, J.J., Dekker, J., Sorbi, M.J. and Bensing, J.M. (1998) Prevalence of chronic benign pain disorder among adults: a review of the literature. *Pain*, **77**, 231–9.
2 Nickel, R. and Raspe, H.H. (2001) Chronischer schmerz: epidemiologie und. inanspruchnahme. *Nervenarzt*, **72**, 897–906.
3 Ospina, M. and Harstall, C. (2002) Prevalence of chronic pain: an overview. Alberta Heritage Foundation for Medical Research, Health Technology Assessment, 28th Report, Edmonton, Canada.
4 Loney, P.L., Chambers, L.W., Bennett, K.L., Roberts, J.G. and Stratford, P.W. (1998) Critical appraisal of the health research literature: prevalence or incidence of a health problem. CDIC Final 19th Report. Health Canada, Ottawa, Canada.
5 Loney, P.L. and Stratford, P.W. (1999) The prevalence of low back pain in adults: a methodological review of the literature. *Physical Therapy*, **79** (4), 384–96.
6 International Association for the Study of Pain (1986) Classification of chronic pain. Descriptions of chronic pain syndromes and definitions of pain terms. *Pain Supplement*, **3**, S1–225.

7 Wolfe, F., Smythe, H.A., Yunus, M.B., Bennett, R.M., Bombardier, C. and Goldenberg, D.L. (1990) The American College of Rheumatology 1990 criteria for the classification of fibromyalgia: report of the multicenter criteria committee. *Arthritis and Rheumatism*, **33**, 160–72.
8 Croft, P., Rigby, A.S., Boswell, R., Schollum, J. and Silman, A. (1993) The prevalence of chronic widespread pain in the general population. *The Journal of Rheumatology*, **20** (4), 710–13.
9 MacFarlane, G.J., Morris, S., Hunt, I.M., Bejamin, S., McBeth, J., Papageourgiou, A.C. *et al.* (1999) Chronic widespread pain in the community: the influence of psychological symptoms and mental disorder on healthcare seeking behavior. *The Journal of Rheumatology*, **26** (2), 413–19.
10 Bowsher, D., Rigge, M. and Sopp, L. (1991) Prevalence of chronic pain in the British population: a telephone survey of 1037 households. *The Pain Clinic*, **4** (4), 223–30.
11 Blyth, F.M., March, L.M., Brnabic, A.J.M., Jorm, L.R., Williamson, M. and Cousins, M.J. (2001) Chronic pain in Australia: a prevalence study. *Pain*, **89** (2–3), 127–34.
12 Helme, R.D. and Gibson, S.J. (1997) *Proceedings of the 8th World Congress on*

Pain, Progress in Pain Research and Management, Vol. **8** (eds T.S. Jensen *et al.*), IASP Press, Seattle WA, pp. 919–44.

13 Birse, E.M. and Lander, J. (1998) Prevalence of chronic pain. *Canadian Journal of Public Health*, **89** (2), 129–31.

14 Brochet, B., Michel, P., Barberger-Gateau, P. and Dartigues, J.-F. (1998) Population-based study of pain in elderly people: a descriptive survey. *Age and Ageing*, **27** (3), 279–84.

15 Buskila, D., Abramov, G., Biton, A. and Neumann, L. (2000) The prevalence of pain complaints in a general population in Israel and its implications for utilization of health services. *The Journal of Rheumatology*, **27** (6), 1521–5.

16 Perquin, C.W., Hazebroek-Kampschreur, A.A., Hunfeld, J.A., Bohnen, A.M., van Suijlekom-Smit, L.W. *et al.* (2000) Pain in children and adolescents: a common experience. *Pain*, **87** (1), 51–8.

17 Elliott, A.M., Smith, B.H., Penny, K.I., Smith, W.C. and Chambers, W.A. (1999) The epidemiology of chronic pain in the community. *Lancet*, **354** (9186), 1248–52.

18 Catala, E., Reig, E., Artes, M., Aliaga, L., Lopez, J.S. and Segu, J.L. (2002) Prevalence of pain in the Spanish population: telephone survey in 5000 homes. *European Journal of Pain*, **6** (2), 133–40.

19 Andersson, H.I., Ejlertsson, G., Leden, I. and Rosenberg, C. (1993) Chronic pain in a geographically defined general population: studies of differences in age, gender, social class and pain localization. *The Clinical Journal of Pain*, **9** (174), 182.

20 Gureje, O., Von Korff, M., Smion, G.E. and Gater, R. (1998) Persistent pain and well-being. A World Health Organization study in primary care. *The Journal of the American Medical Association*, **280** (2), 147–51.

21 Brevik, H., Collett, B., Ventafridda, V., Cohen, R. and Gallacher, D. (2006) Survey of chronic pain in Europe: prevalence, impact on daily life, and treatment. *European Journal of Pain*, **10**, 287–333.

22 Demyttenaere, K., Bruffaerts, R., Lee, S., Posada-Villa, J., Kovess, V., Angermeyer, M.C., Levinson, D., Girolamo, G. et al. (2007) Mental disorders among persons with chronic back or neck pain: results from the world mental health surveys. *Pain*, **129**, 332–42.

23 Millar, W.J. and Chronic, P. (1996) *Health Reports*, **7** (4), 47–53.

24 Moulin, D.E., Clark, A.J., Speechley, M. and Morley-Forster, M.K. (2002) Chronic Pain in Canada – prevalence, treatment, impact and the role of opioid analgesia. *Pain Research and Management*, **7**, 179–84.

25 Boulanger, A., Clark, A.J., Squire, P., Cui, E. and Horbay, G.L.A. (2004) Chronic Pain in Canada: have we improved our management of chronic noncancer pain? *Pain Research and Management*, **12** (1), 39–47.

26 Schopflocher, D.P. (2003) Chronic pain in alberta: a portrait from the 1996 national population health survey and the 2001 canadian community health survey. Alberta Health and Wellness Public Report, Edmonton, Canada.

27 Alberta Health and Wellness (2000) Population projections for alberta and its health regions: 2000–2030. Alberta Health and Wellness Public Report, Edmonton, Canada.

4
The Economics of Chronic Pain

Ceri J. Phillips and Donald Schopflocher

- Chronic pain leads to dramatic increases in health care utilization and therefore direct costs to the health care system. Canadian estimates are an additional Can$3500 per chronic pain sufferer per year, resulting in additional direct system costs over Can$400 000 000/year and rising.

- Societal costs, especially those resulting from work absence and disability, dwarf these direct costs and additional costs to the individual such as those for uninsured treatments, informal care, or intangibles related to loss of quality of life are substantial but difficult to estimate.

- The means to affect a positive impact on the management of patients with chronic pain lies in analyses of effectiveness, efficiency, and equity applied to high quality evidence across the whole health care system.

4.1
Introduction

Pain is a major clinical, social, and economic problem that has exercised generations of healthcare professionals attempting to provide relief and reduce the suffering caused by pain. The prevalence of chronic pain, highlighted in the previous chapter, is high and presents a challenge to those involved in the provision of services. Chronic pain is one of the most widespread and difficult problems the medical community has to face [1], and other symptoms, such as depression, anxiety, physical dysfunction and social isolation, often present with it [2].

The fundamental economic problem confronting us all is that while we have insatiable wants and desires, we only have limited resources (time, energy, expertise and money) at our disposal to satisfy them. Economics is founded on the premise that there will never be enough resources to completely satisfy human desires and the use of resources in one activity means that they cannot be used elsewhere [3]. This problem is particularly evident in relation to health care. The extent of the gap which exists between the demand for services and treatments

Chronic Pain: A Health Policy Perspective
Edited by S. Rashiq, D. Schopflocher, P. Taenzer, and E. Jonsson
Copyright © 2008 WILEY-VCH Verlag GmbH & Co. KGaA, Weinheim
ISBN: 978-3-527-32382-1

and the level of resources available to meet them continues to frustrate politicians, professionals and policy makers alike. The range of healthcare systems that have emerged over time have all had to grapple with the dilemma of allocating limited resources available for the provision of healthcare services in such a way as to maximize the benefits for society; this against a background of increasing expectations within the population as to what can be delivered by healthcare services, continuing advancements in health technology and medical science, and the increasing health needs and demands of an aging population.

The aims of this chapter therefore are,

- to assess the economic impact of chronic pain on the individual, the health system and more widely on the economy; and
- to advocate a coherent and integrated approach to the management of pain, based on the notions of effectiveness, efficiency and equity of service provision.

4.2
Direct Costs of Chronic Pain

The extent of the chronic pain problem poses a significant economic burden for patients, the health services and society as a whole in Canada. In a study that estimated the prevalence of chronic pain in Alberta from the National Population Health Survey (NPHS) [4], described in the previous chapter, various self report measures of health system utilization were also collected. Compared to individuals reporting no pain, those suffering from severe Chronic Pain reported:

- 4 times higher rates of hospitalization in the previous year (24%),
- 4 times the number of consultations with a medical professional in the past 12 months (13.4%),
- 5 times higher rates of unmet health care needs (29%),
- 2 times higher rates of consultation of an alternative care provider (15%),
- 2 times higher rates of attending self-help groups (8%),
- 21% higher proportion reporting the use of over-the-counter pain relievers (88%),
- 6 times higher rates of using narcotic medication (31%),
- 4 times the average number of medications taken (2.9).

In Canada administrative data is collected on each patient visit to hospital, outpatient clinic, or doctor's office. This information is used for payment of medical professionals as well as for statistical purposes. Unfortunately, administrative data sources are unable to provide direct estimates of the prevalence of chronic pain because the International Classification of Diseases ninth Revision (ICD-9-CM) diagnostic system is not organized by symptoms such as pain. Furthermore, chronic pain can be a symptom of a large number of specific diseases (such as arthritis, diabetes, heart disease, endometriosis and hundreds more). As a result, the prevalence of chronic pain is typically estimated from health surveys. However, it was possible to link the health survey responses from the NPHS to administrative data for three years prior to the survey, the year of the survey, and the year after the survey in order to offer partial confirmation of the self-reported measures. The results from this analysis showed that

- in every year there was a gradient in the number of confirmed medical consultations according to the level of chronic pain, with those with the most severe chronic pain averaging 2.5 times as many consultations as those with no pain over the five years;

- in every year there was a gradient in the number of confirmed hospitalizations according to the level of chronic pain, with those with the most severe chronic pain averaging 3.5 times as many consultations as those with no pain over the five years; and

- there was a similar effect for the number of hospital days, with the most severe chronic pain averaging five times as many days as those with no pain over the five years.

Using these data, and average cost figures derived from Alberta Ministry of Health and Wellness, it was estimated that each individual suffering from severe chronic pain costs an additional $3500 (in Canadian year 2000 dollars) per year in direct healthcare costs. Using previous prevalence estimates, as well as Alberta population projections, this translates into a predicted yearly burden of over Cdn$400 million (rising to over Cdn$700 million by 2025) in the absence of effective intervention for individuals with moderate or severe chronic pain.

In other studies, Coyte and colleagues [5] estimated the total cost of musculo-skeletal disorders in Canada in 1994 at US$19.2 billion, with direct medical costs of the total and productivity losses accounting for 71%, while Rapaport and colleagues [6] demonstrated that back pain and arthritis and rheumatism were associated with the greatest impact on healthcare resource utilization.

4.3
Indirect Costs of Chronic Pain

Despite these high levels of resource utilization, there remains what has been called the "crisis of inadequately treated pain" [7], with misconceptions and ignorance among professionals regarding pain, its treatment and effectiveness [8, 9]. At the same time many patients are not being treated, are receiving sub-optimal care, or are not satisfied with the treatment offered [9–11]. In fact, a significant number of people with chronic pain may not consult anyone about their condition, or may choose to self-medicate. A survey of nearly 6000 people across Europe [9] found that up to 27% of respondents had never sought medical help for their pain, and at least 38% of this group were in constant or daily pain. The extent to which people took non-prescription drugs varied between 23% and 59%, while a conservative estimate of over-the-counter medication relating to back pain amounted to US$44 million [12].

As Coyte and colleagues [5] argued, the direct healthcare costs are minor in comparison with the impact on the economy resulting from the consequences of pain. Musculoskeletal conditions are a major cause of pain and disability across the world and among the most common medical cause of long-term sickness absence [13]. The effect of pain, and in particular pain exacerbations, was evident in a study conducted in the US in 2003/04 [14], where it was estimated that the impact of arthritis on lost productive work time amounted to US$7.11 billion, but with 66% of this attributed to the 38% of workers with pain exacerbations.

As well as its impact on absenteeism, pain also has a major impact on worker productivity [15]. For example, a US study found that common pain conditions resulted in lost productivity (also referred to as presenteeism [16]) amounting to $61 billion per year, of which 77% was explained by reduced performance and not work absence [17]. As well as the impact on absenteeism and presenteeism, the odds of quitting one's job because of ill health have been shown to be seven times higher among people with chronic pain problems [12]. The impact of work loss and long-term sickness absence are highly significant and result in several negative outcomes, including poorer recovery and rehabilitation, increased risks of poverty, physical and mental health problems and social exclusion [18]. The longer someone is out of work, the more distant they become from the labor market, the more difficult it is for them to return to work [19], and the greater the impact on overall costs [20, 21]. The exacerbation of problems that result from prolonged periods out of work highlights the need for the development of early and effective interventions to support people to remain at work or to return to work as quickly as possible.

Despite the fact that the economic impact of pain is substantial and imposes a greater economic burden than many other diseases [22], decision makers have tended to concentrate attention on a very minor component of the cost burden, namely prescription drug costs, because they are easy to measure and are therefore an obvious target for restrictions [23]. The acquisition costs of medication are but one very small and insignificant part of a complex and expensive jigsaw, and attempts to focus attention and energies on restricting expenditure in this one area fail to recognize the wide-reaching implications of effective pain management. People whose pain can be effectively managed are less likely to be receiving long-term illness benefits, to be less productive, and to be absent from school and further education. Investment in effective interventions and programs which deliver relief from pain and suffering and reductions in disability levels will generate both economic and social returns that more than re-pay the original investment. In order to develop such a mode of thinking it is essential that "policy makers are fully aware of all aspects associated with the costs of pain and its management," [24] as outlined in the list below [25].

- costs of interventions and therapies for treating pain and securing pain relief (e.g. drug costs and staff costs);
- costs which are incurred as a result of ineffective interventions being provided (e.g. costs of additional GP consultations);

- costs to health service and patients and their families due to lack of appropriate facilities within locality (e.g. costs of accessing alternative therapies);
- costs resulting from inappropriate self-medication and treatment by patients (e.g. costs of treating overdoses);
- costs of treating and preventing adverse events which arise as a result of prescribing decisions (e.g. costs of gastrointestinal bleeds);
- costs of disability claims resulting from people's inability to work;
- costs to the economy of reductions in productivity and absenteeism;
- costs of providing social care and support to people suffering with pain (e.g. costs of home care and respite care);
- costs of informal care provided by families (e.g. loss of earnings);
- costs of intangibles associated with deterioration in the quality of life of patients and their families.

4.4
The Wider Impact of Chronic Pain

The estimates of the burden associated with pain fail to do justice to the extent of suffering and reduced quality of life experienced by patients. It is not merely the economic impact, but rather the tremendous human suffering resulting from chronic pain that warrants pain relief being regarded as a universal human right [8]. Pain affects all of us to varying degrees. For some it may be the briefest of acute sensations, but for others it is a permanent feature of life and has a profound impact on the quality of life. Without adequate treatment, these people are often unable to work or even sometimes to carry out the simplest of tasks. This often leads to problems such as depression or stress which then compound the problems caused by the physical pain. Using the World Health Organization estimate of prevalence of chronic pain [26] of 22%, there would be 2400 million chronic pain days in Canada, which translates into over 250 000 working lives. These "pain days" have a profound impact on patients' quality of life [27–30] and that of other family members, as adjustments are made to adapt to the chronic pain problem [31–35].

4.5
Pain Management Strategies

The burden of suffering that pain imposes on individuals, and the enormous costs which societies have to bear as a result, clearly demonstrate that policy makers and health care decision-makers need to adopt a broad, strategic and coherent perspective in determining issues relating to service provision and resource allocation. Fragmented, budgetary-based interventions and programs based on, at best, inadequate evidence do little to alleviate chronic pain and suffering, and also deprive patients of those services that would have a positive impact. It has therefore been advocated that decisions relating to patient management are made with regard to "the three Es" – effectiveness, efficiency, and equity [24].

4.5.1
Effectiveness

The evidence-base for the effectiveness of interventions and strategies in managing pain is large [36–38] although the issue of what works, where and when remains inconclusive [39–41]. While the evidence base is continuously being updated, incorporating potential new therapeutic areas, interventions and management programs [25, 42–46], questions remain relating to the quality of studies and their relevance for policy and practice [47, 48]. Further, the nature and extent of adverse events associated with some interventions have also resulted in debate as to what actually constitutes "effectiveness" when efficacy and safety are combined. For example, a systematic review of over 5000 patients confirmed that most patients would experience at least one adverse event resulting from opioid use in chronic pain, and that substantial minorities would experience common adverse events of dry mouth, nausea, and constipation, and would not continue treatment because of intolerable adverse events [49].

4.5.2
Efficiency

The term "efficiency" is often misunderstood and confused with the term "economy" and has tended to be used to describe an activity being performed at a given rate at the lowest cost. However, this is too narrow a definition and the economic concept of efficiency embraces both inputs (costs) and outputs or outcomes (benefits). In the context of pain management, it is appropriate to assess the efficiency of interventions from the perspective of cost effectiveness. Cost effectiveness is "designed to compare the costs and benefits of a healthcare intervention to assess whether it is worth doing relative to the resources available" [3] and the need to ensure that interventions are provided which are efficient means that there are three factors to consider, namely,

- to maximise the reduction in pain;
- to minimise the overall cost;
- to minimise the impact of adverse events.

The problem of focusing on acquisition costs was highlighted earlier. It is essential to realize that the cost of treatment is not simply the costs of drugs or medical and nursing time but the total costs of providing the treatment [24]. For example the costs of dealing with adverse events are not insubstantial. In Canada it has been reported that for each dollar spent on non-steroidial anti-inflamatory drugs (NSAIDs), an additional Cdn$0.66 was spent on their side effects [50]. Clinicians, other healthcare professionals and policy makers must carefully weigh up the relative risk and benefit in deciding what are the most effective and efficient approaches to employ in pain management. It has been argued that less emphasis on technological solutions and a shift towards a biopsychosocial model would be an efficient use of limited resources in pain management–"every study published shows that aggressive, multidisciplinary pain management for the most disabled group of chronic patients will produce significant cost savings, to say nothing of the human suffering that will be alleviated" [51]. However, even this claim may be called into question. For example, one systematic review of the effectiveness of multidisciplinary pain treatment of chronic non-malignant pain patients in terms of economic outcomes concluded that "due to serious methodological problems in study designs and outcome measures, it is not possible to draw conclusions on clinical or economical effectiveness" [52].

4.5.3
Equity

The notion of equity has received considerable attention in the literature, as have discussions relating to schemes and policies designed to erode inequities and inequalities in health status and the availability and accessibility of health services.

The availability and accessibility of good quality services for all patients is highly desirable and should form part of the decision-making process. However, a survey of 105 hospitals from 17 European countries showed that only 34% of hospitals had an organized acute pain service, very few hospitals used quality assurance measures and over 50% of anesthesiologists were dissatisfied with post-operative pain management on surgical wards [53]. Similarly, it has been argued that, in selected populations, patients managed through multidisciplinary programs have lower costs, return to work more frequently and experience greater pain control than those who are managed with more traditional methods [43]. However, the availability of such facilities is sketchy and some populations have "no local access to services for patients with long-lasting pain"–a situation likely to deteriorate, as demographic factors intensify the demand for chronic pain services for the fore-seeable future [54].

4.6
Summary and Conclusion

The aims of this chapter therefore have been to assess the economic impact of chronic pain on the health service and more widely on the Canadian economy; and propose a coherent and integrated approach to the management of pain, based on the notions of effectiveness, efficiency, and equity of service provision.

It is clear that chronic pain imposes an enormous burden on individuals, their families and society as a whole. This provides a sound rationale for greater emphasis to be given to pain management. Chronic pain is a complex syndrome that demands a broad strategic perspective for decision-making. Analgesic therapy is only one part of the treatment. It is increasingly evident that, in relation to chronic pain, other factors may contribute to the intensity and persistence of the pain [55] and simple pharmacological interventions, in themselves, are insufficient and need to be located within the context of an overall pain management strategy geared to the needs of the individual patient and constructed on the basis of what we have suggested – namely effectiveness, efficiency and equity. In addition, it has been strongly advocated that society has an obligation to reduce levels of pain and restore normal functioning, based upon both moral principles and economic reality [51].

Yet it is very evident that pain is not given the attention it warrants, based on prevalence rates, its economic cost and the detrimental effects it has on quality of life. Better healthcare does not necessarily require additional resources. Whole systems thinking, based on good quality evidence, rather than an aggregation of narrow, budgetary-focused organizations pursuing their own agendas without regard for the wider perspective, would have a major positive impact on the management of patients with chronic pain. More work is needed to develop a broader, strategic, whole systems agenda in pain management. High quality evidence and health economics techniques and approaches are essential components of this particular agenda. Together they can provide the tools which decision makers can utilize in the drive for reductions in pain and better health and healthcare for our respective communities and societies.

References

1 Latham, J. and Davis, B.D. (1994) The socio-economic impact of chronic pain. *Disability and Rehabilitation*, **16**, 39–44.

2 Rudy, T.E., Kerns, R.D. and Turk, D.C. (1988) Chronic pain and depression, toward a cognitive behavioural model. *Pain*, **35**, 129–40.

3 Phillips, C.J. (2005) *Health Economics, an Introduction for Healthcare Professionals*, Blackwell Publishing, Oxford.

4 Schopflocher, D.P. (2003) Chronic pain in Alberta, a portrait from the 1996 national population health survey and the 2001 Canadian community health survey. Alberta Health and Wellness Public Report, Edmonton, Alberta.

5 Coyte, P.C., Asche, C.V., Croxford, R. and Chan, B. (1998) The economic cost of musculoskeletal disorders in Canada. *Arthritis Care and Research*, **11** (5), 315–25.

6 Rapoport, J., Jacobs, P., Bell, N.R. and Klarenbach, S. (2004) Refining the measurement of the economic burden of chronic diseases in Canada. *Chronic Diseases in Canada*, **25** (1), 13–21.

7 Fishman, S.M., Gallagher, R.M., Carr, D.B. and Sullivan, L.W. (2004) The case for pain medicine. *Pain Medicine*, **5**, 281–86.

8 Cousins, M.J. (2004) Pain relief, a universal human right. *Pain*, **112**, 1–4.

9 Woolf, A.D., Zeidler, H., Hagland, U. *et al.* (2004) Musculoskeletal pain in Europe, its impact and a comparison of population and medical perceptions of treatment in eight European countries. *Annals of Rheumatic Diseases*, **63**, 342–7.

10 Breivik, H., Collett, B., Ventafridda, V., Cohen, R. and Gallacher, D. (2006) Survey of chronic pain in Europe, prevalence, impact on daily life, and treatment. *European Journal of Pain*, **10**, 287–333.

11 Eriksen, J., Jensen, M.K., Sjogren, P., Ekholm, O. and Rasmussen, N.K. (2003) Epidemiology of chronic non-malignant pain in Denmark. *Pain*, **106**, 221–8.

12 Walsh, K., Cruddas, M. and Coggon, D. (1993) Low back pain in eight areas of Britain. *Journal of Epidemiology and Community Health*, **46**, 227–30.

13 Woolf, A.D. and Pfleger, B. (2003) Burden of major musculoskeletal conditions. *Bulletin of the World Health Organization*, **81**, 646–56.

14 Ricci, J.A., Stewart, W.F., Chee, E., Leotta, C., Foley, K. and Hochberg, M.C. (2005) Pain exacerbation as a major source of lost productive time in US workers with arthritis. *Arthritis and Rheumatism*, **53**, 673–81.

15 Burton, W.N., Conti, D.J., Chen, C.Y., Schultz, A.B. and Edington, D.W. (1999) The role of health risk factors and disease on worker productivity. *Journal of Occupational and Environmental Medicine*, **41**, 863–77.

16 Allen, H., Hubbard, D. and Sullivan, S. (2005) The burden of pain on employee health and productivity at a major provider of business services. *Journal of Occupational and Environmental Medicine*, **47**, 658–70.

17 Stewart, W.F., Ricci, J.A., Chee, E., Morganstein, D. and Lipton, R. (2003) Lost productive time and cost due to common pain conditions in the US workforce. *The Journal of the American Medical Association*, **290**, 2443–54.

18 Gannon, B. and Nolan, B. (2007) The impact of disability transitions on social inclusion. *Social Science and Medicine*, **64**, 1425–37.

19 Waddell, G. and Aylward, M. (2005) *The Scientific and Conceptual Basis of Incapacity Benefits*, The Stationery Office, London.

20 Baldwin, M.L. (2004) Reducing the costs of work-related musculoskeletal disorders, targeting strategies to chronic disability cases. *Journal of Electromyography and Kinesiology*, **14**, 33–41.

21 Williams, D.A., Feuerstein, M., Durbin, D. and Pezzullo, J. (1998) Health care and indemnity costs across the natural history of disability in occupational low back pain. *Spine*, **23**, 2329–36.

22 Maniadakis, N. and Gray, A. (2000) The economic burden of back pain in the UK. *Pain*, **84**, 95–103.

23 Smith, I. (2001) Cost considerations in the use of anaesthetic drugs. *Pharmacoeconomics*, **19**, 469–81.

24 Phillips, C.J. (2001) The real cost of pain management. *Anaesthesia*, **56**, 1031–3.

25 Buchbinder, R., Jolley, D. and Wyatt, M. (2001) Population based intervention to change back pain beliefs and disability, three part evaluation. *British Medical Journal*, **322**, 1516–20.

26 Gureje, O., Von Korff, M., Simon, G. *et al.* (1998) Persistent pain and well-being, a World Health Organization study in primary care. *The Journal of the American Medical Association*, **280**, 147–51.

27 Sprangers, M.A.G., de Regt, E.B., Andries, F. *et al.* (2000) Which chronic conditions are associated with a better or poorer quality of life? *Journal of Clinical Epidemiology*, **53**, 895–97.

28 Kerr, S., Fairbrother, G., Crawford, M., Hogg, M., Fairbrother, D. and Khor, K.E. (2004) Patient characteristics and quality of life among a sample of Australian chronic pain clinic attendees. *Internal Medicine Journal*. **34**, 403–9.

29 Ellliott, T.E., Reiner, C.M. and Palcher, J.A. (2003) Chronic pain, depression and quality of life, correlations and predictive value of the SF-36. *Pain Medicine*, **4**, 331–9.

30 Reginster, J.Y. (2002) The prevalence and burden of arthritis. *Rheumatology*, **41** (Suppl. 1), 3–6.

31 Flor, H., Turk, D.C. and Scholz, O.B. (1987) Impact of chronic pain on the spouse, marital, emotional and physical consequences. *Journal of Psychosomatic Research*, **31**, 63–71.

32 Turk, D.C., Flor, H. and Rudy, T.E. (1987) Pain and families. I. Etiology, maintenance and psychological impact. *Pain*, **30**, 3–27.

33 Schwartz, L., Slater, M.A. and Birchler, G.R. (1996) The role of pain behaviors in the modulation of marital conflict in chronic pain couples. *Pain*, **65**, 227–33.

34 Schwartz, L., Slater, M.A., Birchler, G.R. *et al.* (1991) Depression in spouses of chronic pain patients, the role of patient pain and anger, and marital satisfaction. *Pain*, **44**, 61–7.

35 Kemler, M.A. and Furnée, C.A. (2002) The impact of chronic pain on life in the household. *Journal of Pain and Symptom Management*, **23**, 433–41.

36 McQuay, H.J. and Moore, R.A. (1998) *An Evidence Based Resource for Pain Relief*, Oxford University Press, Oxford.

37 Moore, A., Edwards, J., Barden, J. and McQuay, H. (2003) *Bandolier's Little Book of Pain*, Oxford University Press, Oxford.

38 Tramèr, M.R. (ed.) (2003) *Evidence-Based Resource in Anaesthesia and Analgesia*, BMJ Books, London.

39 Eccleston, C., Yorke, L., Morley, S. *et al.* (2003) Psychological therapies for the management of chronic and recurrent pain in children and adolescents. *Cochrane Database of Systematic Reviews*, CD003968.

40 Karjalainen, K., Malmivara, A., van Tulder, M. *et al.* (2003) Multidisciplinary biopsychological rehabilitation for subacute low back pain among working age adults. *Cochrane Database of Systematic Reviews*, CD002193.

41 Karjalainen, K., Malmivara, A., van Tulder, M. *et al.* (2003) Multidisciplinary biopsychological rehabilitation for neck and shoulder pain among working age adults. *Cochrane Database of Systematic Reviews*, CD002194.

42 Campbell, F.A., Tramer, M.R., Carroll, D. *et al.* (2001) Are cannabinoids an effective and safe treatment option in the management of pain? A qualitative systematic review. *British Medical Journal*, **323**, 1–6.

43 Guzmán, J., Esmail, R., Karjalainen, K. *et al.* (2001) Multidisciplinary rehabilitation for chronic low back pain, systematic review. *British Medical Journal*, **322**, 1511–16.

44 Loisel, P., Lemaire, J., Poitras, S. *et al.* (2002) Cost-benefit and cost-effectiveness analysis of a disability prevention model for back pain management, a six-year follow up study. *Occupational and Environmental Medicine*, **59**, 807–15.

45 Stadler, M., Schlander, M., Braeckman, M., Nguyen, T. and Boogaerts, J.G. (2004) A cost-utility and cost-effectiveness analysis of an acute pain service. *Journal of Clinical Anesthesia*, **16**, 159–67.

46 UK Beam Trial Team (2004) United Kingdom back pain exercise and manipulation (UK BEAM) randomised trial, effectiveness of physical treatments for back pain in primary care. *British Medical Journal*, Vol. 329. doi,10.1136/bmj.38282.669225.AE (published 29 November 2004)

47 Moore, R.A. (2001) Pain and systematic reviews. *Acta Anaesthesiologica Scandinavica*, **45**, 1136–9.

48 Eriksen, J., Sjogren, P., Bruera, E., Ekholm, O. and Rasmussen, N.K. (2006) Critical issues on opioids in chronic non-cancer pain, an epidemiological study. *Pain*, **125**(1–2): 172–9.

49 Moore, R.A. and McQuay, H.J. (2005) Prevalence of opioid adverse events in chronic non-malignant pain, systematic review of randomized trials of oral opioids. *Arthritis Research and Therapy*, **7**, R1046–51 (DOI 10.1186/ar1782).

50 Rahme, E., Joseph, L., Kong, S.X. *et al.* (2000) Gastrointestinal health care resource use and costs associated with nonsteroidal anti-inflammatory drugs versus acetaminophen, retrospective cohort study of an elderly population. *Arthritis and Rheumatism*, **43**, 917–24.

51 Loeser, J.D. (1999) Economic implications of pain management. *Acta Anaesthesiologica Scandinavica*, **43**, 957–59.

52 Thomsen, A.B., Sørensen, J., Sjøgren, P. *et al.* (2001) Economic evaluation of multidisciplinary pain management in chronic pain patients, a qualitative systematic review. *Journal of Pain and Symptom Management*, **22**, 688–98.

53 Rawal, N., Allvin, R. and EuroPain Acute Pain Working Party (1998) Acute pain services in Europe, a 17-nation survey of 105 hospitals. *European Journal of Anaesthesiology*, **15**, 354–63.

54 McQuay, H.J., Moore, R.A., Eccleston, C. *et al.* (1997) Systematic review of outpatient services for chronic patient control. *Health Technology Assessment*, **1**(6).

55 Loeser, J.D. and Melzack, R. (1999) Pain, an overview. *Lancet*, **353**, 1607–9.

5
Ethical Considerations in the Treatment of Chronic Pain

Dick Sobsey and Derek Truscott

- The treatment of chronic pain is difficult from the ethical point of view because patient and caregiver are sometimes at odds.

- Pain practitioners wrestle with ethical issues daily, often resorting to fundamental ethics principles on a case-by-case basis.

- Fundamental ethical challenges include undertreatment, the balance between treatment and overtreatment, the subjective nature of the experience of pain, and fears about adverse consequences if opioids are prescribed.

5.1
Scope and Context

Most of us want to do what is right in both the scientific and moral sense. In the scientific sense, doing what is right means acting in a manner most likely to result in a desirable outcome. In the moral sense, doing what is right means determining which outcome is most desirable. Science provides the means and knowledge to determine the probability of outcomes. Ethics provides the means and values to guide our application of science.

Isolated from our values, life is not logically preferable to death, pleasure not logically more desirable than suffering, freedom not logically better than enslavement. Both values and logic are fundamental components of healthcare practice because the health outcomes desired are predicated on values. Science and ethics are so seamlessly intertwined in every healthcare decision that the boundary between them is often difficult to identify, and consideration of ethics tends to be overshadowed by a focus on outcomes. We feel some degree of anxiety when we are unsure that we are taking the best course of action or even when we begin to doubt that any course of action is right.

Chronic pain represents a particularly troubling health circumstance from an ethical perspective. Because pain is inherently subjective, the experience of pain is influenced by the meaning and values attributed to it by the person in pain [1]

Chronic Pain: A Health Policy Perspective
Edited by S. Rashiq, D. Schopflocher, P. Taenzer, and E. Jonsson
Copyright © 2008 WILEY-VCH Verlag GmbH & Co. KGaA, Weinheim
ISBN: 978-3-527-32382-1

and the observer [2]. This may result in ambivalence toward pain on the part of healthcare professionals and society at large, whereby medicine is seen as having a duty to relieve the pain and suffering of patients, while patients are expected to be willing and able to endure pain without complaint [2]. Thus, the values of healthcare staff and policy makers deeply affect the quality and quantity of care provided [3, 4], making ethics of central concern when considering the health policy implications of chronic pain.

The very nature of chronic pain – pain that persists disproportionate to objective disease – puts the person who suffers from it at odds with the healthcare system. Chronic pain is often undertreated because our healthcare system is focused on diagnosis and cure of disease [5]. Yet objective tissue damage is neither necessary nor sufficient for chronic pain. In many of the most prevalent chronic pain syndromes there are no reliable clinical test abnormalities [6]. The curative model of healthcare thus creates an unreceptive environment for care of those with chronic pain [2]. Furthermore, the treatment of chronic pain is often provided through opioids, which can be associated with undesirable outcomes for the patient or others. These may include impairment of physiological or psychological function, masking of symptoms that might assist in diagnosis and treatment of other conditions, promoting addiction, and a host of other known or unanticipated complications of pain treatment. While the mandate to lessen suffering by alleviating or managing pain is fundamental and clear, the ethical issues that arise in the application of this mandate can be numerous and complex. These complexities are particularly prominent in the management of chronic pain because pain that is easily treatable without complications is rarely referred to a specialist.

There is no simple set of values or rules that can be applied to all of the circumstances that arise. There is rarely a single approach or course of action that is clearly superior to all of the other alternatives. Often several basic principles of ethics must be considered in a given situation, and these principles when individually applied appear to suggest conflicting courses of action. As a result, those who seek to treat chronic pain wrestle with ethical issues on a daily basis. While some decisions are simple and easy, many involve choosing among imperfect alternatives with no clearly superior course of action. Some degree of moral distress is desirable, if not inevitable, but when moral discomfort is too high, the resulting stress can be harmful to practitioners and can paralyze the practice. When it is too low or absent, the quality of practice fails to improve as a result of experience and treatment decisions are often inadequately considered. Ethics cannot eliminate this moral distress, but ethics can help us to live with it. Ethics rarely provide simple answers to our moral dilemmas, but they can help us to better understand them and provide a framework for addressing them.

This chapter reviews some essential ethical principles and suggests how they can be applied to health policies that impact persons who suffer from chronic pain through discussion of some examples. While it does not offer simple solutions to difficult problems, it does attempt to offer some practical suggestions for approaching ethical dilemmas in chronic pain.

5.2
Overview of the Literature on Chronic Pain and Ethics

Each professional discipline has its own code of ethics, but the underlying principles are similar [7, 8]. Various authors differ in the number of principles that they discuss and the manner in which they present them, but these differences are generally minor.

5.2.1
Beneficence

The principle of beneficence places a high value on acting for the benefit of others. The primary focus is benefit to patient, but healthcare decisions may also provide benefits to family members, employers, a variety of others, or even society as a whole, provided that they are also in the best interest of the patient. The most obvious and direct benefit of pain management is the alleviation of suffering, and other benefits may include improved physical, psychological, and social function, reduced demands on caregivers, or more productive work life. The ethical obligation to manage pain and relieve the patient's suffering is at the core of the health professional's commitment and a fundamental responsibility [2, 9]. Yet chronic pain tends to be undertreated as a result of focus on diagnosis and cure of disease [5, 10] and fear of opioid addiction [6, 11] and regulatory action [12].

Although efforts to improve treatment of chronic non-malignant pain have focused on increasing access to opioids, this movement has not been matched by attempts to increase access to other effective treatments, such as behavioral, cognitive behavioral, and multidisciplinary treatments [6].

5.2.2
Nonmaleficence

The principle of nonmaleficence requires us to avoid or minimize doing harm or exposing others to unreasonable risk of harm. Of course, many treatments involve some risk or inherently result in at least minor harm. Generally, this means that the likely benefit outweighs any risk or inherent harm, and that it is unlikely that the same benefit could be achieved by some alternative treatment with less potential for harm.

The subjective nature of pain also renders the person who suffers from it more susceptible to unintended harm by healthcare providers. In particular, patients often suffer the certainty of their pain and the uncertainty of others [13]. Challenging the patient's claims concerning the degree of his or her pain or need for help exacerbates suffering [3]. Such a person is tacitly branded as a liar, and comes to be regarded as somehow cowardly, uncooperative and lacking in will power. The patient may come to be labeled as "difficult", "demanding", a "clock watcher", "malingerer" or "manipulator". Once the patient has been so labeled, healthcare providers can feel morally justified in subtly–and sometimes not so subtly– humiliating the patient, disregarding pleas for help, and generally ostracizing him from the community of human moral agents [3]. Such behavior is not only disrespectful; it actually increases the suffering of the patient [13, 14].

Failing to relieve pain or to provide relief when this is available causes harm [15]. In extreme cases it could be regarded as "torture by omission" [16]. Unrelieved pain undermines patients' ability to think or interact socially and contravenes the right of the patient to self-determine his or her healthcare [15].

Moreover, doubt concerning the reality of patients' unrelieved chronic pain has allowed concerns about addiction–which appear to be largely unfounded [17] (though contested [18, 19])–to dominate discussions of treatment, rather than effectiveness [6, 11]. Under the rubric of nonmaleficence, many patients are undermedicated, on the theory that a dosage high enough to produce analgesia will either produce discomfort from side effects that exceeds the discomfort of unrelieved pain, sedation that will undermine quality of life, or respiratory depression that may be life-threatening [2].

5.2.3
Autonomy

The principle of autonomy requires that patients control decisions about their own treatment. It requires health care providers to fully inform patients of the treatment alternatives, including the risks and benefits. Patients must consent before treatment can take place. Since about 1950, there has been increasing recognition of the importance of patient autonomy, to the extent that some ethicists believe that it should be considered the most important ethical principle [7]. In practice, patient autonomy may be the final determiner in refusing treatment, but it cannot normally be used to compel healthcare providers to administer treatments that they believe are harmful or dangerous. Autonomy also may be limited in cases of patients who are considered unable to provide informed consent because of their age or impairment. In such cases, parents, guardians, or even courts may need to act as substitute decision makers. As part of the trend toward patient autonomy, however, there has been increasing recognition that even when patients are unable to provide a competent consent, they should be involved in the decision-making and consent process to the greatest extent possible.

The highly subjective nature of pain has special implications for autonomy. While we can often observe behavior that communicates pain, we have no direct and objective measure of pain. Thus, the healthcare provider may be faced with a dilemma of being requested by the patient for relief from pain–thereby respecting the patient's autonomy–while not believing the patient or desiring to diagnose the root cause of the pain. In such a circumstance the provider risks harming the patient by providing treatment for a non-existent condition with its inherent side-effects–compromising the principle of nonmaleficence–and possibly failing to provide treatment for or even masking whatever condition the patient actually has, if any–compromising beneficence.

The anxieties generated by being totally at the mercy of others with regard to pain relief are often overwhelming and these anxieties tend to magnify the patient's suffering. The other side of this coin is that such anxieties and amplifications may be largely avoided by allowing the patient to make his own properly informed decisions concerning the quantity and timing of pain relief medication. The possibility of patient abuse in this area is considerably less than many have feared, especially since being in control diminishes pain. Among ethics scholars there is agreement that pain—chronic pain in particular—is undertreated [2, 20].

5.2.4
Justice and Social Responsibility

The principle of justice and social responsibility requires healthcare providers to act fairly and consider the impacts of medical decisions on society as a whole. This principle requires consideration of the impact beyond the individual patient and health care provider. The World Health Organization and International Association for the Study of Pain have declared that pain is an important factor in health status, regardless of its etiology [21], and that pain relief is a universal human right [15, 22]. Nevertheless, there are significant disparities in access to high-quality treatment to relieve pain associated with race and ethnicity [23], geographic location [22], socioeconomic status [24], cognitive abilities, and a number of other factors. While some healthcare disparities are more extreme in privately funded healthcare systems [25], publicly funded systems also exhibit disparities.

Fair opportunity says that no persons should receive services on the basis of undeserved advantageous attributes (because no persons are responsible for having these attributes) and that no persons should be denied services on the basis of undeserved disadvantageous attributes (because they similarly are not responsible for having these attributes). Given the positive value attached to courage, cooperativeness, resignation, and will power, those patients with chronic pain who fail to measure up to these stoic attributes are very likely to be underserved. Conversely, those patients who bear their pain without complaint and unquestioningly follow medical advice are very likely to be granted special moral standing and professional and personal favors [3, 4]. Patients with chronic pain who also have a history of substance abuse are particularly prone to being denied effective pain treatment [17].

5.2.5
Integrity

Integrity includes truthfulness, respect for the law and cultural norms, respect for standards of professional practice, and other aspects of trustworthiness. While integrity can be viewed as a personal trait, it can also be complicated by specific situations. Conflicts of interest and dual relationships often challenge integrity.

A conflict of interest exists when a team member has other interests that could conflict with doing what is best for the patient. For example, someone with a financial interest in a drug company or the development of a specific medication would be in a conflict of interest when deciding which drug would be best for a patient.

A dual relationship exists when a team member relates to the patient in more than one role. For example, a physician–patient relationship and a researcher–research participant relationship may interfere with each other.

5.3
Interacting Principles

In practice, it is rare that any of these principles can be applied independently of the others. A close relationship exists between beneficence and nonmaleficence, and some ethicists simply combine the two into a single principle, viewing benefit and harm as two ends of the same continuum. In this model, benefit and harm can balance each other out or neutralize each other. However, this view is simplistic because it fails to recognize the obvious difference between doing great harm while producing great benefit and doing nothing at all.

5.4
Pain, Public Policy, and Ethics

The main health policy needs arising out of this discussion are (i) protocols for determining appropriate treatment and compensation in light of the inherently subjective nature of pain, (ii) a need for guidance as to the responsible use of opioids for treating chronic pain, and (iii) recognition of relief of suffering due to chronic pain as an important healthcare goal.

Patient-centered principles ought to guide efforts to relieve chronic pain, including accepting all patient pain reports as valid while negotiating treatment goals early in care, avoiding harming patients, and incorporating opioids as one part of the treatment plan if they improve the patient's overall health-related quality of life [6].

References

1 Cassell, E.J. (2004) Pain and suffering, in *Encyclopedia of Bioethics*, 3rd edn, Vol. **4** (ed. S. Post), Macmillan Reference, New York, pp. 1961–9.

2 Rich, B.A. (2000) An ethical analysis of the barriers to effective pain management. *Cambridge Quarterly of Healthcare Ethics*, **9**, 54–70.

3 Edwards, R.B. (1984) Pain and the ethics of pain management. *Social Science and Medicine*, **18**, 515–23.

4 Fagerhaugh, S.Y. and Strauss, A. (1977) *Politics of Pain Management*, Addison-Wesley, Menlo Park, CA.

5 Ruddick, W. (1997) Do doctors undertreat pain? *Bioethics*, **11**, 246–55.

6 Sullivan, M. and Ferrell, B. (2005) Ethical challenges in the management of chronic non-malignant pain: negotiating through the cloud of doubt. *The Journal of Pain*, **6**, 2–9.

7 Gillon, R. (2003) Ethics needs principles–four can encompass the rest–and respect for autonomy should be "first among equals". *Journal of Medical Ethics*, **29**, 307–12.

8 Macklin, R. (2003) Applying the four principles. *Journal of Medical Ethics*, **29**, 275–80.

9 Jansen, L.A. (2001) Deliberative decision making and the treatment of pain. *Journal of Palliative Medicine*, **4**, 23–30.

10 Emanuel, L. (2001) Ethics and pain management: an introductory overview. *Pain Management*, **2**, 112–16.

11 Weinman, B.P. (2003) Freedom from pain: establishing a constitutional right to pain relief. *Journal of Legal Medicine*, **24**, 495–539.

12 Reidenberg, M.N. and Willis, O. (2007) Prosecution of physicians for prescribing opioids to patients. *Clinical Pharmacology and Therapeutics*, **81**, 903–6.

13 Blacksher, E. (2001) Hearing from pain: using ethics to reframe, prevent, and resolve the problem of unrelieved pain. *Pain Medicine*, **2**, 169–75.

14 Bedell, S.E., Graboys, T.B., Bedell, E. and Lown, B. (2004) Words that harm, words that heal. *Archives of Internal Medicine*, **164**, 1365–8.

15 Cousins, M.J., Brennan, F. and Carr, D.B. (2004) Pain relief: a universal human right. *Pain*, **112**, 1–4.

16 Cousins, M.J. (1999) Pain: the past, present and future of anesthesiology. The EA Rovenstine Memorial Lecture. *Anesthesiology*, **91**, 538–51.

17 Cohen, M.J.M., Jasser, S., Herron, P.D. and Margolis, C.G. (2002) Ethical perspectives: opioid treatment of chronic pain in the context of addiction. *The Clinical Journal of Pain*, **18**, 99–107.

18 Ballantyne, J.C. and Mao, J. (2003) Opioid therapy for chronic pain. *New England Journal of Medicine*, **349**, 1943–53.

19 Vasudevan, S.V. (2005) Commentary: empower and educate patients diagnosed with chronic nonmalignant pain. *The Journal of Pain*, **6**, 10–11.

20 Ferrell, B.R., Novy, D., Sullivan, M.D. *et al.* (2001) Ethical dilemmas in pain management. *The Journal of Pain*, **2**, 171–80.

21 Breivik, H. (2002) International association for the study of pain: update on WHO-IASP activities. *Journal of Pain and Symptom Management*, **24**, 97–101.

22 Green, C., Todd, K.H., Lebovits, A. and Francis, M. (2006) Disparities in pain: ethical issues. *Pain Medicine*, **7**, 530–3.

23 Cintron, A. and Morrison, R.S. (2006) Pain and ethnicity in the United States: a systematic review. *Journal of Palliative Medicine*, **9**, 1454–73.

24 Lebovits, A. (2005) The ethical implications of racial disparities in pain: are some of us more equal? *Pain Medicine*, **6**, 3–4.

25 Sanmartin, C., Berthelot, J.M., Ng, E., Murphy, K., Blackwell, D.L., Gentleman, J.F. *et al.* (2006) Comparing health and health care use in Canada and the United States. *Health Affairs*, **25**, 1133–42.

6
Pathophysiology of Chronic Pain

James L. Henry

- Pain is a complex phenomenon that combines information from the nervous system with thoughts, emotions and social context.

- Laboratory research has taught us a great deal about the way in which the nervous system creates and processes the necessary signals for pain, but each discovery raises a whole host of further questions.

- Better funding for such research would accelerate the development of further rational, evidence-based treatments for pain, but it is not possible to predict how long this might take.

To the person in pain, its existence is self-evident and utterly convincing. However, the scientist who tries to demonstrate how the pain comes about knows that he or she is taking on a far more complex challenge than the study of other biological phenomena. What can science tell us about what is happening in the cells and tissues of the body during this uniquely human experience?

Scientists study pain at several levels of the nervous system. At the site of injury special nerve cell endings called nociceptors, which have been waiting for exactly this moment, respond to one of a myriad of unpleasant stimuli such as heat, pressure and inflammation, by sending a rapid and urgent signal that something is wrong. Shortly afterwards potent chemicals are released by crushed and broken cells which are detected by other nerves and amplify the pain signal. This whole process is called nociception, the change in the senses induced by a noxious stimulus. Activation of nociceptors can often lead to pain, but this is not always the case. A person may perceive a given event as merely annoying, mildly painful, or excruciating, based not only on the size of the trauma, but also on their thoughts, attitudes and context. An elite athlete or soldier in battle, for example, may be so focused on the task at hand that he or she does not even notice his broken finger until the moment has long passed. Pain is thus the result of integrated neural input. It is highly individual and subjective in nature, often making pain difficult to define scientifically.

Chronic Pain: A Health Policy Perspective
Edited by S. Rashiq, D. Schopflocher, P. Taenzer, and E. Jonsson
Copyright © 2008 WILEY-VCH Verlag GmbH & Co. KGaA, Weinheim
ISBN: 978-3-527-32382-1

From an experimental perspective, pain can be broken down into three types, each mediated by different mechanisms. Nociceptive pain results from activation of nociceptors in peripheral tissues. Neuropathic pain results from injury or irritation to the nerves themselves, such as in shingles or diabetic neuropathy. Inflammatory pain arises from inflamed joints or other tissues.

On a day-to-day basis pain subsides after the recovery from tissue injury, such as a burn, a cut or even a broken bone. However, in some cases pain does not subside, even despite healing of the injury. This is the pathologic condition known as chronic or persistent pain. In this case the individual may experience one or more of the following: spontaneous pain (pain for no apparent reason), hyperpathia (more pain than would be expected after a painful event), hyperalgesia (increased intensity of pain to a further noxious stimulus), secondary hyperalgesia (spreading of sensitivity or pain to nearby, uninjured tissue) and allodynia (sensation of pain from a normally innocuous stimulus).

6.1
Nociception: How It Works

Almost all parts of the body are covered with nerve endings that are each programmed to respond to a specific kind of unpleasant sensation. They require a certain intensity of stimulation before they react and will lie silent until this level is reached. When stimulated they create an electrical pulse known as an action potential, which is transmitted along the attached nerve on the first stage of a long but very rapid journey to the spinal cord and then to the brain. The nerves are composed of thousands of tiny filaments called axons, lying alongside each other like strands of copper wire in a domestic electrical cable. Axons are classified according to the type of receptor they connect to and their diameter. Some are "myelinated", that is, covered with an insulating sheath made from a fatty material called myelin. This protects and nourishes the axon and keeps the electrical pulse within it, resulting in faster transmission of the signal. Large diameter axons are heavily myelinated and these are associated mainly with receptors to movement. There is then a range of axons of diminishing diameter and myelination, dedicated to serving a range of sensory impressions including hair movement, pressure, touch, temperature and nociception.

Subgroups of nerve endings each specialize in distinct sensory modalities such as nociception, or the ability to detect heat, cold or light touch. They differ in size, their destinations in the brain and spinal cord, their degree of myelination and the type of neurotransmitters (chemicals that cause a nerve cell to act in a certain way) that they respond to and secrete. Pain receptors on the skin have been studied more intensely than deeper receptors and are remarkably diverse: we can distinguish more than 20 different types of stimulus for which there is a specific nerve ending programmed and ready to respond to it and each has its own specific trigger. Researchers use specific compounds such as capsaicin (the substance that makes chilli peppers spicy), menthol, camphor, and wasabi to stimulate specific types of these receptors in experimental settings. The conditions required for a given receptor to fire may be highly specific. Thermal nociceptors, for instance, are only activated by temperatures above 45 °C or below 5–10 °C, especially when applied to the skin for durations of greater than one minute. Mechanical nociceptors are only activated by strong physical stimuli, especially when applied over a small surface area. On the other hand, some peripheral receptors may respond to several different types of stimulus, including strong mechanical and thermal stimuli, and are often sensitized in time by repeated application of stimuli. These so-called polymodal nociceptors may also be sensitive to chemical stimuli, such as low pH. It is believed that some of these types of receptor are also located in deeper tissues. Since we know that some receptors can be made more likely to be activated by a number of mechanisms, including the chemical environment, it is theorized that some types of chronic pain may arise from this so-called peripheral sensitization.

As indicated above, information is transmitted from the periphery to the spinal cord and brain by a variety of axon types with myelin sheaths of varying degrees of thickness. The more myelinated axons are thought likely to be the most sensitive to changes in myelination resulting from disease processes or injury. In such cases, when myelination is compromised there is thought to be a dysfunction of the mechanisms that transmit action potentials along axons due to the protective and nourishing function of the myelination. This dysfunction can result in extra, unnecessary electrical activity arising from within the nerve itself (ectopic discharge) and amplification of the nociceptive signal. Such mechanisms are thought to be the basis of some types of neuropathic pain.

The bodies of the nerve cells that transmit pain are located in the dorsal root ganglia, small specialized clumps of nerve tissue that run along the length of the spine, very close to the spinal cord. This is where the vastly complex genetic infrastructure and metabolic machinery of each of these nerve cells reside. The genetic component may be of crucial importance in the understanding of pain states. For example, if a normally non-pain transmitting neurone begins to make a chemical known as substance P, activity in this type of neurone may lead to the perception of pain, even in the absence of a noxious stimulus. Another way in which gene activity might affect pain is in the expression or distribution of sodium channels. These are portals in the wall of the nerve cell through which sodium ions pass in order to generate the action potential. It has recently been reported that a rare inherited disorder in which the person is simply unable to feel pain is associated with a mutation that causes inactivation of a certain specific sodium channel [1]. Interestingly, other mutations that modify the kinetics of this channel, but do not render it inoperable, are associated with the presence of an inherited form of a rare painful disease called erythromelalgia [2, 3] and with other painful illnesses [4]. Thus, a change in the way genes direct the cell to produce and control sodium channels may lead to the perception of persistent pain, again even in the absence of a noxious stimulus. This is a potentially revolutionary area for scientific research.

The axons carrying the nociceptive signal continue on into the spinal cord and relay onto neurones projecting to the brain. They arrive and are distributed in highly organized ways to specific areas of each. The spinal cord is not, as was once thought, merely a passive relay station for sensory information on its way to the brain, but is the site of a great deal of processing of the nociceptive signal. The dorsal horn is the area of the spinal cord where this takes place.

Nociception, the transmission of a signal to the central nervous system in response to tissue damage, is broadly the same in all creatures and is therefore amenable to study in the laboratory. Pain, however, is a uniquely human experience, the result of not only the neurophysiological process but also of many additional factors that generally lie beyond inspection from a neurophysiological point of view, such as personality, circumstance and emotional state. We are beginning to understand how these domains might interact. For example, there are various changes in the spinal cord that have been suggested as leading to chronic pain, known as central sensitization, wind-up, and microglial activation. *Central sensitization* results from low frequency and high frequency nociceptive stimulation, and can lead to increased excitability of dorsal horn neurones, manifested as increased spontaneous discharge, increased receptive field size (possibly a cellular basis of secondary hyperalgesia) and increased responsiveness to innocuous stimulation of the peripheral receptive field (possibly a cellular basis of allodynia). *Wind-up* is initiated by high threshold, C-fiber strength, stimuli delivered at 3 Hz or more to induce cumulative depolarization. Wind-up does not persist following the conditioning stimulus.

The target areas for pain neurones as they arrive in the spinal cord are highly specific. Their organization is thought to serve both the very site specific sensations associated with some painful stimuli (for example, a pinprick in the finger can be located precisely by the sufferer, even without looking), yet a dull poorly-localized sensation associated with other stimuli (e.g. the widespread, dull pain that heralds the onset of appendicitis). This organization may also account for referred pain, the perception of pain in an area other than that where the injury occurs, such as the shoulder-tip pain that is commonly seen when the diaphragm is irritated and which is a common complaint of patients with degenerative joint disease.

6.2
Chemical Mediators of Pain

The nervous system is rich in potent chemical substances that are released, detected and broken down in an ornate sequence that scientists are only beginning to understand. The excitement for researchers in this area is that thoroughly understanding which compounds are in action in a pain state raises the possibility of delivering drugs that affect their action and result in pain relief. An example of this is the amino acid glutamate, which is thought to be part of the sequence of events that results in neuropathic pain. The discovery by laboratory scientists that the action of glutamate can be blocked at a receptor known as NMDA by existing medications resulted in their more widespread use in the treatment of this debilitating condition and has set in motion efforts to develop new drugs that might do so more effectively. Nociceptors also release other types of compounds that act as transmitters or as modulators of nerve excitability. Perhaps the best documented and understood is substance P, an 11 amino acid peptide identified first by Gaddum and von Euler as an extract in 1932, and termed substanz P, for powder. Substance P is released from one type of pain transmitting nerve fiber in response to intense peripheral stimulation and leads to selective amplification of pain pathways in the spinal cord. Other compounds including amino acids such as GABA and glycine, peptides such as enkephalin and dynorphin, and the nucleoside adenosine act to turn down the nociceptive signal.

These are but examples of what is a chemical correlate to the intricate anatomical and morphological complexities that contribute to nociception. Slight disturbances or prolonged synaptic input can generate long-term or even permanent plastic changes in these neurones and these changes may account for some types of chronic pain.

What have been described thus far are events that take place in nerve tissues, but important scientific discoveries about pain transmission have been made by focussing down further, to events taking place within and around individual nerve cells. Calcium plays a major role in short-term and long-term events triggered by synaptic activation. It regulates the amount of other neurotransmitters sent from one pain nerve to the next in sequence in the dorsal horn of the spinal cord. Intracellular calcium participates in activation of and in regulation of gene expression.

6.3
Contribution of Non-Neuronal Cells to Nociception

Recently, we have become aware that direct neuronal signaling to other neurones is not the only process that leads to long-term changes. Immune cells within the spinal cord, which were long thought to only have a protective and nutritive role, have also recently been found to contribute to alterations in neuronal hyperexcitability [5–6]. These immune cells, mainly microglia and astrocytes, are activated from their normal quiescent state by injury to peripheral nerves and also by inflammation or injury within the spinal cord or brain. In fact, rather than increasing excitability of spinal nociceptive neurones the processes thought to occur are through removal of an inhibitory influence, or disinhibition [7].

6.4
Basic Science Contributions to Pain Management

The data that emerge from basic science research on pain mechanisms are yielding ever clearer and simplified explanations for what we see in humans and animal models. The landmark Gate Control Theory of Ronald Melzack (Canada's pre-eminent scientific pain researcher) and Patrick Wall in 1965 [8] set the stage for theories of pain transmission, and this spurred much research into understanding the nature of pain. Yet, these hypotheses oversimplify the complex mechanisms involved and overlook important facts. For example, we now know that most spinal dorsal horn neurones can be excitatory or inhibitory and receive inputs from higher brain centers and other local interneurones in addition to inputs from the periphery. The precise roles of these interneurones in information processing, particularly processing of nociceptive information, are far from being understood. Several different classifications of dorsal horn neurones have been published along morphological, chemical and electrophysiological lines, yet no such classification scheme has been able to claim clearly established links to clinical conditions. Due to their very small size, these neurones are difficult to record from electrophysiologically and thus have not been characterized as clearly. Efforts to synthesize what we know from the morphological, electrophysiological and neurochemical points of view about dorsal horn neurones are ongoing, but slow to yield convincing results. For example, neurones in one part of the dorsal horn (Lamina I) are so few, numbering only in the hundreds rather than in the tens of thousands in the dorsal horn, that it is hard to accept that the full range of perceptions of pain, including allodynia and hyperalgesia, can be served by so few neurones. Clearly, much further research is necessary if we are to gain the depth of understanding needed.

Establishing a functional role for any of these classifications has not been achieved. We have drawn only tenuous conclusions about the possible roles of the different morphological cell types; the various discharge characteristics or the different chemical mediators of synaptic transmission and intracellular signal transduction. Nevertheless, there is considerable hope that this knowledge will yield novel therapeutic approaches. The GluR-A glutamate receptor, a novel form of the glycine receptor containing $\alpha 3$, the prostaglandin E_2 pathway and the Kv4.2-type potassium channel are just some of the parts of the complex processes within the nervous system that are under intense scientific scrutiny for answers to the pain mystery.

Thus, while we know much and there are emerging leads that may be useful in developing novel approaches to management of chronic pain, there remains much to learn. Oversimplified theories of sensory processing mechanisms impede our ability to understand fully these mechanisms. We see the pain system now as multiple pathways, with multiple synaptic junctions and an elaborate network of local and remote control systems acting at each junction and a fathomless capacity for neuroplastic change [9]. If we are to begin to match mechanisms with clinical conditions and to exploit knowledge of mechanisms to develop novel approaches to pain management, we must commit ourselves to the long and difficult task of accumulating a vast amount of scientific knowledge.

6.5
Impact of Policy on Knowledge Generation

Laboratory research has yielded important information about the many aspects of the mechanisms of pain generation. Examples of note include understanding of transduction mechanisms in peripheral sensory nerve terminals and the complexity of mechanisms controlling peripheral sensitivity, the varied nature of the ion channels mediating conduction of action potentials to the central nervous system, the chemical basis of synaptic transmission at the first sensory synapse, recruitment of non-neural cells into pain pathways, the plethora of intracellular signal transduction mechanisms and the resultant forms of altered excitability and, finally, the identification, cloning and regulation of expression of critical proteins, whether receptors, ion channels, transporters or synthesizing enzymes.

These innovations have helped some patients but have not addressed the needs of many others who may be suffering a different type of pain. Innovation can be accelerated, but this would require a different approach to that we have had in the past, and this is governed by policy. Innovation comes most rapidly from the best-funded research efforts attracting the most innovative minds. Therefore, we need to accept that if we intend to ramp up innovation we need to increase its funding.

There is ample reason to be optimistic that these developments and others will eventually lead to new types of treatment for pain, given adequate support and time. In this context, it is sad to note that the Canadian Institutes of Health Research does not have a review process that specifically addresses pain, either for the basic or for the clinical sciences, and that funding for knowledge generation in pain is not commensurate with its impact on the individual and society. The European parliament has declared chronic pain to be a disease in and of itself. This highlights the need for increased knowledge generation. The U.S. House of Representatives has declared this to be the Decade of Pain Care and Research. This was backed up by a 10-step approach and by a major influx of funding for pain care and research. Benefits of this policy change will not be seen immediately, yet in the long run Americans will benefit from this influx of funding.

As Canadians, we justifiably pride ourselves on our healthcare services. However, those people who suffer chronic pain and those others who are engaged in its treatment and research currently see less to be proud of. Bringing research funding up to a level meeting the medical needs imposed by chronic pain would be a valuable first step in addressing the current imbalance.

References

1 Fertleman, C.R., Baker, M.D., Parker, K.A. *et al.* (2006) SNC9A mutations in paroxysmal extreme pain disorder: allelic variants underlie distinct channel defects and phenotypes. *Neuron*, **52**, 767–74.

2 Drenth, J.P.H., Morsche, R.H.M., Guillet, G., Taiev, A., Kirby, R.L. and Jansen, J.B.M.J. (2005) SCN9S mutations define primary erythermalgia as a neuropathic disorder of voltage gated sodium channels. *The Journal of Investigative Dermatology*, **124**, 1333–8.

3 Yang, Y., Wang, Y., Li, S. *et al.* (2004) Mutations in SCN9A, encoding a sodium channel alpha subunit, in patients with primary erythermalgia. *Journal of Medical Genetics*, **41**, 171–4.

4 Hayden, R. and Grossman, M. (1959) Rectal, ocular and submaxillary pain. *American Journal of Diseases of Children*, **97**, 479–82.

5 Marchand, F., Perretti, M. and McMahon, S.B. (2005) Role of the immune system in chronic pain. *Nature Reviews. Neuroscience.*, **6**, 521–32.

6 Tsuda, M., Inoue, K. and Salter, M.W. (2005) Neuropathic pain and spinal microglia: a big problem from molecules in "small" glia. *Trends in Neurosciences*, **28**, 101–7.

7 Sherman, S.E. and Loomis, C.W. (1994) Morphine insensitive allodynia is produced by intrathecal strychnine in the lightly anesthetized rat. *Pain*, **56**, 17–29.

8 Melzack, R. and Wall, P.D. (1965) Pain mechanisms: a new theory. *Science*, **150**, 971–9.

9 Craig, A.D. (2003) Pain mechanisms: labelled lines versus convergence in central processing. *Annual Review of Neuroscience*, **26**, 1–30.

Part Two
Treatment

7
Why Does Pain Become Chronic?

Joel Katz, Colin J. L. McCartney, and Saifudin Rashiq

- Every chronic pain was once an acute pain.
- The more severe the acute pain, the more likely it is to become chronic.
- We know of several things that can be done before surgery to minimize pain, and to help prevent that pain becoming chronic.
- Advanced surgical techniques are less likely to cause severe and chronic pain than traditional operations.
- Depression, anxiety and catastrophizing are associated with the development of chronic pain after injury or surgery.

7.1
Introduction

A fact that is rarely appreciated by clinicians or researchers is that every chronic pain was once acute. The traditional, dominant focus on studying already established chronic pain misses important cues that may help us predict who will go on to develop it and who will recover uneventfully. Though much is known about the biomedical and psychosocial factors that predict the transition of acute pain, whether from illness, injury or surgery to chronic pain, much remains to be learned. In this chapter, we review what is known about the factors that are associated with this transition. Surgery is the only scenario in which the time of onset of injury and pain in an individual can be predicted in advance, and therefore offers a unique opportunity for observational studies of pain, and for research into preventative strategies. The epidemiology of chronic pain following various surgical procedures and the surgical, patient-related, anesthetic, and psychosocial factors that appear to confer a greater risk of developing such pain are therefore reviewed. Outside the context of surgery, we draw upon the results of epidemiologic studies (mostly of acute neck and back pain) to describe what is known about the factors that increase the risk of acute non-surgical pain becoming chronic. Finally, we suggest clinical practice and policy changes that would mitigate the risk of acute pain becoming chronic pain.

Chronic Pain: A Health Policy Perspective
Edited by S. Rashiq, D. Schopflocher, P. Taenzer, and E. Jonsson
Copyright © 2008 WILEY-VCH Verlag GmbH & Co. KGaA, Weinheim
ISBN: 978-3-527-32382-1

7.2
Chronic Post-surgical Pain

The majority of patients who undergo surgery recover uneventfully and within weeks typically resume their normal daily activities. However, chronic postsurgical pain (CPSP) develops in a considerable proportion of patients. The magnitude of the problem is evidenced by recent studies that document the epidemiology and growing awareness of CPSP in the surgical community [1–4]. For example, a prospective study of approximately 5000 postsurgical patients estimated the incidence of acute neuropathic pain in the days after surgery to be between 1 and 3%. Follow-up showed that 56% continued to have ongoing pain one year later [5]. Other prospective studies suggest the incidence of CPSP is considerably higher. A conservative estimate is that up to 10% of patients report severe, intractable CPSP one year after surgery. We know next to nothing about pain beyond the one year mark. These statistics are alarming, especially when one considers the total number of patients worldwide who undergo surgery each year. It comes as no surprise then to see that almost 25% of more than 5000 patients referred to chronic pain treatment centers had CPSP [6].

7.2.1
Epidemiology of CPSP

Table 7.1 shows the incidence of CPSP following a variety of surgical procedures. The one-year incidence ranges from a low of approximately 10%–15% following modified radical mastectomy [22] to a high of 61%–70% for thoracotomy [13] and amputation [16] (Some of the variability in these estimates can be accounted for by differences in the pain intensity used to classify patients: the more stringent the criteria, the lower the estimate). Longer term follow-up studies performed in hernia repair indicate that pain persists in 8.1%–19% of patients for up to six years, with severe or very severe pain occurring in 1.8%. Two years after amputation [17, 23] approximately 60% of amputees report phantom limb pain and 21%–57% report stump (residual limb) pain. All of these rates are unacceptably high.

7.2.2
Surgical Factors Associated with CPSP

CPSP is more likely to develop in certain circumstances. Long operations [24, 25], operations performed in centres with a low annual procedure volume [22], open (vs. laparoscopic) procedures [26] and operations involving nerve damage [5, 26] are known to pose a greater risk. There are also specific surgical techniques that increase the risk [26, 27].

All of these factors appear to be associated with heightened surgical trauma and particularly with intraoperative nerve injury. Nerve injury is likely a major culprit in producing both acute and chronic post-surgical pain. This premise is supported by a large body of animal research. In fact, the most commonly used animal

Table 7.1 Incidence of CPSP following various surgical procedures.

Authors (Ref.)	Surgical procedure	Total number of patients with data in study	Follow-up time (post-op)	Incidence of CPSP
Tasmuth [7]	Modified radical mastectomy (MRM) or breast conserving surgery (BCT) with axillary clearance	93 MRM = 53 BCT = 40	1 year	Breast region MRM = 17% BCT = 33% Ipsilateral arm MRM = 13% BCT = 23%
Aasvang [8]	Hernia repair	694	1 year	Inguinal pain = 56.6% Ejaculation pain = 18.3% Pain in: testes = 39.7% shaft = 5.4% glans = 4.5% thigh = 11.6%
Grant [9]	Hernia repair		5 years	19% groin pain 1.8% severe or very severe groin pain 16.1% testicular pain
Bay-Nielsen [10] Aasvang [11]	Hernia repair	1166	1 year	1 year–pain in groin area = 28.7% Pain interfering with work or leisure = 11%
			6.5 years	6.5 years–8.1% chronic inguinal pain
Katz [12]	Thoracotomy	23	1.5 years	52% with dull, aching or burning pain on a daily to weekly basis, mean pain intensity 3.3/10
Pertunnen [13]	Thoracotomy	67	1 year	61% posthoracotomy pain 3–5% with severe pain
Gotoda [14]	Thoracotomy	91	1 year	Postthoracotomy pain Pain free = 59% Slight pain = 39% Moderate pain = 2% Severe pain = 0%
Nikolajsen [15]	C-section	220	~1 year	12.3% abdominal scar pain
Nikolajsen [16]	Amputation	60	1 year	~70% phantom limb pain
Jensen [17]	Amputation	58	2 years	59% phantom limb pain 21% stump pain
Hanley [18]	Amputation	57	2 years	62% phantom limb pain intensity >0/10 30% phantom limb pain intensity >3/10 57% stump pain intensity >0/10 27% stump pain intensity >3/10

Table 7.1 Continued

Authors (Ref.)	Surgical procedure	Total number of patients with data in study	Follow-up time (post-op)	Incidence of CPSP
Borly [19]	Open cholecystectomy	80	1 year	26% pain
Meyerson [20]	Sternotomy for cardiac surgery (CABG)	318	1 year	28% post-sternotomy pain 13% with pain intensity ≥30/100
Nikolajsen [21]	Hip replacement	1048	12–18 month prevalence	28.1 % chronic hip pain 12.1% pain limited activities to a moderate, severe or very severe degree

models of neuropathic pain involve intentional damage to peripheral nerves [28–30] or spinal nerve roots [31]. Thus, one very useful preventive measure that can be taken against CPSP in humans is to avoid intraoperative nerve damage as far as possible. Clearly, this cannot be done for surgeries such as amputation that necessitate cutting of major nerves. However, the practice of intentionally transecting nerves for surgical convenience can be avoided [26] and doing so may contribute to a reduced incidence of neuropathic CPSP [1, 32].

7.2.3
Patient-Related Factors

The most robust finding to emerge from the surgical pain literature is that pain intensity immediately after the operation predicts the intensity and duration of future pain [1, 4, 33, 34]. This appears to be true days [17, 22, 35–37] and weeks [38] after surgery, as well as for the duration of post-surgical convalescence [17, 18, 26, 35, 38–41]. In addition, the severity of acute postoperative pain in the days and weeks after surgery is also a risk factor for development of CPSP [5, 7, 12, 18, 42–46]. No other patient factor is as consistently related to the development of future pain problems as is current pain. Younger age [38, 46, 47] and female gender [38, 47] predict CPSP but not with the consistency or magnitude with which pain predicts pain. What is yet to be determined, however, is the aspect(s) of pain that is predictive. There are several non-mutually exclusive possibilities for why pain predicts pain including those that propose a causal or associative role: (1) intraoperative nerve damage/postoperative ectopic activity increases the intensity of acute postoperative pain as well as the incidence and intensity of chronic neuropathic pain [12]; (2) central sensitization induced by surgery is maintained by peripheral input [48, 49], (3) perioperative nociceptive activity induces structural changes in the CNS (i.e. centralization of pain) [49, 50]; (4) heretofore unidentified pain genes that confer increased risk of developing intense acute pain and CPSP [51, 52]; (5) consistent response bias over time [12] and psychosocial factors including catastrophizing, social support and social environmental factors [23, 53].

It is important to ascertain the precise mechanisms that underlie the relationship between pain at one time (e.g. preoperative pain or acute post-operative pain) and pain at another (e.g. pain one year after surgery). The idea that acute pain becomes in some way etched into the brain by a memory-like mechanism [54–56] has been at the heart of efforts to halt the transition to chronicity by blocking noxious perioperative impulses from reaching the central nervous system using a pre-emptive approach.

7.2.4
Preemptive/Preventive Analgesia

As we saw in Chapter 6, the old concept that pain is the end-product of a passive system that faithfully transmits a peripheral "pain" signal from receptor to a "pain centre" in the brain like a series of electrical cables is now thoroughly discredited [57]. Basic science and clinical data show instead that cutting tissue, nerve, and bone induces major changes in central neural function that persist well after the cutting has stopped and the injury has healed [48, 50]. The allied idea of it being acceptable to wait to treat surgical pain until after it is well established is, besides being inhumane, now scientifically untenable. Even to this day, in Canadian hospitals, some patients arrive in the post-anesthetic care unit after surgery in severe pain and then receive multiple doses of opioids in an effort to bring the pain down to a tolerable level.

Fortunately, this practice is slowly being supplanted by a preventive approach that aims to block the transmission of the primary afferent injury barrage and the inflammatory response [58–61]. The intent behind this approach is not simply that it reduces nociception and stress during surgery – although these are obviously worthwhile goals – but that it prevents some of the mechanisms that promulgate CPSP from coming into play in the first place. By definition, [59, 62, 63] for studies that compare an active agent with a placebo control condition, the demonstration of preventive analgesia requires that the reduction in analgesic consumption and pain be observed at a point in time that exceeds the clinical duration of action of the target agent used preventively. This is meant to rule out the possibility that the effect is merely an analgesic effect. The longer the time from the administration of the analgesic agent(s), the greater the probability that other factors contribute to the long term effects.

The weight of evidence in this field favors the use of multimodal analgesia, the deployment of pain preventing drugs from more than one class with or without nerve blocking techniques at the same time [62, 64]. This is necessary because there is currently no single intervention that prevents pain signals from reaching the central nervous system for more than a few hours. Benefits of the multimodal approach include improved efficacy, lower doses and fewer adverse effects compared to single-modality drug or other treatment [65].

Recent studies have demonstrated that preventive analgesia is associated with clinically important reductions in CPSP incidence and intensity. For example, the use of bone graft from the hip is a necessary expedient in certain types of reconstructive surgery, but CPSP at the graft removal site is reported in up to 37% of patients at one year and 19% at two years [66–68]. Reuben et al. [69] showed that infiltration of morphine into the bone graft harvest site reduced acute and chronic donor site pain sixfold. Another study [70] demonstrated that chronic donor site pain was reduced from 30% to 10% simply by giving a non-steroidal anti-inflammatory drug one hour prior to surgery and for five days afterwards. Pre-treatment of a similar nature in patients having outpatient anterior cruciate ligament repair [71] was associated with fewer complications and higher activity levels when measured six months postoperatively, an example of how a simple preventative measure taken at the time of surgery can have important and long-lasting beneficial effects. Gabapentin, an anti-epilepsy medication that is commonly used for the treatment of neuropathic pain, can, when taken preventatively by patients before surgery, reduce pain, and narcotic consumption in a variety of types of operations [63]. Why should this be so? The typical explanation offered for the prolonged effect seen in such studies is that the intervention prevented or reduced peripheral and/or central sensitization and thereby reduced long term pain. Unfortunately, there is very little evidence that this is the case since we do not have accurate measures of sensitization in humans and even if we did this still would not indicate that reduction in sensitization is responsible for the long term reductions in pain incidence and/or intensity. Another hypothesis, invoked by Katz and Cohen [72] in discussing the effectiveness of perioperative epidural analgesia versus a sham epidural in reducing pain disability scores, but not pain intensity scores three weeks after surgery [73], is that the reduced hyperalgesia and rate of morphine consumption within the first two days after surgery afforded the epidural groups a "head start" in terms of comfort level and recovery compared with the sham epidural group possibly increasing their self-efficacy in dealing with pain or in mobilization. A similar finding has been reported after radical retropubic prostatectomy [74].

Taken together, the results of these studies are promising. They provide evidence that for some patients, long term pain problems can be prevented or minimized by perioperative multimodal treatment. However, there are several related issues that we still must address:

1. Preventive analgesia does not work for everyone and we do not know for whom such an approach is effective. We do not know why preventive analgesia does not work in some patients. One possibility is that pre-operative pain interferes with the effectiveness of preventive analgesia because central sensitization has already been established [16, 75–77].

2. We do not understand the mechanism(s) by which CPSP is reduced when preventive analgesia is effective, although we are probably correct in attributing the early effects to the pharmacological action of the agents used preventively;

3. The main outcome measures in the vast majority of RCTs are pain intensity or presence/absence of pain and analgesic use. It is rare to find a study that is more comprehensive in the outcome measures assessed. Recommendations for assessment of core measures and domains in clinical pain trials [78] include relevant psychological, emotional and physical variables in addition to those routinely assessed.

Assessment of additional domains of functioning may help to shed light on the predictors of severe acute postoperative pain, the processes involved in recovery from surgery, and the risk factors for developing CPSP [72]. Several psychosocial predictors of CPSP [19, 25, 79] or CPSP disability [23, 53] have been identified including increased preoperative state anxiety, [39], an introverted personality [19], catastrophizing, social support and solicitous responding in the week after amputation [23, 53], fear of surgery [25], and "psychic vulnerability" [79] a construct similar to neuroticism [4].

7.3
Chronic Pain outside the Surgical Context

The undoubted importance of CPSP notwithstanding, most chronic pain occurs in the absence of any previous surgical intervention. Nevertheless, the aphorism that 'every chronic pain was once acute' still applies. Chronic pain can be seen after all forms of injury and disease. If the disease is chronic (such as rheumatoid arthritis, for example), ongoing and visible pathologic changes plausibly explain the symptoms described (although there is still wide variation in the degree of pain, suffering and disability expressed by each patient). Much more perplexing are the cases in which there is no such evidence. The common factor in these cases is that the incident or injury to which the onset of the pain is attributed was either very minor or took place so long ago that both patient and therapist wonder why the natural healing process has not ameliorated the symptoms.

The presence of pain without a demonstrable tissue diagnosis is a deeply unsatisfying scenario. Strenuous efforts are often made to remedy this situation either with diagnostic tests in the individual case, or by group advocacy for the recognition of an organic etiology in new illnesses (fibromyalgia, for example). The inability to name a culpable disease is thought by some patients, their relatives and some clinicians to cast doubt on the veracity of the pain complaint. It also makes the process of obtaining disability-related concessions, financial compensation and damages much more difficult. For this reason, the most intensive research of the link between unpredictable acute pain and its transition to chronicity has been conducted in situations where the financial consequences of disability are most evident; in injured workers seeking wage replacement, and following motor vehicle collisions. The pain experience of injured workers is described in Chapter 15. The pain experience of victims of motor vehicle collision is the subject of several large studies using databases of those involved in collisions usually recruited shortly after the event, and most often for insurance purposes. Though this type of study has attracted strong methodological criticism from some quarters, it provides, uniquely outside the surgical context, the ability to determine the effect of putative risk factors prospectively. Canadian investigators have led the world in this area. Other research approaches include the analysis of population-based health survey data, either collected specifically with pain in mind, or of a more general nature.

What factors, then, predispose an acute non-surgical pain problem to become chronic? Most, but not all researchers find that, in cross-sectional studies at least, increasing age and female gender are predictive of the transition to chronicity. In common with CPSP, it also seems likely that acute pain intensity predicts chronicity to some degree [80, 81]. This immediately poses the question of whether aggressive steps to reduce pain intensity after injury (as distinct from *prior* to the injury as is only possible in surgery), besides being the compassionate thing to do, might reduce the likelihood of developing chronic pain, a question that cannot be answered at this time. Acute pain intensity may, of course, simply be a marker for other factors that predict both intensity and chronicity.

Depression and anxiety are known to be common and strong covariates of chronic pain. The debate about whether these are causes of or a consequence of chronic pain has occupied researchers and clinicians for many years, but recent prospective studies seem to suggest that for some populations, the psychological changes take place first [82]. A trial in elderly people showed that the treatment of depression alone significantly improved pain and pain-related disability [83]. The question of whether better treatment of depression on a population basis will lessen the risk of chronic pain has not been answered.

Pain catastrophizing is characterized by unrealistic beliefs that the current situation will lead to the worst possible pain outcome [84], negative thoughts about the future and self [85] and "an exaggerated negative 'mental set' brought to bear during actual or anticipated pain experience" [86, p. 53]. Pain catastrophizing is a multi-dimensional construct comprising elements of rumination, magnification, and helplessness [86–88]. One of the few consistent findings in the pain literature is that chronic pain patients who do not catastrophize fare better than patients who do [89].

The biopsychosocial model of chronic pain holds that other people in the patient's life make a material contribution to his or her pain experience. Solicitous responding by others, far from helping with pain actually makes pain and pain related disability worse [23, 53]. The operant conditioning model [90, 91] predicts that in offering pain contingent help (e.g. taking over household jobs) in response to expressions (grimacing) and behaviours (guarding) indicative of pain, well-intentioned spouses unwittingly positively reinforce the patient's pain behaviours leading to an increase in their frequency. On the other hand, the lack of any other person at all to provide love and care in a person's life is a risk factor for chronic pain (S. Rashiq, unpublished data). Other known, independent associations with chronic pain are physical inactivity, cigarette smoking, failing to graduate from high school, low income, being overweight and most chronic medical conditions (such as diabetes and heart disease).

7.4
Summary

Taken together, the studies reviewed above indicate that the transition of acute pain to chronic pain is a complex and poorly understood process. Figure 7.1 illustrates this in the surgical context. The noxious effects of surgery (e.g. incision, retraction, inflammatory response, ectopic activity following nerve injury) in conjunction with the competing, beneficial effects of perioperative, preventive multimodal analgesia interact with pre-existing and concurrent pain, psychological and emotional factors (e.g. catastrophizing, pain coping strategies) as well as the social environment (e.g. solicioutousness, social support) to determine the nature, severity, frequency and duration of CPSP. We are a long way from being able to predict with certainty who will recover uneventfully and who will go on to develop debilitating, chronic pain.

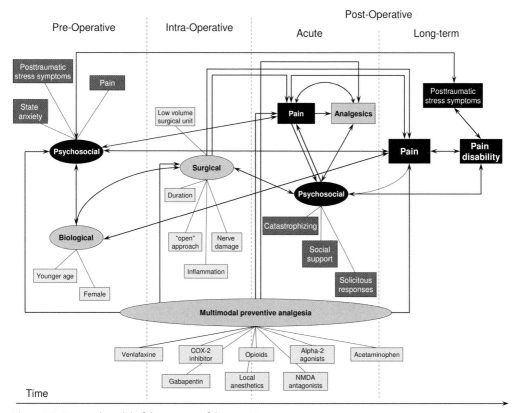

Figure 7.1 Proposed model of the process of the transition of acute post surgical pain to chronic pain.

7.5
Conclusions

The data reviewed in the present chapter suggest the following conclusions:

1. A conservative estimate is that severe CPSP occurs in up to 10% of patients. Procedures with the greatest incidence of CPSP are associated with intentional or unintentional intraoperative nerve injury such as postamputation phantom limb pain and chronic postthoracotomy pain.

2. Avoiding nerve damage will undoubtedly reduce the incidence of CPSP.

3. The most robust predictor of CPSP is the presence of and/or intensity of prior pain, experienced either preoperatively or in the days and weeks after surgery. One of the most important issues to resolve, however, concerns the nature of the relationship(s) between prior pain and CPSP; are they causal, associative, or a combination of the two?

4. Perioperative, multimodal, preventive analgesia regimens have been shown to reduce the incidence and/or the intensity of CPSP.

5. Psychological and social-environmental factors are associated with the development of chronic pain of all types.

7.6
Health Policy Implications and Recommendations

The following health policy recommendations are based in large part on recommendations made by the U.S. National Cancer Policy Board Committee on Cancer Survivorship [92] and draw heavily on this document since there is considerable overlap in the post-discharge needs and experiences of cancer patients and post-surgical patients; indeed many cancer patients will have undergone surgery and developed CPSP. The growing awareness of the problem of CPSP highlights the importance of emphasizing post-hospital discharge surveillance and management.

1. Patients are lost in the transition from acute, in-patient care to their homes. Pain management must be better coordinated post-discharge as surgery is currently considered the important event, not recovery.

2. We must raise awareness among surgeons, family doctors, patients, and the public at large about the needs and optimal care of patients before and after surgery. For example, there is the need for increased support for pre-operative patient education regarding pain and analgesia (e.g. how and when to communicate about pain and the need for analgesics). Education of allied health professionals such as nurses, physiotherapists, psychologists and pharmacists regarding importance of postoperative pain control is currently not regarded as a priority within training programs leading to poor awareness in those professions. All health practitioners need to understand the benefits of good

perioperative pain control in order that effective treatment is instituted and maintained until resolution of pain.

3. There is a belief among both patients and professionals that taking medication for pain after surgery is unnecessary and even harmful. Patients and physicians worry about addiction. Education of both practitioners and patients is needed regarding the following benefits of good pain control after surgery: improved function, faster recovery, reduction in complications and the rarity of addiction in previously opioid naïve individuals.

4. Upon discharge, patients should be given a clear and comprehensive care plan that describes expected time lines for recovery, follow up appointments with surgeons and community health care providers. Where appropriate, communication with social agencies regarding discharge planning is essential so that postoperative analgesic regimes are followed according to plan.

5. Health care providers and physicians should use up to date, evidence-based assessment instruments and practices (e.g. pain assessment, psychological assessment, pain management) to assess and manage CPSP.

6. The Canadian Pain Society (in conjunction with professional associations such as the Canadian Anesthesiologists' Society, the Canadian Psychological Association, and others) are urged to provide educational opportunities to health care providers regarding the transition of acute to persistent post surgical pain.

7. National funding agencies including CIHR, SSHRC, NCIC are urged to develop specific funding initiatives designed to promote basic science and clinical research into the transition of acute to chronic postsurgical pain. In particular more research is required into the debilitating physical and psychological effects of acute and chronic postsurgical pain and how effective pain control can optimize short, medium, and long-term function following surgery.

References

1 Kehlet, H., Jensen, T.S. and Woolf, C.J. (2006) Persistent postsurgical pain: risk factors and prevention. *Lancet*, **367**, 1618–25.

2 Macrae, W.A. (2001) Chronic pain after sternotomy. *Acta Anaesthesiologica Scandinavica*, **45**, 927–8.

3 Macrae, W.A. (2001) Chronic pain after surgery. *British Journal of Anaesthesia*, **87**, 88–98.

4 Perkins, F.M. and Kehlet, H. (2000) Chronic pain as an outcome of surgery. A review of predictive factors. *Anesthesiology*, **93**, 1123–33.

5 Hayes, C., Browne, S., Lantry, G. and Burstal, R. (2002) Neuropathic pain in the acute pain service: a prospective study. *Acute Pain*, **4**, 45–8.

6 Crombie, I.K., Davies, H.T. and Macrae, W.A. (1998) Cut and thrust: antecedent surgery and trauma among patients attending a chronic pain clinic. *Pain*, **76**, 167–71.

7 Tasmuth, T., Kataja, M., Blomqvist, C., von Smitten, K. and Kalso, E. (1997) Treatment-related factors predisposing to chronic pain in patients with breast cancer – a multivariate approach. *Acta Oncologica*, **36**, 625–30.

8 Aasvang, E.K., Mohl, B., Bay-Nielsen, M. and Kehlet, H. (2006) Pain related sexual dysfunction after inguinal herniorrhaphy. *Pain*, **122**, 258–63.

9 Grant, A.M., Scott, N.W. and O'Dwyer, P.J. (2004) Five-year follow-up of a randomized trial to assess pain and numbness after laparoscopic or open repair of groin hernia. *The British Journal of Surgery*, **91**, 1570–4.

10 Bay-Nielsen, M., Perkins, F.M. and Kehlet, H. (2001) Pain and functional impairment 1 year after inguinal herniorrhaphy: a nationwide questionnaire study. *Annals of Surgery*, **233**, 1–7.

11 Aasvang, E.K., Bay-Nielsen, M. and Kehlet, H. (2006) Pain and functional impairment 6 years after inguinal herniorrhaphy. *Hernia*, **10**, 316–21.

12 Katz, J., Jackson, M., Kavanagh, B.P. and Sandler, A.N. (1996) Acute pain after thoracic surgery predicts long-term post-thoracotomy pain. *The Clinical Journal of Pain*, **12**, 50–5.

13 Perttunen, K., Tasmuth, T. and Kalso, E. (1999) Chronic pain after thoracic surgery: a follow-up study. *Acta Anaesthesiologica Scandinavica*, **43**, 563–7.

14 Gotoda, Y., Kambara, N., Sakai, T., Kishi, Y., Kodama, K. and Koyama, T. (2001) The morbidity, time course and predictive factors for persistent post-thoracotomy pain. *European Journal of Pain (London, England)*, **5**, 89–96.

15 Nikolajsen, L., Sorensen, H.C., Jensen, T.S., and Kehlet, H. (2004) Chronic pain following Caesarean section. *Acta Anaesthesiologica Scandinavica*, **48**, 111–6.

16 Nikolajsen, L., Ilkjaer, S., Christensen, J.H., Kroner, K. and Jensen, T.S. (1997) Randomised trial of epidural bupivacaine and morphine in prevention of stump and phantom pain in lower-limb amputation. *Lancet*, **350**, 1353–7.

17 Jensen, T.S., Krebs, B., Nielsen, J. and Rasmussen, P. (1985) Immediate and long-term phantom limb pain in amputees: incidence, clinical characteristics and relationship to pre-amputation limb pain. *Pain*, **21**, 267–78.

18 Hanley, M.A., Jensen, M.P., Smith, D.G., Ehde, D.M., Edwards, W.T. and Robinson, L.R. (2007) Preamputation pain and acute pain predict chronic pain after lower extremity amputation. *The Journal of Pain*, **8**, 102–9.

19 Borly, L., Anderson, I.B., Bardram, L., Christensen, E., Sehested, A., Kehlet, H. et al. (1999) Preoperative prediction model of outcome after cholecystectomy for symptomatic gallstones. *Scandinavian Journal of Gastroenterology*, **34**, 1144–52.

20 Meyerson, J., Thelin, S., Gordh, T. and Karlsten, R. (2001) The incidence of chronic post-sternotomy pain after cardiac surgery–a prospective study. *Acta Anaesthesiologica Scandinavica*, **45**, 940–4.

21 Nikolajsen, L., Brandsborg, B., Lucht, U., Jensen, T.S. and Kehlet, H. (2006) Chronic pain following total hip arthroplasty: a nationwide questionnaire study. *Acta Anaesthesiological Scandinavica*, **50**, 495–500.

22 Tasmuth, T., Blomqvist, C. and Kalso, E. (1999) Chronic post-treatment symptoms in patients with breast cancer operated in different surgical units. *European Journal of Surgical Oncology*, **25**, 38–43.

23 Hanley, M.A., Jensen, M.P., Ehde, D.M., Hoffman, A.J., Patterson, D.R. and Robinson, L.R. (2004) Psychosocial predictors of long-term adjustment to lower-limb amputation and phantom limb pain. *Disability and Rehabilitation*, **26**, 882–93.

24 Kalso, E., Mennander, S., Tasmuth, T. and Nilsson, E. (2001) Chronic post-sternotomy pain. *Acta Anaesthesiologica Scandinavica*, **45**, 935–9.

25 Peters, M.L., Sommer, M., de Rijke, J.M., Kessels, F., Heineman, E., Patijn, J. et al. (2007) Somatic and psychologic predictors of long-term unfavorable outcome after surgical intervention. *Annals of Surgery*, **245**, 487–94.

26 Liem, M.S., van Duyn, E.B., van der Graaf, Y. and van Vroonhoven, T.J. (2003) Recurrences after conventional anterior and laparoscopic inguinal hernia repair: a randomized comparison. *Annals of Surgery*, **237**, 136–41.

27 Cerfolio, R.J., Price, T.N., Bryant, A.S., Sale Bass, C. and Bartolucci, A.A. (2003) Intracostal sutures decrease the pain of thoracotomy. *The Annals of Thoracic Surgery*, **76**, 407–11. discussion 11-2.

28 Wall, P.D., Devor, M., Inbal, R., Scadding, J.W., Schonfeld, D., Seltzer, Z. *et al.* (1979) Autotomy following peripheral nerve lesions: experimental anaesthesia dolorosa. *Pain*, **7**, 103–11.

29 Bennett, G.J. and Xie, Y.K. (1988) A peripheral mononeuropathy in rat that produces disorders of pain sensation like those seen in man. *Pain*, **33**, 87–107.

30 Seltzer, Z., Dubner, R. and Shir, Y. (1990) A novel behavioral model of neuropathic pain disorders produced in rats by partial sciatic nerve injury. *Pain*, **43**, 205–18.

31 Kim, S.H. and Chung, J.M. (1992) An experimental model for peripheral neuropathy produced by segmental spinal nerve ligation in the rat. *Pain*, **50**, 355–63.

32 Rasmussen, S. and Kehlet, H. (2007) Management of nerves during leg amputation–a neglected area in our understanding of the pathogenesis of phantom limb pain. *Acta Anaesthesiologica Scandinavica*, **51**, 1115–16.

33 Katz, J. (1997) Pain begets pain– predictors of long-term phantom limb pain and post-thoracotomy pain. *Pain Forum*, **6**, 140–4.

34 Dworkin, R.H. (1997) Which individuals with acute pain are most likely to develop a chronic pain syndrome? *Pain Forum*, **6**, 127–36.

35 Nikolajsen, L., Ilkjaer, S., Kroner, K., Christensen, J.H. and Jensen, T.S. (1997) The influence of preamputation pain on postamputation stump and phantom pain. *Pain*, **72**, 393–405.

36 Caumo, W., Schmidt, A.P., Schneider, C.N., Bergmann, J., Iwamoto, C.W., Adamatti, L.C. *et al.* (2002) Preoperative predictors of moderate to intense acute postoperative pain in patients undergoing abdominal surgery. *Acta Anaesthesiologica Scandinavica*, **46**, 1265–71.

37 Scott, L.E., Clum, G.A. and Peoples, J.B. (1983) Preoperative predictors of postoperative pain. *Pain*, **15**, 283–93.

38 Thomas, T., Robinson, C., Champion, D., McKell, M. and Pell, M. (1998) Prediction and assessment of the severity of post-operative pain and of satisfaction with management. *Pain*, **75**, 177–85.

39 Harden, R.N., Bruehl, S., Stanos, S., Brander, V., Chung, O.Y., Saltz, S. *et al.* (2003) Prospective examination of pain-related and psychological predictors of CRPS-like phenomena following total knee arthroplasty: a preliminary study. *Pain*, **106**, 393–400.

40 Brander, V.A., Stulberg, S.D., Adams, A.D., Harden, R.N., Bruehl, S., Stanos, S.P. *et al.* (2003) Predicting total knee replacement pain: a prospective, observational study. *Clinical Orthopaedics and Related Research*, 27–36.

41 Poobalan, A.S., Bruce, J., King, P.M., Chambers, W.A., Krukowski, Z.H. and Smith, W.C. (2001) Chronic pain and quality of life following open inguinal hernia repair. *The British Journal of Surgery*, **88**, 1122–6.

42 Lau, H., Patil, N.G., Yuen, W.K. and Lee, F. (2003) Prevalence and severity of chronic groin pain after endoscopic totally extraperitoneal inguinal hernioplasty. *Surgical Endoscopy*, **17**, 1620–3.

43 Callesen, T., Bech, K. and Kehlet, H. (1999) Prospective study of chronic pain after groin hernia repair. *The British Journal of Surgery*, **86**, 1528–31.

44 Senturk, M., Ozcan, P.E., Talu, G.K., Kiyan, E., Camci, E., Ozyalcin, S. *et al.* (2002) The effects of three different analgesia techniques on long-term postthoracotomy pain. *Anesthesia and Analgesia*, **94**, 11–15.

45 Romundstad, L., Breivik, H., Roald, H., Skolleborg, K., Romundstad, P.R. and Stubhaug, A. (2006) Chronic pain and sensory changes after augmentation mammoplasty: long term effects of preincisional administration of methylprednisolone. *Pain*, **124**, 92–9.

46 Poleshuck, E.L., Katz, J., Andrus, C.H., Hogan, L.A., Jung, B.F., Kulick, D.I. *et al.* (2006) Risk factors for chronic pain following breast cancer surgery: a prospective study. *The Journal of Pain*, **7**, 626–34.

47 Kalkman, C.J., Visser, K., Moen, J., Bonsel, G.J., Grobbee, D.E. and Moons, K.G. (2003) Preoperative prediction of severe postoperative pain. *Pain*, **105**, 415–23.

48 Coderre, T.J. and Katz, J. (1997) Peripheral and central hyperexcitability: differential signs and symptoms in persistent pain.

The Behavioral and Brain Sciences, **20**, 404–19. (discussion 35–513).

49 Devor, M., Basbaum, A.I., Bennett, G.J., Blumberg, H., Campbell, J.N., Dembowsky, K.P. *et al.* (1991) Group report: mechanisms of neuropathic pain following peripheral injury, in *Towards a New Pharmacotherapy of Pain* (eds A.I. Basbaum and J.M. Besson), John Wiley & Sons, New York, NY, pp. 417–40.

50 Woolf, C.J. and Salter, M.W. (2000) Neuronal plasticity: increasing the gain in pain. *Science*, **288**, 1765–9.

51 Devor, M., del Canho, S. and Raber, P. (2005) Heritability of symptoms in the neuroma model of neuropathic pain: replication and complementation analysis. *Pain*, **116**, 294–301.

52 Devor, M. and Raber, P. (1990) Heritability of symptoms in an experimental model of neuropathic pain. *Pain*, **42**, 51–67.

53 Jensen, M.P., Ehde, D.M., Hoffman, A.J., Patterson, D.R., Czerniecki, J.M. and Robinson, L.R. (2002) Cognitions, coping and social environment predict adjustment to phantom limb pain. *Pain*, **95**, 133–42.

54 Katz, J. and Melzack, R. (1990) Pain "memories" in phantom limbs: review and clinical observations. *Pain*, **43**, 319–36.

55 Mitchell, S.W. (1872) *Injuries of Nerves and Their Consequences*, J.B. Lippincott, Philadelphia.

56 Nathan, P.W. (1962) Pain traces left in the central nervous system, in *The Assessment of Pain in Man and Animals* (eds C.A. Keele and R. Smith), Livingstone, Edinburgh, pp. 129–34.

57 Melzack, R. and Wall, P.D. (1996) *The Challenge of Pain*, Basic Books, New York.

58 Katz, J. (1995) Pre-emptive analgesia: evidence, current status and future directions. *European Journal of Anaesthesiology. Supplement*, **10**, 8–13.

59 Katz, J. and McCartney, C.J.L. (2002) Current status of pre-emptive analgesia. *Current Opinion in Anaesthesiology*, **15**, 435–41.

60 Kissin, I. (1994) Preemptive analgesia: terminology and clinical relevance. *Anesthesia and Analgesia*, **79**, 809.

61 Kissin, I. (2000) Preemptive analgesia. *Anesthesiology*, **93**, 1138–43.

62 Katz, J. and Clarke, H. (2008) Preventive analgesia and beyond: current status, evidence, and future directions, in *Clinical Pain Management: Acute Pain* (eds D.J. Rowbotham and P.E. Macintyre), Arnold, London, in press.

63 McCartney, C.J., Sinha, A. and Katz, J. (2004) A qualitative systematic review of the role of N-methyl-D-aspartate receptor antagonists in preventive analgesia. *Anesthesia and Analgesia*, **98**, 1385–400.

64 Reuben, S.S. and Buvanendran, A. (2007) Preventing the development of chronic pain after orthopaedic surgery with preventive multimodal analgesic techniques. *The Journal of Bone and Joint Surgery*, **89**, 1343–58.

65 Gilron, I. and Max, M.B. (2005) Combination pharmacotherapy for neuropathic pain: current evidence and future directions. *Expert Review of Neurotherapeutics*, **5**, 823–30.

66 Fernyhough, J.C., Schimandle, J.J., Weigel, M.C., Edwards, C.C. and Levine, A.M. (1992) Chronic donor site pain complicating bone graft harvesting from the posterior iliac crest for spinal fusion. *Spine*, **17**, 1474–80.

67 Goulet, J.A., Senunas, L.E., DeSilva, G.L. and Greenfield, M.L. (1997) Autogenous iliac crest bone graft. Complications and functional assessment. *Clinical Orthopaedics and Related Research*, 76–81.

68 Kurz, L.T., Garfin, S.R. and Booth, R.E. Jr. (1989) Harvesting autogenous iliac bone grafts. A review of complications and techniques. *Spine*, **14**, 1324–31.

69 Reuben, S.S., Vieira, P., Faruqi, S., Verghis, A., Kilaru, P.A. and Maciolek, H. (2001) Local administration of morphine for analgesia after iliac bone graft harvest. *Anesthesiology*, **95**, 390–4.

70 Reuben, S.S., Ekman, E.F., Raghunathan, K., Steinberg, R.B., Blinder, J.L. and Adesioye, J. (2006) The effect of cyclooxygenase-2 inhibition on acute and chronic donor-site pain after spinal-fusion surgery. *Regional Anesthesia and Pain Medicine*, **31**, 6–13.

71 Reuben, S.S. and Ekman, E.F. (2007) The effect of initiating a preventive multimodal analgesic regimen on long-term patient outcomes for outpatient anterior cruciate ligament reconstruction surgery. *Anesthesia and Analgesia*, **105**, 228–32.

72 Katz, J. and Cohen, L. (2004) Preventive analgesia is associated with reduced pain disability 3 weeks but not 6 months after major gynecologic surgery by laparotomy. *Anesthesiology*, **101**, 169–74.

73 Katz, J., Cohen, L., Schmid, R., Chan, V.W. and Wowk, A. (2003) Postoperative morphine use and hyperalgesia are reduced by preoperative but not intraoperative epidural analgesia: implications for preemptive analgesia and the prevention of central sensitization. *Anesthesiology*, **98**, 1449–60.

74 Gottschalk, A., Smith, D.S., Jobes, D.R., Kennedy, S.K., Lally, S.E., Noble, V.E. *et al.* (1998) Preemptive epidural analgesia and recovery from radical prostatectomy: a randomized controlled trial. *The Journal of the American Medical Association*, **279**, 1076–82.

75 Bach, S., Noreng, M.F. and Tjellden, N.U. (1988) Phantom limb pain in amputees during the first 12 months following limb amputation, after preoperative lumbar epidural blockade. *Pain*, **33**, 297–301.

76 Klasen, J., Haas, M., Graf, S., Harbach, H., Quinzio, L., Jurgensen, I. *et al.* (2005) Impact on postoperative pain of long-lasting pre-emptive epidural analgesia before total hip replacement: a prospective, randomised, double-blind study. *Anaesthesia*, **60**, 118–23.

77 Aida, S., Fujihara, H., Taga, K., Fukuda, S. and Shimoji, K. (2000) Involvement of presurgical pain in preemptive analgesia for orthopedic surgery: a randomized double blind study. *Pain*, **84**, 169–73.

78 Dworkin, R.H., Turk, D.C., Farrar, J.T., Haythornthwaite, J.A., Jensen, M.P., Katz, N.P. *et al.* (2005) Core outcome measures for chronic pain clinical trials: IMMPACT recommendations. *Pain*, **113**, 9–19.

79 Jorgensen, T., Teglbjerg, J.S., Wille-Jorgensen, P., Bille, T. and Thorvaldsen, P. (1991) Persisting pain after cholecystectomy. A prospective investigation. *Scandinavian Journal of Gastroenterology*, **26**, 124–8.

80 Dufton, J.A., Kopec, J.A., Wong, H., Cassidy, J.D., Quon, J., Mcintosh, G. and Koehoorn, M. (2006) Prognostic factors associated with minimal improvement following acute whiplash-associated disorders. *Spine*, **31**, E759–65.

81 Holm, L.M., Carroll, L.J., Cassidy, D., Skillgate, E. and Ahlbom, A. (2007) Widespread pain following whiplash-associated disorders; incidence, course and risk factors. *The Journal of Rheumatology*, **34**, 193–200.

82 Carroll, L.J., Cassidy, J.D. and Cote, D. (2004) Depression as a risk factor for onset of an episode of troublesome neck and low back pain. *Pain*, **107**, 134–9.

83 Lin, E.H.B., Katon, W., Von Korff, M. *et al.* (2003) Effect of improving depression care on pain and functional outcomes among older adults with arthritis: a randomized controlled trial. *The Journal of the American Medical Association*, **290**, 2428–9.

84 Keefe, F.J., Brown, G.K., Wallston, K.A. and Caldwell, D.S. (1989) Coping with rheumatoid arthritis pain: catastrophizing as a maladaptive strategy. *Pain*, **37**, 51–6.

85 Turk, D.C. (2002) A diathesis-stress model of chronic pain and disability following traumatic injury. *Pain Research and Management*, **7**, 9–19.

86 Sullivan, M.J.L., Thorn, B., Haythornthwaite, J.A., Keefe, F., Martin, M., Bradley, L.A. et al. (2001) Theoretical perspectives on the relation between catastrophizing and pain. *The Clinical Journal of Pain*, **17**, 52–64.

87 Osman, A., Barrios, F.X., Gutierrez, P.M., Kopper, B.A., Merrifield, T. and Grittmann, L. (2000) The Pain Catastrophizing Scale: further psychometric evaluation with adult samples. *Journal of Behavioral Medicine*, **23**, 351–65.

88 Osman, A., Barrios, F.X., Kopper, B.A., Hauptmann, W., Jones, J. and O'Neill, E. (1997) Factor structure, reliability, and validity of the Pain Catastrophizing Scale. *Journal of Behavioral Medicine*, **20**, 589–605.

89 Jensen, M.P., Turner, J.A., Romano, J.M. and Karoly, P. (1991) Coping with chronic pain: a critical review of the literature. *Pain*, **47**, 249–83.

90 Fordyce, W.E. (1989) The cognitive/behavioral perspective on clinical pain, in

Managing the Chronic Pain Patient New York (eds J.D. Loeser and K.J. Egan), Raven Press, NY, pp. 51–64.

91 Fordyce, W.E. (1976) *Behavioral Methods for Chronic Pain and Illness*, Mosby, St. Louis, MO.

92 Hewitt, M., Greenfield, S. and Stovall, E. (eds) (2006) *From Cancer Patient to Cancer Survivor: Lost in Transition*, The National Academies Press, Washington, DC.

8
What Chronic Pain Looks Like to the Clinician

Chris Spanswick

- There are many distinct conditions subsumed by the term chronic pain, each with its own complexities and challenges for optimal assessment and treatment.

- Patients and family physicians initiate consultations with multidisciplinary pain clinics for a great variety of reasons that may be obscure at the time of consultation but are often crucial determiners of clinical outcome and satisfaction.

- Typical patients referred to multidisciplinary pain clinics have been in pain for many years, have multiple chronic illnesses, have failed to improve with community-based single modality treatment, are physically de-conditioned, are in considerable distress and are likely to have financial challenges.

8.1
Introduction

The purpose of this chapter is to describe patients with chronic pain referred to a specialty pain clinic. The first sections describe the typical patient of a typical pain clinic and their complaints. The next section describes special challenges surrounding the emotional state of the patient on presentation to the clinic. The final section discusses the reasons why patients consult and why their family doctors refer.

8.2
The Typical Patient

What does the "average" patient attending a pain clinic look like?

She is likely to be female (70%), reflecting the distribution of pain disorders in the general population and the tendency for women to consult more often than men with healthcare problems (Annual Report of the Calgary Health Region

Chronic Pain: A Health Policy Perspective
Edited by S. Rashiq, D. Schopflocher, P. Taenzer, and E. Jonsson
Copyright © 2008 WILEY-VCH Verlag GmbH & Co. KGaA, Weinheim
ISBN: 978-3-527-32382-1

Chronic Pain Centre, 2006). She will most likely be in her late forties. She will most likely have completed high school; nearly half will have completed further education. She is likely married or in a common law relationship (60%).

On average, she will have a history of pain of eight to ten years prior to being referred to a pain clinic. Her pain will score between 6 and 7 on a 0 to 10 severity scale; where 0 is no pain and 10 is the worst pain one could imagine. She will report a low quality of life on self report health related quality of life scales such as the SF-36 [1].

8.3
Common Pain Conditions

Pain may be divided into different types, depending on the underlying mechanisms. It may be musculoskeletal if it arises from muscles, bones, or joints. It may be visceral if it arises from hollow organs within the body, such as the gut. It may be neuropathic if it arises from damaged nerves or nerves that are not functioning normally. In practice, most pains are a mixture often involving more than one mechanism. This may be seen in spinal pain, which often presents with both a musculoskeletal pain and a neuropathic pain.

8.3.1
Spinal Pain

Spinal pain is the commonest reason that patients are referred to a chronic pain clinic, accounting for 75 to 80% of referrals [2]. *Low back pain* is much more common than *neck or cervical pain* although it is quite common for patients to present with more than one spinal pain disorder.

8.3.2
Neck Pain

Chronic disabling neck pain is said to affect up to 5% of the general population [3]. Patients with persistent neck pain may have pain that radiates into the arms. This may be a function of the severity of their pain or may be due to irritation of the nerves in the neck that go to the arms. Usually, if the nerves are affected the patient will describe pins and needles in the upper limbs and may also have numbness. The nerve supply to the back part of the head is shared with the upper part of the neck. Patients with neck problems may therefore also experience headaches. However, it is also possible for neck problems to be a trigger for migraine. Patients with persistent neck pain problems have significant restriction of movements. The pain and stiffness has a profound impact on activities of daily living. Even the act of holding a book may be impossible for some. Patients often complain of problems with sleep, concentration and a sense of weakness in the neck such that they feel they need some form of support to take the weight off.

8.3.3
Whiplash

Neck pain is often associated with motor vehicle collisions. About 10 to 20% of patients with a *Whiplash Associated Disorder* following a motor vehicle collision can expect to have prolonged pain that may never settle [3]. Whiplash describes the mechanism of neck injury, usually due to a motor vehicle collision. It involves a rapid hyperextension and/or hyperflexion of the cervical spine due to sudden changes in acceleration of the body. Many of these may occur as a result of low speed (<10 mph) collisions.

In the clinic the term "whiplash associated disorder" is used to describe neck pain and other symptoms following an injury involving the mechanism outlined above. The Quebec Task Force on whiplash associated disorders (WAD) published a monograph in 1995 in order to try and clarify assessment and management of whiplash disorders [4].

Since the neck pain follows a collision, there is usually some form of litigation or insurance claim, which compounds the clinical presentation and usually contributes to the level of distress. Epidemiological studies indicate a wide variety in frequency of neck pain in claimants following WAD in different countries [5]. This indicates the complex nature of the problem that includes factors other than the injury, including socio-economic factors, fear of recurrent pain, social norms and peer behavior. Some patients go on to develop widespread pain and up to a fifth of patients with whiplash associated disorder will develop fibromyalgia syndrome (FMS) [6].

Patients often arrive in great distress. By the time they are referred to a pain clinic most "standard" treatments have not worked, jobs may be in jeopardy and there may be depression and significant problems with sleep. Patients are not only distressed but often confused by conflicting clinical opinions from different experts. They are unsure what to do next.

8.3.4
Low Back Pain

Low back pain may be associated with leg pain. This may simply be pain radiating from the spine or if it extends to the foot may be neuropathic and due to irritation of nerves in the spine. A thorough history, examination and, if indicated, special tests including MRI scanning, may be necessary to exclude treatable causes such as a prolapsed disc or narrowing of the spinal canal through which the spinal cord or nerves pass.

Many patients, however, have mechanical low back pain and do not have any nerve involvement. Most often MRI scanning does not show any specific cause [7]. The relationship between MRI scans and symptoms is not straightforward and often MRIs do not distinguish between individuals with pain and those without. Similarly, X-rays are often of little help. As a result, there is no specific explanation for the patient's pain and no specific treatment to offer. Patients, not unreasonably, tend to want to repeat such procedures in the hope of finding a specific explanation and a specific treatment that can provide long term benefit.

8.3.5
Sciatica

Sciatica is a term used to describe leg pain due to a nerve in the lumbar spine (lower back) being compressed or irritated. The lay public often use the term inaccurately to indicate pain from the back radiating into the leg. The sciatic nerve is a large nerve that serves sensation and motor power to the leg. It takes its nerve supply from a number of nerve roots that exit from the spine at the lower back.

Patients typically experience pain radiating down the back of the leg. The pain is commonly described as a shooting pain and is made worse on coughing or sneezing. The pain usually extends down to the toes. It may be associated with pins and needles in the leg, and more so in the foot, as well as a burning sensation in one side of the foot. If the motor nerves are also affected the patient may complain of very specific muscle weakness, for example inability to lift the foot, leading to tripping over curb stones or on stairs.

The commonest cause is irritation or compression of one of the nerves as it exits the spinal canal. The commonest cause of this is a prolapsed, bulging or even ruptured intervertebral disc pressing on the nerve. The sciatic nerve itself may also be squeezed by spasm of muscles around it in the buttock. This disorder is called Piriformis Syndrome.

Careful clinical examination (searching for changes in sensation, reflexes and motor power) will enable the professional to determine the most likely cause and special tests (including MRI scanning) will help confirm the cause.

In some cases, further surgery may be needed if the previous surgery was either inadequate or at the wrong level. However the outcome following second and subsequent surgery is less good [8] and indeed may make the situation worse.

8.3.6
Failed Back Surgery Syndrome

Failed back surgery syndrome (FBSS) is the general term used to describe cases in which back surgery performed for back pain has left the patient with pain and those in which decompressing a nerve has failed to resolve nerve pain in the leg. The patient will typically have had symptoms for many months and several imaging studies and surgical opinions. There is usually activity-dependent pain, which may limit activities of daily living and work, a constriction of social life and pleasurable activity and often a considerable burden of side-effects from drugs given to ease the symptoms. FBSS patients are often reluctant to exercise, believing that this may worsen their predicament, and this often becomes a self-fulfilling prophecy as lack of exercise leads to physical deconditioning and yet more disability.

FBSS is not really a diagnosis but a description. In the US more than 500 000 spinal surgeries were performed in 2001, approximately twice the number of back operations per capita compared with Canada [9]. The rate of FBSS is reported to be between 10 and 40% [10].

Assessing patients with FBSS can be challenging for a number of reasons. First the patient is usually distressed about having undergone significant invasive surgery without achieving symptom reduction or resolution. Not only are they left with the ongoing pain for which the surgery was initially offered, but it may seem that there are no other options left for them to pursue. Second, the diagnosis (either initial or subsequent) may not be clear and therefore further treatment may be difficult to plan. It is difficult to know whether the back has failed, the surgery has failed or whether the surgery was appropriate.

There are a number of different types of spinal surgery for differing clinical problems. These include decompressive surgery for nerve or spinal cord entrapment and fusion where the purpose is to prevent movement of part of the spine. The place of fusion surgery is still hotly debated.

Psychological distress often clouds the picture for the clinician, which complicates the assessment and treatment process even further. It is essential to have the expertise of a number of professions and specialties in order to address the multifaceted nature of the problems that patients with FBSS present.

8.3.7
Arthritis

Arthritis is common and has a significant impact on physical functioning, work, quality of life and healthcare use. It is anticipated that up to 6 million people in Canada will have some form of arthritis by 2035 and that 15% of the Canadian population will have been diagnosed as having a long-term problem with arthritis [11]. Even these figures may be too conservative [12].

The term arthritis includes a number of different diseases, which are usually painful and have a significant effect on activities of daily living. Osteoarthritis is perhaps the best known to the lay public, but there are a number of other arthritides and disease processes that give rise to pain and inflammation of joints, these include rheumatoid arthritis, ankylosing spondylitis, arthritis associated with psoriasis and gout. Most of these diseases are connective tissue diseases and also affect other organs within the body. The mechanism of "injury" is known to be autoimmune. In other words the body's immune system attacks the body itself. This leads to inflammation and ultimately destruction of tissue. In the case of rheumatoid arthritis this includes joints and the tissues surrounding the joints.

The chronic pain of arthritis is due in part to the ongoing disease process. Much of the management of inflammatory arthritides involves using disease modifying drugs, anti-inflammatory medications, and steroids, in addition to appropriate analgesics [13]. It is for this reason that family practitioners refer patients in the first instance to rheumatologists rather than pain clinics. A large number of patients are, however, managed in the community by their family physician without referral to a rheumatologist [14]. Smaller numbers of patients with arthritis are seen in pain clinics. Many patients attending pain clinics have multiple pain problems and it is not uncommon for arthritis to be part of the overall clinical picture.

8.3.8
Fibromyalgia and Widespread Pain

Many patients have more than one pain problem and frequently develop pains in other parts of their body and may ultimately develop widespread pain.

Although *fibromyalgia* [15] is present in between 3 and 6% of the general population [16], it makes up a larger proportion of the pain clinic population. The condition is very variable and may be very severe. It is commonly associated with fatigue, non-restorative sleep, irritable bowel syndrome and clinical depression. Patients with fibromyalgia may also have concurrent disease, for example arthritis, which can complicate their situation.

Many patients do not have symptoms and physical findings that are compatible with a diagnosis of fibromyalgia, but still present with widespread pain. These patients are just as challenging to assess and treat. It is vital that patients are screened medically to check for other known causes of widespread symptoms that may be treatable.

For many patients, previous attempts at improving levels of physical function have been unsuccessful because of large increases in either pain and/ or fatigue. This leads to a sense of frustration and can lead to hopelessness. It is common for patients with fibromyalgia to be on a large number of different medications. These may include painkillers, including strong opioids, tranquilizers and sleeping tablets. The patient's problems may indeed be made worse by inappropriate or unhelpful medications. Pain clinic physicians often help patients modify their drug usage. Helping patients with these problems requires considerable skill and patience.

8.3.9
Neuropathic Pain

Neuropathic pain affects a small proportion of the population [17]. About 30% of patients referred to a pain clinic have symptoms of neuropathic pain. Neuropathic pain usually has one or more recognizable symptoms which can help make the diagnosis. These include shooting pain, burning pain and other sensory symptoms such as tingling. It may be initiated by trauma, surgery, infections (e.g. shingles) or be part of a peripheral neuropathy such as diabetic neuropathy. Careful assessment is needed to check for possible underlying causes.

Some patients display skin sensitivity, which can be so bad that they are unable to tolerate anything touching their skin (even clothing). These patients always have major problems at night and their sleep is badly disrupted.

Neuropathic pain is often a difficult diagnosis for the non-specialist to make and may be mistaken for a musculoskeletal pain. This can lead to unsuccessful therapy and frustration on the part of the patient and physician. Neuropathic pain responds very variably to treatment, even for the most experienced practitioners.

8.3.10
Diabetic Neuropathy

Diabetes is the commonest cause of neuropathy in the West [18]. It is most commonly seen in diabetic patients over the age of 50 and affects up to 50% of patients. It can be present in many different forms depending upon the particular part of the peripheral nervous system affected.

In diabetes, damage and loss of peripheral nerve fibers may occur. This may be due to poor circulation to the nerves. Most patients are seen within diabetic clinics and services, especially if they have acute symptoms. Most if not all services regularly monitor their patient population for the complications of diabetes, of which the development of peripheral neuropathy is an important one. A number of patients with chronic or persistent pain associated with their neuropathy may be referred to a pain clinic.

Patients may present with acute or chronic symptoms. These include altered sensation, tingling, numbness, burning pain and shooting pains. A common presentation is with peripheral burning pins and needles in the feet. This is usually associated with sensory changes. The most common [18] is a Distal Symmetric Sensorimotor Polyneuropathy. This can initially present with similar symptoms with distressing numbness and tingling in the feet. These sensations may spread up the leg and can eventually spread to hands and trunk. Patients often report both burning sensations and shooting pains. Patients show loss of touch, pain and vibration sense. Some patients notice changes in their balance because of a lack of normal position sense from the legs and feet, not because of any involvement of the balance mechanism in the inner ear.

Diabetes can also cause single nerve neuropathy or radiculopathy. Pain is often experienced in the low back or hip and radiates down the leg. Patients describe aching pain with superimposed shooting or stabbing pain. There is often some muscle weakness and changes in reflexes but few sensory changes. Most patients recover but recovery may not be complete.

8.3.11
Trigeminal Neuralgia (Tic Douloureux)

Trigeminal neuralgia is a unilateral (single-sided) facial pain, which is characteristically a recurrent severe, brief, stabbing pain. It is usually experienced in the area served by the trigeminal nerve. The underlying cause is not fully understood, although there does seem to be an association with a blood vessel being in close contact with the nerve within the skull [19]. More recent surgical treatments are aimed at decompressing or moving the vessel away from the nerve.

Trigeminal neuralgia is more common in the elderly with a peak incidence between 50 and 70 [20]. Patients describe bouts of severe debilitating pain with an electric shock quality. The pain can be so severe as to render the patient suicidal. The episodes of pain can be provoked by minimal stimulation of the face or by chewing. Commonly the patients have difficulty eating and male patients may stop shaving for fear of provoking an episode. Some patients have extensive dental extractions when the problem is mistakenly attributed to problems with their teeth.

Trigeminal neuralgia is usually intermittent, with the patient suffering bouts of attacks with periods of (relative) freedom from pain for weeks or months at a time. Assessment and management include a careful history, clinical examination and institution of further tests (e.g. MRI of the head) if other causes are being sought or if surgery is being contemplated. The history is usually unmistakable. Clinical examination usually reveals no sensory loss. Fortunately trigeminal neuralgia usually responds well to anticonvulsant medications. Side effects of these drugs can be troublesome, especially in elderly patients.

8.3.12
Post Herpetic Neuralgia

Post herpetic neuralgia (PHN) is a condition following an infection with the varicella virus (the chickenpox virus). The acute infection is commonly called *shingles*. The virus remains dormant in the nervous system following the usual childhood infection (chickenpox). It can re-emerge later in life, more commonly in people over the age of 50, as shingles. The precise reasons for this are not fully understood, but it is thought to be due to decreases in immunity [21]. The infection tends to affect a single nerve and causes a rash and considerable pain. The pain often precedes the rash and can lead to a misdiagnosis until the rash appears later. The acute infection is short lived and in most people the associated pain will settle over time.

The occurrence of PHN increases with age along with the increase in shingles. PHN is defined as pain persisting after the healing of the rash associated with the acute infection. The healing of the rash commonly happens in a month, but since pain may continue to improve after this, it is common to describe PHN as occurring after three months.

The pain is experienced in the distribution of the nerve affected and occurs in an estimated 14% of people suffering acute shingles. The likelihood of developing PHN following acute shingles increases with age [22]. The pain has all the characteristics of nerve injury pain with patients describing burning and shooting pain. Often patients describe intense sensitivity of the skin. They are sometimes unable to tolerate the touch of clothes. All of these symptoms can persist indefinitely. Patients are usually distressed and sometimes frankly depressed, which magnifies their suffering.

Treatments include drugs used in the management of neurogenic pain including: antidepressants, anticonvulsant drugs and opioids. Topical agents have been used but the outcomes are variable and there is little evidence of efficacy with the exception of topical Lidocaine (Lidoderm) although it should be said that the benefit is only modest [23]. Surgery (cutting the nerve) appears to be unhelpful and can make the situation worse. Early use of antiviral agents does not seem to make a great difference in the subsequent occurrence of PHN. There is now evidence that vaccinating the older population (over 70) not only reduces the frequency of acute shingles in that age group but also reduces the frequency of PHN [24].

8.3.13
Complex Regional Pain Syndrome

The International Association for the Study of Pain has attempted to clarify the diagnosis of complex regional pain syndrome (CRPS). Two types are described; type 1, without evidence of nerve injury and type 2 with evidence of nerve injury [25].

This syndrome has been known by a multitude of names (reflex sympathetic dystrophy, Sudek's atrophy, hand-shoulder syndrome, causalgia to name a few). It is a very distressing and poorly understood condition.

CRPS Type 1 may occur following an apparently innocuous injury, fractures, heart attack, and strokes. It may even occur without a previous injury. Because of difficulties in diagnosis and lack of uniformity of cases it is difficult to estimate the incidence of either type 1 or type 2 CRPS, but it is rare. Most texts indicate rates varying between 1% following fractures to as much as 25% following strokes [26].

CRPS is characterized by chronic pain, usually in a limb, associated with local changes in sweating, skin color, skin sensitivity, skin temperature, swelling, muscle spasm, wasting of muscles, and changes in hair and nail growth. Patients do not necessarily present with all the symptoms and signs.

However, the predominant feature is severe intractable pain, usually described as a burning sensation although other descriptors are also used. The pain is commonly accompanied by painful sensitivity of the skin. The pain and other symptoms and clinical findings extend outside the territory of any individual nerve, hence the term "regional" pain syndrome. Perhaps the most striking features are the level of dysfunction and the loss of use of the limb and profound psychological distress.

CRPS Type 2 is associated with a nerve injury. The incidence is thought to be approximately 1%. It is seen more commonly during war in association with high velocity projectile injuries to nerves in the upper or lower limb. The presentation is very similar to CRPS Type1.

There are no definitive objective diagnostic tests for CRPS. Nerve conduction studies may be normal in CRPS Type 1. X-rays may show patchy localized osteoporosis and bone scans may show abnormalities, but the absence of these abnormalities does not exclude a diagnosis of CRPS. CRPS is therefore essentially a clinical diagnosis.

The underlying mechanisms have been the subject of hot debate for decades. Past theories led to beliefs that abnormal behavior in the sympathetic nervous system was the main cause of the ongoing pain and other symptoms. This theory led to a host of treatments that were standard treatment until proper randomized trials found them to be ineffective [27]. Until recently there has been no animal model other than reproducing nerve injuries [28]. There is still a poor understanding of the underlying mechanisms and therefore current treatments have tended to be generalized rather than specific. Unfortunately, there is currently no cure and many patients have persistent pain and disability and often require intensive long term support.

8.3.14
Co-morbidities

There is a close association between chronic disease and pain [29]. It is not surprising that the more chronic diseases one has, the more likely one is to experience chronic pain. As noted above, this includes diseases like arthritis and diabetes, but pain may also accompany multiple sclerosis, depression, and other chronic illnesses. Many of the diseases of older age are systemic, meaning that they are not confined to one part of the body. Someone with osteoarthritis, for example, may present with pain in more than one part of the body, but will also likely have become quite deconditioned from lack of use of the affected parts and perhaps generally deconditioned if they have been inactive.

Most patients, therefore, do not simply present with one pain problem. They present with a number of co-morbidities. Their pain may or may not be due to their other illnesses. More importantly, the presence of other concurrent health problems makes the assessment and treatment planning and management much more complex [30].

The presence of co-morbid diseases also means that the patient will likely be consulting other disciplines and healthcare professionals. Some of the treatments and management plans for their pain may impact on the management of their concurrent diseases and vice versa. Effective care for these individuals requires close coordination of care between and among all treating professionals, the patient and often their family.

8.4
How Do Patients Present?

Perhaps the most striking observation about patients referred to a chronic pain clinic is their level of distress. They often display a number of emotions in addition to their obvious discomfort with their pain. These may include helplessness, worry, sadness, anger, fatigue, fear and, sometimes, even hopelessness.

8.4.1
Worry/Anxiety

Pain and anxiety often go together [31]. When pain is severe and continuous it is not surprising that patients become concerned about the meaning of the pain. Most will begin to entertain worrying thoughts about the pain and its consequences. (Does this herald cancer or serious disease? How am I going to cope with this? Is this ever going to stop? How am I going to survive financially? What will happen to the family?)

These worries will not only influence how patients cope with their pain but also directly influence how much pain they experience. It is well known that anxiety augments pain experienced from any injury.

8.4.2
Sadness/Depression

Patients with chronic pain suffer losses. Their life is no longer "normal". They may no longer obtain enjoyment out of life. They may lose contact with their workmates; they may lose their place as an active contributor to family life; they may lose intimacy with their partner; they may lose self-worth; they may lose the future they had planned.

With loss comes sadness. With lack of positive reward from family, friends, workmates, mood tends to drift down. It is common for patients to progress to clinical depression. Patients who are depressed experience more pain than those without depression [32].

Patients often feel that no one believes them, understands them, can find anything wrong or can do anything to help them. The more they try to convince people that their pain is real, it seems the more people question the very existence of their pain. None of the previous treatments have helped and it seems that they face the rest of their life on their own trying to cope with their pain with no hope of future relief.

Some patient's sense of despair can become so great that they even contemplate suicide. Suicidal ideation is not uncommon, although suicidal intent is less common [31].

8.4.3
Anger/Frustration

Patients become frustrated with themselves for not being able to cope with their pain, especially if they have overcome previous injuries. They become frustrated with the limitations that the pain appears to place on them. It seems that the harder they try to keep active the more they hurt. It is as though no matter what they do, they end up in severe pain.

Frustration can turn to anger. This is more commonly seen among patients who are either in dispute with the benefit system or are involved in litigation [33]. Anger directed at the "person" who caused their injury can be pervasive and very destructive. It can lead to states of intense physical arousal and can contribute to muscle tension and thus pain.

8.4.4
Fatigue/Exhaustion/Sleep Disorders

Patients use up significant amounts of energy to cope with their pain. Physically, muscle tension will be increased around the painful site, which restricts and limits movement. In addition, coping with pain is emotionally draining. This reduces any reserves of energy that they have and commonly leads patients to experience significant tiredness and fatigue. This can be accentuated by fatigue from lack of restorative sleep, which usually accompanies chronic pain.

Almost invariably patients with chronic pain give a history of significant interference with sleep [34]. This is usually both initiation as well as maintenance of sleep. Inactivity and depressed mood can also contribute to poor sleep. Non-restorative sleep itself can lead to chronic fatigue, difficulties in concentration, irritability and may also increase pain. Patients have usually tried various tranquillizers and sleep medications. At best, these provide short-term help, but rarely do they help in the long term. In addition, many of the currently available tranquillizers have a significant risk of addiction. Thus sleep is probably one of the most difficult secondary problems to treat successfully.

8.4.5
Physical/Social Functioning

Most patients begin to experience some financial strain. Frequently, the partner will have to increase their working hours to make up any reduction in family income, as well as take on or continue with the responsibility of running the home. This will often put a strain on the relationship. The enjoyment soon disappears from family life. The children and partners often become irritable and the patient frequently feels guilty and misunderstood.

Chronic pain patients, as a group, are known to be some of the most disabled patients [35]. Although limiting activity does help control the pain a little, the patient becomes progressively unfit and deconditioned. This leads to the patient becoming less physically capable of performing physical activities, irrespective of their pain, and to being more prone to minor injuries when they attempt to improve their activity level.

Although limiting activity helps to some extent, the price is dear. Patients usually become frustrated with their limitations and will often attempt to perform some tasks in an effort to get back towards normal. However, these activities are now outside their physical capabilities and often provoke an acute episode of pain that may last for hours or days. For the patient, this simply reinforces the idea that they should keep their activities at a low level for fear of permanently increasing their pain.

8.4.6
Coping

For most patients with chronic pain there is no cure, so they have to cope with their pain. Some patients seem to cope well despite their pain problem and others clearly do not. Those who cope well may not consult the healthcare system very much, if at all. It is those who appear not to cope well who consult and require frequent support and reassurance. There are a number of factors that differentiate good copers from poor copers. Coping skills are of major importance and the factors that influence coping are an important component of a full assessment.

8.5
Why Do Patients Consult?

Although chronic pain is very common, not everyone with chronic pain consults his or her family physician. Many seem to manage with over-the-counter medications and minimal use of healthcare. Some patients have undoubtedly been bruised by their previous experiences of the healthcare system, particularly if they have not been believed or have been treated dismissively.

Patients who do consult do so for a number of reasons. The severity of pain might be the most obvious and important factor, but it is not necessarily the most important. Understanding the patient's agenda will help guide the consultation and ensure that issues important to the patient are explored. If these issues are not addressed, the patient will not be satisfied and will go on to repeated and costly consultations with others in the healthcare system.

8.6
Why Do Family Physicians Refer Patients to Specialists?

8.6.1
Diagnosis

The family physician (FP) may need some assistance with diagnosis or ensuring that he/she has not missed anything of importance. Often reassurance that there is no simple diagnosis and that the full spectrum of potential medical diagnoses have been checked is all that may be required. The FP may use the pain clinic to reassure the patient that his/her pain problem has been taken seriously and investigated appropriately.

8.6.2
Treatment

Some treatments are specialized and may need either sanction of or delivery by a specialist. Some FPs are uncomfortable commencing particular treatments without the express support of a specialist. This may include starting strong opioids.

Some treatments are highly specialized and need to be performed in specialized clinics. Others require the input from a multi- or interdisciplinary team and therefore cannot be easily delivered in primary care.

8.6.3
High Levels of Distress

It is difficult to deal with very high levels of distress within the time constraints of the FP's office. Often patients' symptoms do not respond to medication and multimodal therapy is required.

8.6.4
Problem Medication Use

FPs recommend and prescribe a wide range of medications for both acute and chronic pain. Some of these have the potential for addiction. Although the risk of addiction is small, a number of patients develop problem medication use. This is exemplified by an apparent loss of control. The patient may begin to use more than the recommended dose and often escalates the dose without any major benefit (see Chapter 15). Helping these patients manage this situation requires skill and time, often with assistance from a number of professionals in a specialized clinic.

8.6.5
Help with Patient Management

Patients with chronic pain can be intellectually and emotionally challenging, even for the most seasoned physicians. Sometimes the family physician needs moral support if they continue to manage the patient in the community. It is important to offer such support to primary care physicians to prevent burnout.

Some patient's problems need the skills of more than one profession and referral to an interdisciplinary clinic is not only appropriate but essential if progress is to be made.

8.6.6
Other Obstacles to Managing Pain in the Community

Some FPs simply don't like working with patients with chronic pain. There is no simple answer to the patient's problem. The physician finds him/herself constantly under pressure from the patient. Each consultation seems to be a battle, either to achieve the impossible (cure the pain) or to try and retain some control over prescribing. The patient demands steadily more medication and the physician worries about coming into conflict with regulatory bodies about his prescribing. Trying to deal with patients with complex problems in the time available may seem impossible. There are also significant problems with reimbursement. In many areas there is no additional fee for looking after patients with chronic pain or funding mechanisms to allow more time to be spent with the patient. Some jurisdictions allow additional fees if the patient has already been assessed in a multidisciplinary pain clinic.

Finally some physicians recognize the need for a more broad-based approach to managing patients with chronic pain but they are frustrated by the lack of community-based resources. For example it is common for there to be major restrictions on the availability of rehabilitative physiotherapy or clinical psychology in the community. Not unsurprisingly, physicians feel uncomfortable in managing such patients knowing what is required and not being able to access appropriate help.

References

1 Sjolund, B. (2007) Biomedical and pain assessment in secondary and tertiary care settings, in *Pain Management: Practical Applications of the Biopsychosocial Perspective in Clinical and Occupational Settings* (eds C.J. Main, M.J.L. Sullivan and P. Watson), Churchill Livingstone Elsevier, Edinburgh, pp. 135–43.

2 Wilkes, L.M., Castro, M., Mohan, S., Sundaraj, S.R. and Noore, F. (2003) Health status of patients with chronic pain attending a pain center. *Pain Management Nursing*, **4** (2), 70–6.

3 Bogduk, N. (2003) Neck pain and whiplash, in *Clinical Pain Management – Chronic Pain* (eds T.S. Jensen, P.R. Wilson and A.S.C. Rice), Arnold, London, pp. 505–19.

4 Spitzer, W.O., Skovron, M.L., Salmi, L.R. *et al.* (1995) Scientific monograph of the Quebec Task Force on whiplash-associated disorders: redefining "whiplash" and its management. *Spine*, **20** (Suppl 1), 1–74.

5 Ghatan, S. and Godkin, R. (2001) Neck pain, in *Bonica's Management of Pain* (eds D.J. Loeser, S.H. Butler, C.R. Chapman and D.C. Turk), Lippincott, Williams and Wilkins, Philadelphia, pp. 1003–18.

6 Russell, I.J. (2001) Fibromyalgia syndrome, in *Bonica's Management of Pain* (eds D.J. Loeser, S.H. Butler, C.R. Chapman and D.C. Turk), Lippincott, Williams and Wilkins, Philadelphia, pp. 543–56.

7 Peat, G.M. (2000) Evaluation of outcome, in *Pain Management: An Interdisciplinary Approach* (eds C.J. Main and C.C. Spanswick), Churchill Livingstone, Edinburgh, pp. 363–85.

8 Waddell, G.W. (2004) The biopsychosocial model, in *The Back Pain Revolution*, Churchill Livingstone, Edinburgh, pp. 265–85.

9 Waddell, G.W. (2004) US health care for back pain, in *The Back Pain Revolution*, Churchill Livingstone, Edinburgh, pp. 419–37.

10 Oaklander, A.L. and North, R.B. (2001) Failed back surgery syndrome; low back pain, in *Bonica's Management of Pain* (eds D.J. Coeser, S.H. Butler, C.R. Chapman and D.C. Tunk), 3rd edn, Lippincott, Williams and Wilkins, Philadelphia, pp. 1540– 48.

11 Badley, E.M. and Wang, P.P. (1998) Arthritis and the aging population: projections of arthritis prevalence in Canada 1991 to 2031. *The Journal of Rheumatology*, **25** (1), 138–44.

12 Perruccio, A.V., Power, J.D. and Badley, E.M. (2006) Revisiting arthritis prevalence projections – it's more than just the aging of the population. *The Journal of Rheumatology*, **33** (9), 1856–62.

13 Cohen, P.L. (2003) Systemic autoimmunity, in *Fundamental Immunology* (ed. W.E. Paul), 4th edn, Lippincott, Williams and Wilkins, Philadelphia, pp. 1371–1400.

14 Lacaille, D., Anis, A.H., Guh, D.P. and Esdaile, J.M. (2005) Gaps in care for rheumatoid arthritis: a population study. *Arthritis and Rheumatism*, **53** (2), 241–8.

15 Wolfe, F. *et al.* (1990) The American College of Rheumatology 1990 criteria for the classification of fibromyalgia. Report of the Multicenter Criteria Committee. *Arthritis and Rheumatism*, **33** (2), 160–72.

16 Wolfe, F. and Cathey, M.A. (1983) Prevalence of primary and secondary fibrocitis. *The Journal of Rheumatology*, **10**, 965–8.

17 Hall, G.C., Carroll, D., Parry, D. and McQuay, H.J. (2006) Epidemiology and treatment of neuropathic pain: the UK primary care perspective. *Pain*, **122**, 156–62.

18 Shembalkar, P. and Anand, P. (2003) Peripheral neuropathies, in *Clinical Pain Management; Chronic Pain* (eds T.S. Jensen, P.R. Wilson and A.S.C. Rice), Arnold, London, pp. 367–81.

19 Zakrzewska, J.M. and Harrison, S.D. (2003) Facial pain, in *Clinical Pain Management; Chronic Pain* (eds T.S. Jensen, P.R. Wilson and A.S.C. Rice), Arnold, London, pp. 481–504.

20 Loeser, D.J. (2003) Cranial neuralgias, in *Bonica's Management of Pain* (eds D.J. Loeser, S.H. Butler, C.R. Chapman and D.C. Turk), Lippincott, Williams and Wilkins, Philadelphia, pp. 855–66.

21 Brisson, M., Edmunds, W.J., Law, B., Gay, N.J., Walld, R., Brownell, M., Roos, L. and De Serres, G. (2001) Epidemiology of varicella zoster virus infection in Canada and the United Kingdom. *Epidemiology and Infection*, **127**, 305–14.

22 Hope-Simpson, R.E. (1975) Postherpetic neuralgia. *The Journal of the Royal College of General Practitioners*, **25**, 571–5.

23 Watson, C.P.N. (2003) Post herpetic neuralgia, in *Clinical Pain Management; Chronic Pain* (eds T.S. Jensen, P.R. Wilson and A.S.C. Rice), Arnold, London, pp. 451–8.

24 Brisson, M., Pellissier, J.M. and Levin, M.J. (2007) Cost-effectiveness of herpes zoster vaccine: flawed assumptions regarding efficacy against postherpetic neuralgia. *Clinical Infectious Diseases*, **45**, 1527–9.

25 Boas, R.A. (1996) Complex regional pain syndromes: symptoms, signs and differential diagnosis, in *Progress in Pain Research Management* (eds W. Janig and M. Stanton-Hicks), IASP Press, Seattle, WA, pp. 79–92.

26 Sandroni, P., Dotson, R. and Low, P.A. (2003) Complex regional pain syndromes, in *Clinical Pain Management; Chronic Pain* (eds T.S. Jensen, P.R. Wilson and A.S.C. Rice), Arnold, London, pp. 383–401.

27 Ramamurthy, S. and Hoffman, J. (1995) Intravenous regional guanethidine in the treatment of reflex sympathetic dystrophy/causalgia: a randomized, double blind study. *Anesthesia and Analgesia*, **81**, 718–23.

28 Coderre, T.J., Xanthos, D.N., Francis, L. and Bennett, G.J. (2004) Chronic post-ischemia pain (CPIP): a novel animal model of complex regional pain syndrome-type I (CRPS-I; reflex sympathetic dystrophy) produced by prolonged hindpaw ischemia and reperfusion in the rat. *Pain*, **112** (1–2), 94–105.

29 Schopflocher, D. (2003) *Chronic Pain in Alberta: A portrait from the 1996 National Population Health Survey and the 2001 Canadian Community Survey*, Health Surveillance, Alberta Health, Edmonton, Alberta. http://www.health.gov.ab.ca/public/dis_chronicpain.pdf.

30 Spanswick, C.C. and Million, R. (2000) Medical assessment, in *Pain Management; an Interdisciplinary Approach* (eds C.J. Main and C.C. Spanswick), Churchill Livingstone, Edinburgh, pp. 139–61.

31 King, S. (2003) Psychiatric diagnosis and chronic pain, in *Clinical Pain Management – Chronic Pain* (eds T.S. Jensen, P.R. Wilson and A.S.C. Rice), Arnold, London, pp. 631–5.

32 Currie, S.R. and Wang, J. (2004) Chronic back pain and major depression in the general Canadian population. *Pain*, **107**, 54–60.

33 Main, C.J., Sullivan, M.J.L. and Watson, P. (2007) Tertiary pain management programs, in *Pain Management: Practical Applications of the Biopsychosocial Perspective in Clinical and Occupational Settings*, Churchill Livingstone Elsevier, Edinburgh, pp. 241–319.

34 Lavigne, G., Sessle, B.J., Choiniere, M. and Soja, P.J. (eds) (2007) *Sleep and Pain*, IASP Press, Seattle.

35 Benjamin, S. and Barnes, D. (1987) The selfcare assessment schedule (SCAS) – II reliability and validitiy. *Journal of Psychosomotic Research*, **31** (2), 203–14.

9
Drug Treatment for Chronic Pain

Mary Lynch

- There are effective drug treatments for many kinds of chronic pain, but most of these treatments have significant limitations.
- Many Canadian physicians do not know how best to treat chronic pain using currently available medicines.
- There is a reluctance to prescribe some drugs because of fear of addiction, diversion and censure.
- It is difficult or impossible in many jurisdictions to obtain timely consultation with a pain specialist.
- The lack of a national pain drug strategy leads to much pain drug research and education being funded by pharmaceutical manufacturers.

9.1
Scope and Context of the Issue

For millennia mankind has sought relief from pain by using medicines. Ancient texts refer to the use of many herbal remedies for pain. In the first century Emperor Claudius' personal physician recorded detailed instructions on the preparation of opium and Galen recommended the juice of the poppy for the relief of pain in older persons. Cannabis was a remedy with known pain relieving effects in use in Ancient China [1], Persia, India, Egypt, Greece and Rome. The bark of the willow (from which aspirin is derived) was one of the many herbal remedies used by Sumerian, Egyptian, Greek and Roman physicians. In Hippocratic writings it was praised for combating the pains of childbirth and fever [1].

Chronic Pain: A Health Policy Perspective
Edited by S. Rashiq, D. Schopflocher, P. Taenzer, and E. Jonsson
Copyright © 2008 WILEY-VCH Verlag GmbH & Co. KGaA, Weinheim
ISBN: 978-3-527-32382-1

9.2

The Evidence for Prescription Drug Use in the Management of Chronic Pain

Despite significant developments in pharmacological technology over time, very little has changed with regards to drug options for the management of pain. Derivatives of the poppy and the bark of the willow continue to play a prominent role. Cannabinoids have continued to be used as traditional remedies in many parts of the world and, with increasing evidence for analgesic and other therapeutic effects, are now receiving increased attention in Europe and North America [2–5]. The main additions to the pain therapists' drug armamentarium include members of the antidepressant and anticonvulsant groups of drugs, and topical preparations (applied directly to the painful part), the latter supported by evidence identifying peripheral mechanisms of chronic pain.

Recent reviews have identified several key groups of medication for which there is high quality evidence supporting efficacy in the management of chronic pain [6–9]. This evidence has been used to develop recommendations and treatment algorithms for pharmacologic management of chronic neuropathic pain [6, 9] and chronic pain in general [7, 8]. The classes of agents for which there is well-established evidence of analgesic (pain-relieving) efficacy include the non-steroidal anti-inflammatories (NSAIDs), the tricyclic antidepressants (TCAs), specific anticonvulsants and the opioids (narcotic or morphine-like drugs). The cannabinoids (cannabis derivatives) have good support to justify their use as second or third line treatment and there is growing evidence that specific topical preparations are effective as sole agents in mild to moderate pain with potential to be used in combination with systemic therapy in moderate to severe pain [8].

Clinical experience suggests that the drug treatment of chronic pain is best placed within an overall plan of pain management that includes consideration of interdisciplinary, integrative, and active participatory self-management strategies reviewed in other chapters in this volume.

A generally similar approach to selecting medication for chronic pain is used by most practising clinicians: Once the physician has established the working diagnosis and has identified that analgesic medication is necessary, the usual approach is to start with a non-opioid analgesic such as an NSAID or acetaminophen for mild to moderate pain. If this is inadequate, or if there is an element of sleep difficulty, the next step is to add an antidepressant with analgesic qualities. If there is a component of neuropathic pain, then a trial of one of the anticonvulsant analgesic agents is appropriate. If these steps are inadequate, then an opioid analgesic may be added. Cannabinoids and topical agents may also be appropriate as single agents or in combination. In an individual patient, more than one pain mechanism may be at play and more than one agent may be necessary for pain control. There is also significant individual variation in response to medications. For this reason it is important to take an individualized approach to each patient and adjust doses according to treatment response and side effects. It is also appropriate to use a combination of agents with different mechanisms of action in an effort to obtain adequate pain control. This combination approach has been supported by a randomized, double blind, placebo controlled trial which found that gabapentin and morphine combined achieved better analgesia at lower doses than when the agents were used alone [10].

9.3
Specific Drugs

9.3.1
NSAIDs and Acetaminophen

Acetylsalicyclic acid (aspirin, ASA), ibuprofen and acetaminophen are available over the counter in North America and are widely used for pain control. ASA and ibuprofen, members of the NSAID group, exhibit analgesic, anti-pyretic (fever-reducing) and anti-inflammatory effects. Acetaminophen is a different agent and does not exhibit significant anti-inflammatory effects but does act as a good analgesic and anti-pyretic [8].

There are at least 10 different drug classes represented within the NSAID group. Their analgesic action has been established for some time. Most are as effective as ASA but several are better. There is poor correlation between an NSAID's anti-inflammatory activity and its analgesic efficacy, leading to the conclusion that relief of pain from NSAIDs involves additional mechanisms [8]. Unfortunately these agents generally provide only mild relief and can be associated with potentially serious adverse effects including gastric ulceration and gastric bleeding and, with the exception of ASA, an increased risk of hypertension and cardiovascular events such as myocardial infarction and stroke. For these reasons their usefulness in more than mild chronic pain is limited.

Acetaminophen is as potent as ASA in most types of pain except inflammatory arthritis. In head-to-head patient preference studies comparing acetaminophen with NSAIDs, over twice as many patients preferred NSAIDs. However, given its good safety profile, acetaminophen is still considered the first-line drug for patients with osteoarthritis [11]. Precisely how acetaminophen works to relieve pain is still unknown but there is growing evidence supporting a central serotonergic mechanism [8] and a possible cannabinoid/vanilloid receptor connection [12].

9.3.2
Tricyclic Antidepressants

This class of medications was originally developed for the treatment of depression, for which there are now better choices available. However, several systematic reviews and meta-analyses have concluded that tricyclic antidepressants are effective in relieving pain in a number of chronic pain conditions, in doses that are much lower than would have been used for mood elevation [8, 13–15]. The number needed to treat (NNT)[1] is generally in the range of 2.1–2.6 (Table 9.1). TCAs are

---

1) Number needed to treat refers to the number of patients who must be exposed to the treatment in order to experience a significant reduction in pain, defined in this case as 50% reduction in pain or more.

Table 9.1 Average of number needed to treat among placebo-controlled trials examining tricyclic and SNRI antidepressants for neuropathic pain for benefit (50% reduction of pain), minor and major harm. Reprinted with permission of Pulses Publishing (Lynch and Watson 2006 [8]).

Agent	NNT Benefit	NNT Minor harm	NNT "Major harm"[a]	Number of studies[b]
Amitriptyline	2.4	20.4	30.5	6
Imipramine	2.1	1.4	13.7	4
Desipramine	2.4	12.4	15.2	3
Nortriptyline	2.6	1.4	8	3
Clomipramine	2.1	no dichotomous data available	8.7	1
Average TCAs	2.3	8.9	17	
Venlafaxine	4.0			2
SSRIs	6.7			3

a Major harm consists of withdrawal from the study due to adverse effects.
b This column refers to the number of studies for which there was adequate information with which to calculate an average NNT.
SSRIs = serotonin specific reuptake inhibitors.
Please note that these figures derive from studies using different methodology, different data analysis, with different numbers of patients. There are few comparative trials and the external validity may be poor because of the selection that goes into trials. Thus the NNT data is a rough guide only.

effective in diabetic neuropathy, postherpetic neuralgia, tension headache, migraine, atypical facial pain, fibromyalgia and low back pain. In neuropathic pain, TCAs relieve brief lancinating pain, constant dysesthetic pain, allodynia and spontaneous pain. The pain relief from TCAs is generally moderate in degree. Side effects such as sedation, postural hypotension, dry mouth and constipation are common. The sedative property of this class of drugs is often exploited when pain is accompanied by sleep difficulty by administering them in the evening. TCAs with a balanced inhibition of serotonin (5-hydroxytryptamine or 5-HT) and norepinephrine (NA) reuptake, such as amitriptyline, imipramine and clomipramine, as well as agents with greater NA reuptake inhibition such as desipramine and nortriptyline, appear to be effective analgesics. The mechanism of action is unknown. It is probably multi-factorial as these agents are capable of multiple actions of relevance to pain transmission [8].

9.3.3
Other Antidepressants

There are several additional chemical groups of anti-depressants. To date, only the serotonin norepinephrine reuptake inhibitors (SNRIs) have exhibited potential as analgesics in randomized controlled trials. The two antidepressants available from this group include venlafaxine (Effexor) and duloxetine. Both are effective anti-depressants with strong inhibition of 5-HT and NA reuptake. This balanced neu-rotransmitter profile is similar to that found with the TCAs.

There are six randomized controlled trials examining the analgesic effects of venlafaxine. The largest trial of 244 patients with painful diabetic neuropathy found a significant reduction in pain with a NNT of 4.5 for 50% pain reduction at week 6 [16]. Several smaller studies demonstrated mixed results [17–21]. At present there is initial evidence indicating that venlafaxine in a dose range of 150–225 mg/day may exhibit some analgesic effect in painful diabetic neuropathy, however the NNT in order to obtain an analgesic effect is higher than with the TCA group. Further controlled trials are necessary. One potential advantage of venlafaxine is its less troublesome side effect profile compared to TCAs.

To date there are five published randomized controlled trials examining the efficacy of duloxetine in the treatment of pain. In three multidose trials involving a total of 1139 patients with diabetic neuropathy, duloxetine 60 and 120 mg/day significantly reduced pain severity as compared with placebo [22–24]. In fibromy-algia, results have been mixed in two randomized controlled trials [25, 26].

9.3.4
Anticonvulsants

There is good evidence that several anticonvulsants (seizure-preventing drugs) relieve neuropathic pain, based on their ability to reduce neuronal excitability. The most well studied agents include gabapentin, carbamazepine, lamotrigine, and pregabalin. Table 9.2 presents further detail regarding anticonvulsants exhibiting analgesic potential, proposed mechanisms of action and dosing [8].

9.3.4.1 Gabapentin
Gabapentin has been the first-choice anticonvulsant for pain purposes for more than a decade. Several large randomized controlled trials show that gabapentin is effective in postherpetic neuralgia [30, 31] diabetic neuropathy [32–34] and mixed diagnoses of neuropathic pain [35]. There is also support for it being an effective analgesic in spinal cord injury [36], trigeminal neuralgia [35], Guillain–Barre Syn-drome [37], phantom limb pain [38] and in prophylaxis of chronic daily headache [39]. Overall, gabapentin is well tolerated and the most common side events are mild to moderate dizziness and somnolence most of which occur in the early phase of treatment. Less common additional side effects include ataxia (difficulty walking) and confusion. There are only two comparative trials comparing gabap-entin with other analgesics, one head to head trial with the TCA amitriptyline [34] which found both agents to exhibit similar efficacy but with different side effects, and one placebo controlled trial examining gabapentin alone and in combination with morphine which demonstrated the combination to be superior to either agent alone [10]. A newer and potentially very exciting use of this drug is to help reduce

Table 9.2 Anticonvulsants with documented analgesic effects. From: Management issues of trigeminal neuropathic pain from a medical perspective [27], A treatment algorithm for neuropathic pain [28], Oxcarbazepine, topiramate, zonisamide and levetriacetum: potential use in neuropathic pain [29] also see text for specific agents and more detailed references.

Agent	Dose range (mg/day)	Mechanism of action	Indications supported by at least one RCT[a]	Side effects	Comments
Gabapentin (Neurontin)	1200–3600	N-type Ca^{2+} channel blocker	PHN DN mixed neuropathic pain	sedation dizziness ataxia confusion	Does not require metabolism in liver so is a better choice in liver dysfunction, clearance will be diminished in renal dysfunction.
Pregabalin (Lyrica)	150–600	α_2-δprotein of voltage gated Ca^{2+} channels	PHN DN	dizziness somnolence	Analgesic effect is seen within first 3 days, does not require liver metabolism, renal excretion primary route of elimination
Carbamazepine (Tegretol)	200–2000[b]	Na^+ channel blockade	TN	sedation dizziness ataxia diplopia hepatitis rash hyponatremia	CBC, electrolytes and liver function studies pre-treatment and every 2 weeks for 3 months then less frequently[b] (the hyponatremia may result in a confusional state). Most serious potential side effects are aplastic anemia, hepatitis, serious dermatologic reactions[c]
Lamotrigine (Lamictal)	200–400	Na^+ channel blockade	TN DN post stroke pain spinal cord injury	mild rash to serious dermatologic reactions[d]	CBC and liver function studies pre-treatment and at 4 weeks
Oxcarbazepine (Trileptal)	600–1200	Na^+ channel blockade	TN	sedation headache dizziness rash[c] vertigo ataxia nausea diplopia hyponatremia	CBC, electrolytes and liver function studies pre-treatment and at 4 weeks Serious dermatologic reactions and multi-organ hypersensitivity reactions have been reported

Table 9.2 Continued

Agent	Dose range (mg/day)	Mechanism of action	Indications supported by at least one RCT[a]	Side effects	Comments
Topiramate (Topamax)	50–200 mg	Na$^+$ channel blockade ⇈ GABA inhibition ⇊ glutamate excitation Modulates Ca^{2+} channels	Migraine prophylaxis	paresthesia fatigue nausea anorexia weight loss changes in taste	Effect is modest, topiramate was associated with approximately 1 less migraine per month than placebo in 3 large RCTs

TN (trigeminal neuralgia), PHN (post herpetic neuralgia), DN (diabetic neuropathy), RCT (randomized placebo-controlled trial)

a For details see section relating to the specific agent in the text.
b Doses of up to 2000 mg/day may be required in trigeminal neuralgia.
c Life threatening dermatologic reactions such as Stevens–Johnson syndrome, toxic epidermal necrolysis and lupus, may be serious and require discontinuation of carbamazepine and oxcarbazepine
d Rash ranging from simple morbilliform type to potentially serious rashes including Stevens-Johnson syndrome and toxic epidermal necrolysis have been reported.

or prevent the pain that follows surgery. Healthy, pain free subjects require less narcotic analgesics after certain kinds of surgery if they are given one or more doses of gabapentin before the operation [37].

9.3.4.2 Pregabalin

Large randomized controlled trials have found that pregabalin exhibits significant analgesic efficacy in post-herpetic neuralgia [40–43] and painful diabetic peripheral neuropathy [44–46]. There is also support for efficacy in spinal cord injury pain [47]. Pregabalin is generally well tolerated, the most common side effects being dizziness and somnolence. The overall benefit of pregabalin is that it seems to offer approximately 30% advantage over placebo, a similar effect to that of gabapentin. It is significantly more costly than gabapentin. To date there are no head to head trials comparing these two drugs.

9.3.4.3 Carbamazepine

Controlled trials have demonstrated analgesic effects in trigeminal neuralgia, diabetic neuropathy, and migraine prophylaxis [27, 48–50]. Carbamazepine remains the most successful first line approach in treatment of trigeminal neuralgia.

9.3.4.4 Oxcarbazepine

Oxcarbazepine (Trileptil) is an analogue of carbamazepine. A recent review found three randomized controlled trials demonstrating an analgesic effect in trigeminal neuralgia, and one controlled trial found comparable analgesia between amitriptyline and oxcarbazepine in cancer-related neuropathic pain with fewer adverse events in the oxcarbazepine group [29].

9.3.4.5 Lamotrigine

Randomized controlled trials have demonstrated a significant analgesic effect for lamotrigine as compared to placebo in patients with trigeminal neuralgia [51], diabetic neuropathy [52], central poststroke pain [53] and incomplete spinal cord injury pain [54]. In HIV-neuropathy there was initial evidence of greater reduction in pain scores for patients on lamotrigine as compared with placebo [55]. A larger trial found no difference in average pain score between HIV patients on lamotrigine as compared with placebo when looking at the whole sample, but when subgroups were examined, patients receiving antiretroviral therapy exhibited significantly reduced pain with lamotrigine as compared to placebo. A trial examining a dose of 200 mg/day of lamotrigine in a group of patients with mixed diagnoses of neuropathic pain did not demonstrate greater analgesic effect than placebo [56].

In summary, carbamazepine remains an established first line option in the treatment of trigeminal neuralgia. There is good evidence supporting gabapentin and pregabalin for the treatment of post herpetic neuralgia and painful diabetic neuropathy, and growing evidence for lamotrigine and oxcarbazepine in neuropathic pain. Carbamazepine may also be used in neuropathic pain if the pain is predominantly electric shock-like or if it is caused by multiple sclerosis. Large trials have identified that topiramate is not analgesic in painful diabetic neuropathy, and the efficacy of topiramate in other types of pain remains to be established. Thus, it is reasonable to use gabapentin, pregabalin and carbamazepine first (in the case of liver disease carbamazepine should be avoided) and then to move to lamotrigine or oxcarbazepine, if there is no response or the patient is unable to tolerate side effects. In the case of trigeminal neuralgia, baclofen provides an additional option [8].

The evidence to date supports similar analgesic efficacy between the antidepressant and anti-convulsant agents in treatment of neuropathic pain, and the clinician may be guided primarily by the side effect profile and co-morbidities [8].

9.3.5
Opioids

There is a growing body of evidence that controlled release opioid analgesics have a role to play in a subset of patients with chronic pain. A recent meta-analysis of 41 randomized controlled trials involving 6019 patients found that opioids were more effective than placebo for both pain and functional outcomes in both nociceptive and neuropathic pain [57]. Guidelines for the use of opioid analgesics in chronic pain have been established by the Canadian Pain Society [58] and details regarding treatment using opioids for chronic pain are presented in a recent review [8]. The main message is that opioids are a reasonable and efficacious treatment for people with chronic pain [8]. The average duration of trials was only 5 weeks (range 1–16 weeks) and there is a need for longer term trials examining efficacy and safety parameters. Recommended front line agents include codeine, hydromorphone, morphine, oxycodone, and tramadol used orally on a time contingent basis. Additional options include the fentanyl patch where the oral route is not a reasonable option (malabsorption, vomiting) or has failed, and methadone if the previous conventional opioids have failed [59].

Unfortunately there are side effects. A systematic review of over 5000 patients confirmed that most patients would experience at least one adverse event resulting from opioid use in chronic pain, and that many would experience common adverse events of dry mouth, nausea, and constipation, and would not continue treatment because of intolerable adverse events [60].

9.3.6
Cannabinoids

The potent anti-nociceptive and antihyperalgesic effects of cannabinoid agonists in animal models of acute and chronic pain, the presence of cannabinoid receptors in pain-processing areas of the brain, spinal cord and periphery and evidence supporting endogenous modulation of pain systems by cannabinoids, provide support that cannabinoids exhibit significant potential as analgesics. Fifteen of eighteen randomized controlled trials examining cannabinoids in the treatment of pain have demonstrated a significant analgesic effect. Table 9.3 presents further detail. Cannabinoid agents tested included synthetic analogs as well as cannabis and cannabis-based extracts. These agents were tested in a number of pain conditions.

Taken together, the evidence supports that cannabinoids exhibit a moderate analgesic effect in neuropathic pain and cancer pain with preliminary evidence for action in other types of pain such as spinal pain and headache. In Canada there are four cannabinoid agents available. These include the naturally occurring agent cannabis available under the Medical Marihuana Access Regulations (MMAR), a cannabis buccal spray (Sativex), a synthetic THC analog nabilone (Cesamet) and dronabinol (synthetic Δ-9-THC in sesame oil sold under the trade name Marinol). Guidelines for the use of cannabinoids available in Canada for chronic pain management have been established [61] and recently updated [62]. Based on current evidence supporting that cannabinoids are analgesic and safe it is reasonable to use a cannabinoid as a second or third line agent, either as a single agent or in combination with other agents exhibiting a different mechanism of action. In patients exhibiting a constellation of symptoms including nausea, anorexia or spasticity one might consider introducing a cannabinoid earlier as there is evidence that cannabinoids exhibit antiemetic and antispasticity action [63]. Currently available cannabinoid agonists should be avoided in patients with a history of psychosis.

Table 9.3 Randomized controlled trials examining cannabinoid agents in chronic pain.

Author and date	Agent	Population (N) design	Results	Outcome summary
Abrams (2007)	Cannabis	HIV neuropathy (50)	Significant decrease in pain	+
Wissel (2006)	Nabilone 1 mg/day	Spasticity related pain in UMNS (11) crossover	Significant decrease in pain but not spasticity	+
Pinsger (2006)	Nabilone 0.25–1 mg/day	Chronic pain (30)	Significant reductions in spinal pain and headache	+
Rog (2005)	Cannabis based buccal spray extract (Sativex)	Central pain in MS (64) parallel group	Significant reductions in pain (NRS, NPS) and sleep disturbance (NRS)	+
Berman (2004)	Cannabis based buccal spray extract Sativex and another with primarily THC	Neuropathic pain brachial plexus avulsion (48) crossover	Significant reductions in pain (NRS) and sleep disturbance (NRS) but not to the full 2 point reduction (NNT ≥ 1 = 3) NNT ≥ 2 = 7.5)	~
Svendsen (2004)	dronabinol	Central pain in MS (24) crossover	Significant reductions in pain (NRS) modest reductions 1 point on a 0–10 point scale NNT for 50% relief = 3.45	+
Zajicek (2003) 14 wks Rx	Oral cannabis extract or THC	MS (211/206/213) parallel group	Significant reductions in subjective pain and spasticity on category rating scales No change Ashworth scale for spasticity	+
Zajicek (2006) 52 wks Rx	Oral cannabis extract or THC	MS (172/154/176) parallel group	Significant reduction in pain *Small but significant reductions in Ashworth scale for spasticity*	+
Wade (2004)	Cannabis based buccal spray extract Sativex	MS (160) parallel group	No significant difference in pain scores (VAS) between active and placebo all decreased *There was a significant reduction in spasticity (VAS) scores*	–
Karst (2003)	CT-3 Synthetic analog of THC-11-oic acid	Neuropathic pain with hyperlagesia or allodynia (19) crossover	Significant reduction in pain (VAS and VRS)	+
Notcutt (2004)	Cannabis based buccal sprays extracts THC CBD THC:CBD P	Chronic pain (34)"N of 1" 2 week open/RCT 1 week Rx periods X 2 for each CBME	Significant reduction in pain (VAS) for THC and THC;CBD	+

Table 9.3 Continued

Author and date	Agent	Population (N) design	Results	Outcome summary
Wade (2003)	Cannabis based buccal sprays extracts THC CBD THC:CBD P	Neurogenic symptoms in MS/spinal cord injury/brachial plexus injury/limb amputation (24) "N of 1"	Significant reductions in pain CBD and THC CBME *Reductions in spasm with THC and CBD/THC and spasticity with THC*	+
Noyes (1975)	THC	Cancer (10) *crossover*	Oral THC 15–20 mg significant reduction in pain (4 pt categ scale) *Sedation was a problem*	+
Noyes (1975)	THC	Cancer (36) *crossover*	Oral THC 20 mg significant reduction in pain *Sedation was a problem*	+
Jochimsen (1978)	Benzopyrano-peridine THC cogener	Cancer (37) *crossover*	No more effective than codeine or placebo (VAS)	–
Staquet (1978)	Nitrogen analog THC	Cancer (15) *crossover*	Significantly better than placebo *Dose limiting side effects*	+
Maurer (1990)	THC	Painful spastic paraparesis	5 mg THC significant analgesic and antispasticity effect	+
Holdcroft (1997)	THC	Familial Mediterranean fever "N of (1)"	10 mg 5 × day no difference in VAS but level of morphine use significantly less in THC condition	+

9.3.7
Muscle Relaxants

For most muscle relaxants the indication is for management of acute pain. However, in practice, there are many patients who suffer with chronic pain who are using muscle relaxants in an attempt to find relief. A recent systematic review of controlled trials examining the use of muscle relaxants in nonspecific low back pain identified 30 studies that met inclusion criteria [64]. Four trials studied benzodiazepines, eleven studied non-benzodiazepines and two studied anti-spasticity muscle relaxants in comparison with placebo. Results showed that muscle relaxants are effective for short-term symptomatic relief in patients with acute and chronic low back pain. However the incidence of drowsiness and dizziness and other side effects is high. Caution was recommended and it was left to the discretion of the physician to review the risk-benefit analysis. It was noted that additional large high quality trials are necessary.

9.3.8
Topical Analgesics

The involvement of peripheral mechanisms in the generation of chronic pain supports the use of topical agents in its management.

9.3.8.1 Topical NSAIDs

Systematic reviews of randomized controlled trials have identified that topical NSAIDs are effective in relieving pain in acute (soft tissue trauma and sprains) and chronic pain [65, 66]. Topical NSAIDs exhibited few adverse events; these were primarily cutaneous in nature (rash or itching at the site of application) and gastro-intestinal (GI) adverse events were rare (compared to a 15% rate with oral use). The only topical NSAID commercially available in Canada is diclofenac 1.5% in dimethlysulfoxide (Pennsaid). Data from three randomized controlled trials indicates that this agent is significantly better than placebo in the treatment of osteoarthritis of the knee [67–69] and was as effective as the oral agent and exhibited fewer adverse events.

There are a number of over-the-counter salicylate-containing preparations available in Canada (most contain trolamine salicylate or methyl salicylate). These agents generally fall into the category of rubefacients (or agents that act by counter irritation). A systematic review of topical rubefacients containing salicylates notes that salicylates are difficult to categorize, as they do not seem to work in the same way as other NSAIDs [70]. This review indicated that trials of rubefacients are limited by number, size, quality and validity; the best assessment of limited information suggests rubefacients containing salicylates may be efficacious in acute pain and moderately to poorly efficacious in chronic arthritic and rheumatic pain.

9.3.8.2 Capsaicin

Capsaicin is the active ingredient of chili peppers and similar plants in the capsicum family. Capsaicin is available as a topical cream (Zostrix). A recent systematic review revealed six double blind placebo controlled trials for neuropathic pain conditions with a NNT of 5.7, and three controlled trials examining capsaicin in musculoskeletal pain with a NNT of 8.1 [71]. An earlier meta-analysis identified that capsaicin cream was better than placebo in the treatment of diabetic neuropathy, osteoarthritis and psoriasis [72]. Thus there is evidence that topical capsaicin is better than placebo for treatment of pain from diabetic neuropathy, osteoarthritis and possibly in psoriasis, however the treatment effect is modest and the NNT relatively high. Capsaicin may be beneficial to some patients with neuropathic or arthritic pain as an adjuvant analgesic, but is unlikely to be adequate as the sole analgesic agent.

9.3.8.3 Topical Tricyclic Antidepressants and Ketamine

There are preliminary data regarding the effectiveness of topically applied antidepressants for pain. Two randomized controlled trials demonstrated topical doxepin 3–5% is analgesic in a mixed group of patients with neuropathic pain [73, 74]. In a 3 week randomized placebo controlled trial, the concentration of amitriptyline 2%/ketamine 1% was not better than placebo [75]; however a higher concentration of amitriptyline 4% /ketamine 2% was significantly better than placebo in post herpetic neuralgia [76].

9.3.8.4 Topical Lidocaine Patch

Subequent to initial data suggesting that a topical gel containing 5% lidocaine led to a significant decrease in the pain of post herpetic neuralgia, a topical 5% lidocaine patch has been developed. There are three randomized controlled trials examining the lidocaine patch in post herpetic neuralgia [77–79] and one in patients with a variety of peripheral neuropathic pain syndromes [80], all have found the lidocaine 5% patch provides significantly better analgesia than a vehicle placebo patch. The lidocaine patch is not available in Canada.

9.3.9
Other Analgesic Agents

There are two other agents (or groups of agents) used in specific types of pain that have good support for efficacy. They include baclofen for trigeminal neuralgia and the tryptans (e.g. sumatriptan, Imitrex) for treatment of migraine headaches. Clonidine is also used. Details regarding these agents can be found in a recent review [8].

9.4
Conclusions

There is significant evidence that the NSAIDs and acetaminophen are effective for reducing the intensity of chronic pain but they are usually inadequate to control moderate to severe pain and the NSAIDs can cause serious gastrointestinal side effects that are potentially life threatening. In addition, the COXIB group of NSAIDs has been shown to increase the risk for cardiovascular events such as heart attack and stroke. There is strong evidence that tricyclic antidepressants are analgesic in many types of pain. There is preliminary evidence for the use of duloxetine in the treatment of diabetic neuropathy. There is strong evidence that gabapentin and pregabalin are analgesic in post herpetic neuralgia and diabetic neuropathy. Carbamazepine remains the best first line analgesic for trigeminal neuralgia. Opioids have been shown to be effective analgesics for both somatic and neuropathic pain. There is growing evidence for cannabinoids with support for a couple of currently available agents and significant potential for new agents in development. There is initial data to support topical approaches for some types of neuropathic pain.

Further study is required in many areas. For instance, the effect of duloxetine in pain other than diabetic neuropathy is not well established. For venlafaxine evidence is mixed and there is only one good large RCT in diabetic neuropathy.

Longer-term trials examining safety and efficacy of the anticonvulsants and chronic opioids are needed as well as large multicenter trials regarding the cannabinoids and topicals.

9.5
Implications for Policy Makers, Patients, Families and Communities

One of the biggest barriers to pharmacotherapy facing many Canadians with chronic pain is the problem of access to treatment. Access to care by specialists and teams with up to date knowledge regarding pain management and access to medications are both also problematic.

A recent study has found that wait-times for treatment at publicly funded pain clinics across Canada were unacceptably long (over 1 year at 30% of clinics with a range of up to 5 years) and that there were large regions of Canada where there is no access to a pain clinic [81]. A systematic review identified that patients experience a significant deterioration in health-related quality of life and psychological wellbeing while waiting for treatment for chronic pain during the 6 months from the time of referral to treatment. It is unknown at what point this deterioration begins as results from the 14 trials involving wait-times of 10 weeks or less yielded mixed results with wait-times amounting to as little as 5 weeks, associated with deterioration [82]. It is not known how waiting and deterioration affect outcomes to treatment but it is probable that there is an adverse effect.

This lack of access to care may be compounded by a lack of access to appropriately prescribed medications. On the one hand, many physicians are uncomfortable with prescribing agents such as opioids, cannabinoids and others that may be appropriate and necessary in specific cases to obtain adequate pain control, leaving patients to suffer needlessly. On the other hand, some physicians who do their best in good faith to help patients with complex chronic pain find themselves prescribing greater and greater doses of such medications and may cause overmedication or become victims of dishonest 'patients' seeking to obtain drugs for diversion. Part of the problem is a lack of specialists and services for the management of chronic pain in spite of significant scientific advances in the field [83].

Drug costs are an additional problem. Even if patients are able to access service they may not be able to afford medications that may provide relief. In many cases provincial and insurance formularies do not include some of the analgesic medications reviewed above. In addition the regulatory climate has created a situation where the development of novel agents is a lengthy and very expensive process (average 15 years and billions of dollars in research and development for each agent). The situation is even more difficult for agents that may have, or are perceived to have, abuse potential. This is compounded by the fact that the pharmaceutical industry is profit driven and marketing strategies add additional cost.

Strategies are needed to facilitate the research and development of lower cost drug options. This could be accomplished through strategic funding of basic science and clinical trials targeting development of analgesics through federal granting agencies such as CIHR. Trials of analgesics have not been a priority of funding agencies. At present, grant applications may be rejected if it is thought that the pharmaceutical industry should be funding them. The pharmaceutical industry is involved in analgesic research, but their focus on profit and acute pain limits their interest in treatment of chronic pain, particularly in research involving established agents.

It is up to the Canadian academic community to find solutions that will improve the health, quality of life and level of function of Canadians suffering from persistent pain. At present there is excellent human pain research being done in Canada but researchers are located in separate institutions across the country and do not have a mechanism to connect and integrate pursuit of an overall research program.

Pain drug research is where cancer research was 40 years ago and AIDS research was 10 years ago. There are international (e.g. Pediatric Oncology Group-POG) and national (Canadian HIV Trials Network-CTN) research networks that have made rapid progress toward treatment of cancer and AIDS. A similar approach is needed for pain. Canadian pain researchers have recognized this and initiated the process. Two CIHR funded consortia; the Canadian Consortium on Pain Mechanisms Diagnosis and Management (Cure Pain) and the Canadian Consortium for the Investigation of Cannabinoids (CCIC) have been created. Both identify pain research as a primary focus (see www.curepain.ca, www.ccicnewsletter.com). Both consortia have identified that a formal national clinical trials network is needed. Recently the Canadian Foundation for Innovation (CFI) has provided seed funding to provide support for development of the Canadian Pain Trials Network (CPTN) a clinical trials network dedicated to pain research. The vision of the CPTN is to integrate Canadian pain labs in a way that will allow pursuit of an organized, aggressive, human pain research program where Canada as a whole becomes the laboratory. This would allow more efficient development of new treatments, enable examination of demographics of pain with more power, and provide the clinical arm for a national training program in pain research. CFI provided seed funding for space and equipment, what is required now is funding for operations similar to what was done in the case of the CTN a decade ago.

In summary, in order to enable better care for pain, to improve the health of patients, families and communities, a multi-pronged approach targeting access to care and access to medications is critical. If this is done then significant relief of chronic pain is available through a combination of careful pharmacological management and other approaches reviewed in this volume. It will be necessary to support these and other innovative solutions as chronic pain continues to be an escalating public healthcare problem with significant human and economic costs.

References

1 Dormandy, T. (2006) *The Worst of Evils, The Fight Against Pain*, Yale University Press, New Haven and London.

2 Russo, E. (1998) Cannabis for migraine treatment: the once and future prescription? An historical and scientific review. *Pain*, **76**, 3–8.

3 Joy, J.E., Watson, S.J. *et al.* (1999) *Marijuana and Medicine: Assessing the Science Base*, National Academy Press, Washington DC.

4 Lynch, M.E. (2005) Preclinical science regarding cannabinoids as analgesics: an overview. *Pain Research and Management*, **10**, 7A–14A.

5 Ware, M. and Beaulieu, P. (2005) Cannabinoids for the treatment of pain: an update on recent clinical trials. *Pain Research and Management*, **10**, 27A–30A.

6 Finnerup, N.B., Otto, M. *et al.* (2005) Algorithm for neuropathic pain treatment: an evidence based proposal. *Pain*, **118**, 289–305.

7 Griffin, R.S. and Woolf, C.J. (2005) *Pharmacology of Analgesia. Principles of Pharmacology: The Pathophysiologic Basis of Drug Therapy*, Lippincott, Williams and Wilkins, Philadelphia, pp. 229–43.

8 Lynch, M.E. and Watson, C.P. (2006) The pharmacotherapy of chronic pain, a review. *Pain Research and Management*, **11**, 11–38.

9 Moulin, D.E., Clark, A.J. *et al.* (2007) Pharmacologic management of chronic neuropathic pain concensus statement and guidelines from the Canadian Pain Society. *Pain Research and Management*, **12**, 13–21.

10 Gilron, I., Bailey, J.M. *et al.* (2005) Morphine, gabapentin, or their combination for neuropathic pain. *The New England Journal of Medicine*, **352**, 1324–34.

11 Tannenbaum, H., Bombardier, C. *et al.* (2005) An evidence based approach to prescribing NSAIDs, the third Canadian consensus conference, **33**, 140–57.

12 Hogestatt, E.D., Jonsson, B.A.G. *et al.* (2005) Conversion of acetaminophen to bioactive N-acylphenolamine AM404 via fatty acid amide hydrolase-dependent arachidonic acid conjugation in the nervous system. *The Journal of Biological Chemistry*, **280**, 31405–12.

13 McQuay, H.J., Tramer, M. *et al.* (1996) A systematic review of antidepressants in neuropathic pain. *Pain*, **68**, 217–27.

14 McQuay, H.J. and Moore, R.A. (1997) Antidepressants and chronic pain. *BMJ*, **314**, 763–4.

15 Lynch, M.E. (2001) Antidepressants as analgesics. A review of random controlled trials examining analgesic effects of antidepressant agents. *Journal of Psychiatry and Neuroscience*, **26**, 30–6.

16 Rowbotham, M., Goli, V. *et al.* (2004) Venlafaxine extended release in the treatment of painful diabetic neuropathy: a double-blind, placebo-controlled study. *Pain*, **110**, 697–706.

17 Tasmuth, T., Hartel, B. *et al.* (2002) Venlafaxine in neuropathic pain following treatment of breast cancer. *European Journal of Pain*, **6**, 17–24.

18 Sindrup, S.H., Bach, F.W. *et al.* (2003) Venlafaxine versus imipramine in painful polyneuropathy: a randomized, controlled trial. *Neurology*, **60**, 1284–9.

19 Forssell, H., Tasmuth, T. *et al.* (2004) Venlafaxine in the treatment of atypical facial pain: a randomized controlled trial. *Journal of Orofacial Pain*, **18**, 131–7.

20 Ozyalcin, S.N., Koknel Talu, G. *et al.* (2005) The efficacy and safety of venlafaxine in the prophylaxis of migraine. *Headache*, **45**, 144–52.

21 Yucel, A., Ozyalcin, S. *et al.* (2005) The effect of venlafaxine on ongoing and experimentally induced pain in neuropathic pain patients: a double blind, placebo controlled study. *European Journal of Pain*, **9**, 406–16.

22 Goldstein, D.J., Lu, Y. *et al.* (2005) Duloxetine vs. placebo in patients with painful diabetic neuropathy. *Pain*, **116**, 109–18.

23 Raskin, J., Pritchett, Y.L. *et al.* (2005) A double blind, randomized multicenter trial comparing duloxetine with placebo in management of diabetic peripheral pain. *Pain Medicine*, **5**, 346–56.

24 Wernicke, J.F., Pritchett, Y.L. *et al.* (2006) A randomized controlled trial of duloxetine in diabetic peripheral neuropathic pain. *Neurology*, **67**, 1411–20.

25 Arnold, L.M., Lu, Y. *et al.* (2004) A double-blind, multicenter trial comparing duloxetine with placebo in treatment of fibromyalgia patients with or without major depressive disorder. *Arthritis and Rheumatism*, **50**, 2974–84.

26 Arnold, L.M., Rosen, A. Pritchett, Y.C., D'Souza, D.N., Goldstein, D.J., Ivengar, S., Wernicke, J.F. (2005) A randomized double blind, placebo-controlled trial of duloxetine in the treatment of women with fibromyalgia with and without major depressive disorder. *Pain*, **119**, 5–15.

27 Watson, C.P.N. (2004) Management issues of neuropathic trigeminal pain from a medical perspective. *Journal of Orofacial Pain*, **18**, 366–73.

28 Namaka, M., Gramlich, C.R., Ruhlen, D., Melanson, M., Sutton, I., Major, J. (2004) A treatment algorithm for neuropathic pain. *Clin Ther*, **26**, 951–79.

29 Guay, D.P. (2003) Oxcarbazepine, topiramate, zonisamide, and levetiracetam: potential use in neuropathic pain. *The American Journal of Geriatric Pharmacotherapy*, **1**, 18–37.

30 Rowbotham, M., Harden, N. *et al.* (1998) Gabapentin for the treatment of postherpetic neuralgia. *JAMA*, **280**, 1837–42.

31 Rice, A.S.C., Maton, S. *et al.* (2001) Gabapentin in postherpetic neuralgia: a randomised, double blind, placebo controlled study. *Pain*, **94**, 215–24.

32 Backonja, M., Baydoun, A. *et al.* (1998) Gabapentin for the symptomatic treatment of painful neuropathy in patients with diabetes mellitus. *JAMA*, **280**, 1831–6.

33 Gorson, K.C., Schott, C. *et al.* (1999) Gabapentin in the treatment of painful diabetic neuropathy: a placebo controlled trial. *Journal of Neurology, Neurosurgery and Psychiatry*, **66**, 251–2.

34 Morello, C.M., Leckband, S.G. *et al.* (1999) Randomized double blind study comparing efficacy of gabapentin with amitriptyline on diabetic peripheral

neuropathy pain. *Archives of Internal Medicine*, **159**, 1931–7.

35 Serpell, M.G. and Group, N.P.S. (2002) Gabapentin in neuropathic pain syndromes: a randomised, double blind, placebo-controlled trial. *Pain*, **99**, 557–66.

36 Levendoglu, F., Ogun, C.O. *et al.* (2004) Gabapentin is a first line drug for the treatment of neuropathic pain in spinal cord injury. *Spine*, **29**, 743–51.

37 Pandey, C.K., Singhal, V. *et al.* (2005) Gabapentin provides effective postoperative analgesia whether administered pre-emptively or post-incision. *Canadian Journal of Anaesthesia*, **52** (*8*), 827–31.

38 Bone, M., Critchley, P. *et al.* (2002) Gabapentin in postamputation phantom limb pain: a randomized, double-blind, placebo-controlled, cross-over study. *Regional Anesthesia and Pain Medicine*, **27**, 481–6.

39 Spira, P.J., Beran, R.G. *et al.* (2003) Gabapentin in the prophylaxis of chronic daily headache: a randomized, placebo-controlled study. *Neurology*, **61**, 1753–9.

40 Dworkin, R.H., Corbin, A.E. *et al.* (2003) Pregabalin for treatment of postherpetic neuralgia, a randomized, placebo-controlled trial. *Neurology*, **60**, 1274–83.

41 Sabatowski, R., Galvez, R. *et al.* (2004) Pregabalin reduces pain and improves sleep and mood disturbances in patients with post-herpetic neuralgia: results of a randomized, placebo-controlled clinical trial. *Pain*, **109**, 26–35.

42 Freynhagen, R., Strojek, K. *et al.* (2005) Efficacy of pregabalin in neuropathic pain evaluated in a 12-week, randomised, double-blind, multicenter, placebo-controlled trial of flexible- and fixed-dose regimens. *Pain*, **115**, 254–63.

43 van Seventer, R., Feister, H.A. *et al.* (2006) Efficacy and tolerability of twice-daily pregabalin for treating pain and related sleep interference in postherpetic neuralgia: a 13 week, randomized trial. *Current Medical Research and Opinion*, **22**, 375–84.

44 Lesser, H., Sharma, U. *et al.* (2004) Pregabalin relieves symptoms of painful diabetic neuropathy. *Neurology*, **63**, 2104–10.

45 Rosenstock, J., Tuchman, M. *et al.* (2004) Pregabalin for the treatment of painful diabetic peripheral neuropathy: a double-

blind, placebo-controlled trial. *Pain*, **110**, 628–38.

46 Richter, R.W., Portenoy, R.K. *et al.* (2005) Relief of painful diabetic peripheral neuropathy with pregabalin: a randomized, placebo controlled trial. *The Journal of Pain*, **6**, 253–60.

47 Siddall, P.J., Cousins, M.J. *et al.* (2006) Pregabalin in central neuropathic pain associated with spinal cord injury, a placebo-controlled trial. *Neurology*, **67**, 1792–800.

48 McQuay, H.J. and Carroll, D. (1995) Anticonvulsant drugs for management of pain: a systematic review. *British Medical Journal*, **311**, 1047–52.

49 Berde, C.B. (1997) New and old anticonvulsants for management of pain. *IASP Newsletter Technical Corner*, *Jan/Feb*, 3–5.

50 Backonja, M. and Serra, J. (2004) Pharmacologic management part 1: better studied neuropathic pain diseases. *Pain Medicine*, **5**, S28-47.

51 Zakrzewska, J.M., Chaudhry, Z. *et al.* (1997) Lamotrigine (Lamictal) in refractory trigeminal neuralgia: results from a double-blind placebo controlled trial. *Pain*, **73**, 223–30.

52 Eisenberg, E., Lurie, Y. *et al.* (2001) Lamotrigine reduces painful diabetic neuropathy: a randomized, controlled study. *Neurology*, **57**, 505–9.

53 Vestergaard, K., Andersen, G. *et al.* (2001) Lamotrigine for central poststroke pain, a randomized controlled trial. *Neurology*, **56**, 184–90.

54 Finnerup, N.B., Sindrup, S.H. *et al.* (2002) Lamotrigine in spinal cord injury pain: a randomized controlled trial. *Pain*, **96**, 375–83.

55 Simpson, D.M., Loney, R. *et al.* (2000) A placebo-controlled trial of lamotrigine for painful HIV-associated neuropathy. *Neurology*, **54**, 2115–19.

56 McCleane, G. (1999) 200 mg daily of lamotrigine has no analgesic effect in neuropathic pain: a randomized, double-blind, placebo controlled trial. *Pain*, **83**, 105–7.

57 Furlan, A.D., Sandoval, J.A. *et al.* (2006) Opioids for chronic noncancer pain: meta-analysis of effectiveness and side effects. *CMAJ*, **174**, 1589–94.

58 Jovey, R.D., Ennis, J. *et al.* (2003) Use of opioid analgesics for the treatment of chronic noncancer pain – A consensus statement and guidelines from the Canadian Pain Society, 2002. *Pain Research and Management*, **8**, 3A–28A.

59 Lynch, M.E. (2005) A review of the use of methadone for treatment of chronic non-cancer pain. *Pain Research and Management*, **10**, 133–44.

60 Moore, R.A. and McQuay, H.J. (2005) Prevalence of opioid adverse events in chronic non-malignant pain, systematic review of randomised trials of oral opioids. *Arthritis Research and Therapy*, **7**, R1046–51 (DOI 10.1186/ar1782).

61 Clark, A.J., Lynch, M.E. *et al.* (2005) Guidelines for the use of cannabinoid compounds in chronic pain. *Pain Research and Management*, **10**, 44A–6A.

62 Clark, A.J., Lynch, M.E. *et al.* (2007) Updated guidelines for the use of cannabinoid compunds available in Canada for the treatment of chronic pain. *ICRS Annual Meeting Abstracts*: Abstract 126.

63 Watson, S.J., Benson, J.A.J. *et al.* (2000) Marijuana and medicine: assessing the science base: a summary of the 1999 Institute of Medicine Report. *Archives of General Psychiatry*, **57** (6), 547–52.

64 Van Tulder, M.W., Touray, T. *et al.* (2003) Muscle relaxants for nonspecific low back pain. *Cochrane Database of Systematic Reviews* (4.Art No.:CD0044252.DOI): 10.1002/14651858.CD004252.

65 Moore, R.A., Tramer, M.R. *et al.* (1998) Quantitative systematic review of topically applied non-steroidal anti-inflammatory drugs. *British Medical Journal*, **316**, 333–8.

66 Heyneman, C.A., Lawless-Liday, C. *et al.* (2000) Oral versus topical NSAIDs in rheumatic diseases. *Drugs*, **2000**, 555–74.

67 Bookman, A.A., Williams, K.S. *et al.* (2004) Effect of a topical diclofenac solution for relieving symptoms of primary osteoarthritis of the knee: a randomized controlled trial. *CMAJ*, **171**, 333–8.

68 Roth, S.H. and Shainhouse, J.Z. (2004) Efficacy and safety of a topical diclofenac solution (Pennsaid) in the treatment of primary osteoarthritis of the knee: a randomized, double-blind, vehicle-controlled clinical trial. *Archives of Internal Medicine*, **164**, 2017–23.

69 Tugwell, P.S., Wells, G.A., Shainhouse, J.Z. (2004) Equivalence study of a topical diclofenac solution (Pennsaid) compared with oral diclofenac in symptomatic treatment of osteoarthritis of the knee: a randomized controlled trial. *The Journal of Rheumatology*, **31**, 2002–12.

70 Mason, L., Moore, R.A. *et al.* (2004) Systematic review of efficacy of topical rubefacients containing salicylates for the treatment of acute and chronic pain. *BMJ*, **328** (*7446*), 995, Epub2004March19.

71 Mason, L., Moore, A., Derry, S., Edwards, J.E., McQuay, H.J. (2004) Systematic review of topical capsaicin for the treatment of chronic pain. *BMJ*, doi:10.1136/bmj.33042.506748.EE.

72 Zhang, W.Y. and Li Wan Po, A. (1994) The effectiveness of topically applied capsaicin. A meta-analysis. *European Journal of Clinical Pharmacology*, **46**, 517–22.

73 McCleane, G.J. (2000) Topical application of doxepin hydrochloride, capsaicin and a combination of both produces analgesia in chronic human neuropathic pain: a randomized, double blind, placebo-controlled study. *British Journal of Clinical Pharmacology*, **49**, 574–9.

74 McCleane, G.J. (2000) Topical doxepin hydrochloride reduces neuropathic pain: a randomized, double-blind placebo controlled study. *The Pain Clinic*, **12**, 47–50.

75 Lynch, M.E., Clark, A.J. *et al.* (2005) Topical amitriptyline2% and ketamine 1% in neuropathic pain syndromes: a randomized double blind placebo

controlled trial. *Anesthesiology*, **103**, 140–6.

76 Lockhart, E. (2004) Topical combination of amitriptyline and ketamine for post herpetic neuralgia. *The Journal of Pain*, **5** (*S1*), 82.

77 Rowbotham, M.C., Davies, P.S. *et al.* (1996) Lidocaine patch: double-blind placebo controlled study of a new treatment method for post-herpetic neuralgia. *Pain*, **65**, 39–44.

78 Galer, B.S., Rowbotham, M.C. *et al.* (1999) Topical lidocaine patch relieves postherpetic neuralgia more effectively than a vehicle topical patch: results of an enriched enrollment study. *Pain*, **80**, 533–8.

79 Galer, B.S., Jensen, M.P. *et al.* (2002) The lidocaine patch 5% effectively treats all neuropathic pain qualities: results of a randomized, double blind, vehicle controlled, 3 week efficacy study with use of the neuropathic pain scale. *The Clinical Journal of Pain*, **18**, 297–301.

80 Meier, T., Wasner, G. *et al.* (2003) Efficacy of lidocaine patch 5% in the treatment of focal peripheral neuropathic pain syndromes: a randomized, double-blind, placebo-controlled study. *Pain*, **106**, 151–8.

81 Peng, P., Chouiniere, M. *et al.* (2006) Characterization of multidisciplinary pain treatment facilities (MPTF) in Canada: Stop Pain Project-Study II. *Pain Research and Management 11*, **121**, P-59.

82 Lynch, M.E., Campbell, F.A., Clark, A.J. *et al.* (2007) A systematic review of the effect of waiting for treatment for chronic pain. *Pain*, 2007, Epub ahead of print: PMID:17707589.

83 Bond, M., Breivik, H., Niv, D. Global day against pain, new declaration. http://www.painreliefhumanright.com 2004:1–4.

10
Non-Drug Treatments for Chronic Pain

Alexander J. Clark

- Systematic reviews of non-pharmacological treaments for chronic pain show that the overall quality of the research in this area is modest and that most data are for short-term outcomes only.

- Treatment strategies with some empirical support for their efficacy include:
  - acupuncture for short-term relief of chronic low back and neck pain;
  - individual and group exercise therapy for a variety of chronic pain syndromes;
  - laser therapy for short-term relief of rheumatoid and osteoarthritis and chronic low back pain;
  - massage therapy for chronic low back pain;
  - relaxation therapy for chronic low back pain;
  - transcutaneous electrical nerve stimulation for short-term relief of osteo-arthritis of the knee and chronic low back pain.

10.1
Introduction

Chronic pain sufferers seek treatment from a variety of health care professionals and non-traditional healers. These practitioners administer a variety of treatments, some of which are well known to the general public, including acupuncture, exercise therapies, manipulation, laser therapy, transcutaneous electrical nerve stimulation, ultrasound, psychological treatment and massage therapy. These modalities can be offered as solo treatment provided by a practitioner functioning independently or as a part of a multidisciplinary treatment program. Many of these treatments have been the subject of clinical research. While this research varies in methodological quality, it represents the best available way of assessing the value of these interventions for this patient group.

Chronic Pain: A Health Policy Perspective
Edited by S. Rashiq, D. Schopflocher, P. Taenzer, and E. Jonsson
Copyright © 2008 WILEY-VCH Verlag GmbH & Co. KGaA, Weinheim
ISBN: 978-3-527-32382-1

This chapter will describe the available research evidence for the more commonly used non-pharmacological treatments when used in isolation. Where appropriate evidence exists, the efficacy of these modalities as part of multidisciplinary treatment programs will be discussed. We conclude with recommendations to enhance the benefit and reduce potential risk offered by these approaches to the treatment of chronic pain.

10.2
Acupuncture

Acupuncture is part of the traditional Chinese system of medicine. In recent years it has become a popular treatment modality in Western cultures and is practised by both traditional Chinese medicine healers as well as by Western practitioners who have undergone specific training. In some jurisdictions there are specific requirements for registration as an acupuncturist. Among Western trained practitioners, physicians, physiotherapists and occupational therapists may seek additional training in acupuncture treatment and include this modality in their chronic pain practices. Acupuncture treatment may be provided in the traditional manner through the placement and manipulation of fine needles or may be provided with the additional stimulation of electric currents, electroacupuncture.

Systematic reviews on the efficacy of acupuncture for lateral elbow pain [1], rheumatoid arthritis [2], shoulder pain [3], low back pain [4] and chronic neck pain [5] were identified. From these reviews it can be concluded that acupuncture may provide short-term benefit in pain reduction in patients with chronic neck and low back pain, but that this relief is not better than that obtainable from other therapies. There is no evidence of long-term benefit in pain reduction or improvement of function for any condition. Use of acupuncture in the treatment of chronic pain for these conditions should be short-term and focused on conditions with proven efficacy. More good quality clinical trials are needed to assess efficacy in other conditions associated with chronic pain.

10.3
Cognitive Behavioral Therapy (with Bruce Dick)

Cognitive behavioral therapy (CBT) is a psychological treatment based on the premise that a person's thoughts, mood/affect, and behavior exert a considerable influence upon each other [6]. It follows that any adopted treatment or strategy that affects one of those factors will affect the other two. Traditionally, CBT has been used as a means of improving negative mood. Within the context of chronic pain management, CBT is often administered within a multidisciplinary environment. In such an environment, medical and physical therapies are accompanied by psychological interventions that target negative feelings such as depression and anxiety that co-exist with the pain. The psychological treatment often aims to change negative and unhelpful thought patterns that are related to or are a consequence of the chronic pain. Additional behavioral patterns that are also frequently targeted include the reduction of activity avoidance, reducing pain-related fear and distress, and pain behavior.

The majority of single studies of the efficacy of CBT in pain management are not of sufficient quality to provide evidence of the efficacy of this treatment modality for chronic pain on their own. However, in a systematic review of the efficacy of CBT on chronic pain in adults (excluding headache) [7], data from 25 trials were evaluated. While considerable variability was identified between studies with regard to measures used and experimental design, CBT was found to provide significant benefit in terms of pain experience, positive cognitive coping and appraisal, and reduced pain behavior. CBT was not found to be associated with benefit in the domains of mood/affect, negative cognitive coping, or social role functioning. Overall, it was concluded that CBT is an effective treatment modality for managing chronic pain.

10.4
Exercise

Exercise, in both group and one-to-one programs, is commonly advocated for patients with chronic pain to enhance physical capacity in order to improve function. Systematic reviews of the role of exercise have been conducted in rheumatoid arthritis [8], osteoarthritis of the knee [9, 10], fibromyalgia [11], patellofemoral pain syndrome (knee pain in adolescents/young adults) [12], peripheral neuropathy [13], low back pain [14], and chronic neck pain [15]. In general, various types of exercise seem to reduce pain and improve function, but the long-term efficacy of exercise was not assessed in these reviews. Improvements were observed in aerobic capacity and muscle strength. Pain does not generally get worse over the course of an exercise training program (a fear often expressed by patients before they begin), and in some cases (e.g. low back pain), mild decreases in pain intensity are observed. The combination of exercise with other modalities (e.g. mobilization and manipulation) seems to produce greater benefits than these modalities alone. Group exercise programs appear to be as effective as one-to-one programs in some conditions and should be considered as potentially more cost effective. The benefits of exercise in conjunction with other modalities need further study.

10.5
Laser Therapy

Laser therapies are frequently part of a comprehensive physiotherapy treatment program for musculoskeletal injuries, chronic and degenerative conditions and to heal wounds. The light source is placed in contact with the skin allowing the photon energy to penetrate the tissue, where it is thought to interact with various intracellular biomolecules. Systematic reviews of laser therapy have been undertaken in chronic (mechanical) neck pain [16], rheumatoid arthritis [17], osteoarthritis [18], and chronic low back pain [19]. There is limited evidence that laser therapy may be useful in providing short-term reduction in pain in rheumatoid arthritis, osteoarthritis and chronic low back pain. There is no evidence for a long-term effect or for use in chronic mechanical neck pain. Laser therapy should only be considered for short-term use in chronically painful conditions.

10.6
Manipulation and Mobilization

Manipulation is movement of short amplitude and high velocity that moves a joint beyond where a patient's muscles could move it by themselves but that does not cause ligament rupture. *Mobilization* is movement administered by the clinician within normal joint range in order to increase the overall range of motion. Manipulation is considered to have a higher complication rate compared to mobilization due to the velocity and movement of the joint. There is very limited information about the effectiveness of manipulation and mobilization in conditions associated with chronic pain. Neither mobilization nor manipulation was superior when compared to each other. There is little evidence in mechanical neck disorders that manipulation or mobilization used in isolation or with other passive physical medicine modalities is beneficial (although some benefits when used in association with exercise were seen [20]). In chronic low back pain, manipulation was beneficial, but only reduced pain by less that 10 mm on a 100 mm pain rating scale, a degree not considered clinically significant. There is no evidence that spinal manipulation therapy is superior to other standard therapies such as analgesic drugs, physical therapy, exercises, back school or usual general practitioner care in chronic low back pain [21]. There is no evidence for the use of traction in back pain treatment [22].

10.7
Massage Therapy

Various types of massage (deep transverse friction massage, classical massage and acupressure massage) are advocated for the treatment of chronic pain. Deep transverse friction is a specific type of connective tissue massage and is applied by the finger(s) directly to the painful area, across the direction of the muscle fibers. It can be used after an injury or for mechanical overuse in muscles, tendons and ligaments. In acupressure massage, points on the body are massaged using finger or thumb in a rapid circular motion with medium pressure. Massages last between 5 and 15 minutes. Classical (also known as Swedish) massage includes a variety of techniques specifically designed to relax muscles by applying pressure to them against deeper muscles and bones, and rubbing in the same direction as the flow of blood returning to the heart. It employs five different movements: long gliding strokes, kneading of individual muscles, friction, hacking or tapping, and vibration. Massage therapy has been studied in systematic reviews for tendonitis [23], chronic low back pain [24], and mechanical neck disorders [25]. Massage, particularly acupressure massage, may be beneficial in the treatment of chronic low back pain, especially when combined with exercise and education, although it is not necessarily better than other types of treatment. The evidence for massage therapy in neck pain and tendonitis is lacking. Long-term effects of massage therapy are unclear. Massage therapy in conjunction with other therapies (such as relaxation therapy, acupuncture and self-care education) may be of benefit.

10.8
Occupational and Physical Therapy

Occupational therapy, which aims to facilitate task performance and decrease the impact of illness on daily activities, is a cornerstone in the management of severe chronic disease. Thirty eight studies were included in a systematic review which concluded that occupational therapy has a positive effect on functional ability in patients with rheumatoid arthritis [26]. Physical conditioning programs (also called work conditioning, work hardening and functional restoration/exercise programs) that include a cognitive-behavioral approach plus intensive physical training which is in some way work-related seem to be effective in reducing the number of sick days for some workers with chronic back pain when compared to usual care [27].

10.9
Relaxation Therapy

There are many different types of relaxation techniques which include meditation, mind/body interaction, music- or sound-induced relaxation, mental imagery, and biofeedback. Rhythmic, deep, visualized or diaphragmatic breathing may also be used. Most studies of relaxation therapies are of poor quality and provide conflicting results. There is some evidence of short-term benefit in chronic low back pain for combined cognitive therapy and progressive relaxation therapy [28]. Mindfulness based stress reduction, a learned meditation technique that has been applied to many chronic psychological and physical health conditions, appears to be associated with significant and sustained improvements in pain intensity [29], but has yet to be subjected to adequately sized randomized trials.

10.9.1
Thermotherapy (Application of Heat and/or Cold)

Application of heat and cold, by various means such as hot packs, ice packs, paraffin wax baths and faradic baths, is a common treatment for pain.

In a review of studies of patients with rheumatoid arthritis, no significant effects were found for hot pack, cold pack or faradic bath use on a number of objective measures of disease or pain intensity. Positive results were seen for paraffin wax baths alone in the range of motion, grip strength and pain on non-resisted motion after four weeks of treatment [30]. Balneotherapy (spa therapy), one of the oldest forms of therapy used to relieve pain and improve joint function in rheumatoid arthritis, has been studied in six trials, most of which reported positive findings but all of which were methodologically flawed. It was concluded that the scientific evidence is insufficient to draw any conclusions [31].

In osteoarthritis of the knee, ice massage improved swelling, range of motion, function and knee strength, but not pain, and only in the short term [32]. Hot packs were not useful. A systematic review of nine trials concluded that there is insufficient evidence to support the common practice of superficial heat and cold for low back pain [33]. There is moderate evidence that heat wrap therapy provides a small short-term reduction in pain and disability in acute and sub-acute low back pain and that exercise further reduces pain and improves function in those conditions. In summary, the application of heat or cold may provide limited benefit of short-term duration in some individuals with chronic pain.

10.9.2
Transcutaneous Electrical Nerve Stimulation

Transcutaneous electrical nerve stimulation (TENS) is the application of electrical current across the skin using adhesive electrodes connected to a small battery-powered device. It is thought to produce analgesia by stimulating one type of nerve which reduces impulse transmission in other, pain transmitting nerves. Current can be delivered at different frequencies and intensities, and there is no consensus on the number and configuration of electrodes that are to be applied, making systematic research difficult. The efficacy of TENS has been studied in shoulder pain after stroke [34], osteoarthritis of the knee [35], chronic low back pain [36], rheumatoid arthritis in the hand [37], and chronic pain [38]. TENS is very widely used, but there is only limited evidence that it can lead to short-term reduction of pain in osteoarthritis of the knee and chronic low back pain. There are no systematic reviews of the efficacy of long-term use of TENS although some patients report experiencing benefit. A large systematic review of 19 studies of the efficacy of TENS in chronic pain was inconclusive, as the published trials did not provide information on the stimulation parameters, which parameters were most likely to provide optimum pain relief, nor did they answer questions about long-term efficacy. It concluded that the analgesic efficacy of TENS in chronic pain remains uncertain.

10.10
Ultrasound

Therapeutic ultrasound is a physical therapy modality often administered for the management of chronic pain and loss of function. Systematic reviews of the efficacy of ultrasound have been conducted for patellofemoral syndrome [39], osteoarthritis of the hip and knee [40], rheumatoid arthritis [41], and shoulder pain [27]. There is no evidence that that ultrasound is beneficial for any of the chronic pain conditions studied. The use of ultrasound cannot be supported for treating chronic pain conditions.

10.11
Conclusion

Many different non-drug modalities are used to treat chronic pain. Many have short-term benefit but few have any evidence of long-term benefit and some have no evidence of benefit at all. Often, results appear to be better if modalities are combined and this is particularly so when exercise is combined with other modalities. Generally, most studies are of low quality and small numbers and there is a need to perform high quality studies.

Understandably perhaps, many patients with chronic pain are willing to try modality treatments whether or not there is scientific justification for their use and whether or not they are required to pay out of pocket for doing so. There is a wide range of such treatments available. Many of their practitioners make extravagant claims of success and charge high fees. Research studies need to be undertaken to determine whether modalities not covered by this review should be considered when deciding what care to provide to individuals with conditions that result in chronic pain.

Based on the scientific evidence available, reasonable recommendations would be to suggest that many of the modalities reviewed here could be tried for a short period of time to assess benefit, but should not be used in the long term unless continuing objective improvement is observed. Combinations of modalities may be more efficacious than single modalities.

References

1 Green, S., Buchbinder, R., Barnsley, L., Hall, S., White, M., Smidt, N. and Assendelft, W. (2002) Acupuncture for lateral elbow pain. *Cochrane Database of Systematic Reviews*, 1, John Wiley & Sons, Ltd, Chichester, UK, DOI: 10.1002/14651858.CD003527.

2 Casimiro, L., Barnsley, L., Brosseau, L., Milne, S., Robinson, V.A., Tugwell, P. and Wells, G. (2005) Acupuncture and electroacupuncture for the treatment of rheumatoid arthritis. *Cochrane Database of Systematic Reviews*, 4, John Wiley & Sons, Ltd, Chichester, UK, DOI: 10.1002/14651858.CD003788.pub2.

3 Green, S., Buchbinder, R. and Hetrick, S. (2005) Acupuncture for shoulder pain. *Cochrane Database of Systematic Reviews*, 2, John Wiley & Sons, Ltd, Chichester, UK, DOI: 10.1002/14651858.CD005319.

4 Furlan, A.D., van Tulder, M.W., Cherkin, D.C., Tsukayama, H., Lao, L., Koes, B.W. and Berman, B.M. (2005) Acupuncture and dry-needling for low back pain. *Cochrane Database of Systematic Reviews*, 1, John Wiley & Sons, Ltd, Chichester, UK, DOI: 10.1002/14651858.CD001351.pub2.

5 Trinh, K.V., Graham, N., Gross, A.R., Goldsmith, C.H., Wang, E., Cameron, I. D., Kay, T. and Cervical Overview Group

(2006) Acupuncture for neck disorders. *Cochrane Database of Systematic Reviews*, **3**, John Wiley & Sons, Ltd, Chichester, UK, DOI: 10.1002/14651858.CD004870.pub3.

6 Beck, A. (1993) *Cognitive Therapy and the Emotional Disorders*, Penguin, New York.

7 Morley, S., Eccleston, C. and Williams, A. (1999) Systematic review and meta-analysis of randomized controlled trials of cognitive behavioral therapy. *Pain*, **80**, 1–13.

8 van den Ende, C.H.M., Vliet Vlieland, T.P.M., Munneke, M. and Hazes, J.M.W. (1998) Dynamic exercise therapy for treating rheumatoid arthritis. *Cochrane Database of Systematic Reviews*, **4**, John Wiley & Sons, Ltd, Chichester, UK, DOI: 10.1002/14651858.CD000322.

9 Fransen, M., McConnell, S. and Bell, M. (2001) Exercise for osteoarthritis of the hip or knee. *Cochrane Database of Systematic Reviews*, **2**, John Wiley & Sons, Ltd, Chichester, UK, DOI: 10.1002/14651858.CD004376.

10 Brosseau, L., MacLeay, L., Robinson, V.A., Tugwell, P. and Wells, G. (2003) Intensity of exercise for the treatment of osteoarthritis. *Cochrane Database of Systematic Reviews*, **2**, John Wiley & Sons, Ltd, Chichester, UK, DOI: 10.1002/14651858.CD004259.

11 Busch, A.J., Barber, K.A.R., Overend, T.J., Peloso, P.M.J. and Schachter, C.L. (2002) Exercise for treating fibromyalgia syndrome. *Cochrane Database of Systematic Reviews*, **2**, John Wiley & Sons, Ltd, Chichester, UK, DOI: 10.1002/14651858.CD003786.

12 Heintjes, E., Berger, M.Y., Bierma-Zeinstra, S.M.A., Bernsen, R.M.D., Verhaar, J.A.N. and Koes, B.W. (2003) Exercise therapy for patellofemoral pain syndrome. *Cochrane Database of Systematic Reviews*, **4**, John Wiley & Sons, Ltd, Chichester, UK, DOI: 10.1002/14651858.CD003472.

13 White, C.M., Pritchard, J., Turner-Stokes, L. (2004) Exercise for people with peripheral neuropathy. *Cochrane Database of Systematic Reviews*, **4**, John Wiley & Sons, Ltd, Chichester, UK, DOI: 10.1002/14651858.CD003904.

14 Hayden, J.A., van Tulder, M.W., Malmivaara, A. and Koes, B.W. (2005) Exercise therapy for treatment of non-specific low back pain. *Cochrane Database of Systematic Reviews*, **3**, John Wiley & Sons, Ltd, Chichester, UK, DOI: 10.1002/14651858.CD000335.pub2.

15 Kay, T.M., Gross, A., Goldsmith, C., Santaguida, P.L., Hoving, J., Bronfort, G. and Cervical Overview Group (2005) Exercises for mechanical neck disorders. *Cochrane Database of Systematic Reviews*, **3**, John Wiley & Sons, Ltd, Chichester, UK, DOI: 10.1002/14651858.CD004250.pub3.

16 Gross, A.R., Aker, P.D., Goldsmith, C.H., Peloso, P. (1998) Physical medicine modalities for mechanical neck disorders. *Cochrane Database of Systematic Reviews*, **2**, John Wiley & Sons, Ltd, Chichester, UK, DOI: 10.1002/14651858.CD000961.

17 Brosseau, L., Robinson, V., Wells, G., de Bie, R., Gam, A., Harman, K., Morin, M., Shea, B. and Tugwell, P. (2005) Low level laser therapy (Classes I, II and III) for treating rheumatoid arthritis. *Cochrane Database of Systematic Reviews*, **4**, John Wiley & Sons, Ltd, Chichester, UK, DOI: 10.1002/14651858.CD002049.pub2.

18 Brosseau, L., Robinson, V., Wells, G., de Bie, R., Gam, A., Harman, K., Morin, M., Shea, B. and Tugwell, P. (2007) Low level laser therapy (Classes III) for treating osteoarthritis. *Cochrane Database of Systematic Reviews*, **1**, John Wiley & Sons, Ltd, Chichester, UK, DOI: 10.1002/14651858.CD002046.pub3.

19 Yousefi-Nooraie, R., Schonstein, E., Heidari, K., Rashidian, A., Akbari-Kamrani, M., Irani, S., Shakiba, B., Mortaz Hejri, S., Mortaz Hejri, S. and Jonaidi, A. (2007) Low level laser therapy for nonspecific low-back pain. *Cochrane Database of Systematic Reviews*, **2**, John Wiley & Sons, Ltd, Chichester, UK, DOI: 10.1002/14651858.CD005107.pub2.

20 Gross, A.R., Hoving, J.L., Haines, T.A., Goldsmith, C.H., Kay, T., Aker, P., Bronfort, G. and Cervical overview group (2004) Manipulation and mobilisation for mechanical neck disorders. *Cochrane Database of Systematic Reviews*, **1**, John Wiley & Sons, Ltd, Chichester, UK, DOI: 10.1002/14651858.CD004249.pub2.

21 Assendelft, W.J.J., Morton, S.C., Yu Emily, I., Suttorp, M.J. and Shekelle, P.G. (2004) Spinal manipulative therapy for low-back

pain. *Cochrane Database of Systematic Reviews*, **1**, John Wiley & Sons, Ltd, Chichester, UK, DOI: 10.1002/14651858. CD000447.pub2.

22 Clarke, J.A., van Tulder, M.W., Blomberg, S.E.I., de Vet, H.C.W., van der Heijden, G.J.M.G., Bronfort, G. and Bouter, L.M. (2007) Traction for low-back pain with or without sciatica. *Cochrane Database of Systematic Reviews*, **2**, John Wiley & Sons, Ltd, Chichester, UK, DOI: 10.1002/14651858.CD003010.pub4.

23 Brosseau, L., Casimiro, L., Milne, S., Robinson, V.A., Shea, B.J., Tugwell, P. and Wells, G. (2002) Deep transverse friction massage for treating tendinitis. *Cochrane Database of Systematic Reviews*, **4**, John Wiley & Sons, Ltd, Chichester, UK, DOI: 10.1002/14651858.CD003528.

24 Furlan, A.D., Brosseau, L., Imamura, M. and Irvin, E. (2002) Massage for low-back pain. *Cochrane Database of Systematic Reviews*, **2**, John Wiley & Sons, Ltd, Chichester, UK, DOI: 10.1002/14651858. CD001929.

25 Haraldsson, B.G., Gross, A.R., Myers, C.D., Ezzo, J.M., Morien, A., Goldsmith, C., Peloso, P.M., Bronfort, G. and Cervical overview group (2006) Massage for mechanical neck disorders. *Cochrane Database of Systematic Reviews*, **3**, John Wiley & Sons, Ltd, Chichester, UK, DOI: 10.1002/14651858.CD004871.pub3.

26 Steultjens, E.M.J., Dekker, J., Bouter, L.M., van Schaardenburg, D., van Kuyk, M.A.H. and van den Ende, C.H.M. (2004) Occupational therapy for rheumatoid arthritis. *Cochrane Database of Systematic Reviews*, **1**, John Wiley & Sons, Ltd, Chichester, UK, DOI: 10.1002/14651858. CD003114.pub2.

27 Schonstein, E., Kenny, D.T., Keating, J. and Koes, B.W. (2003) Work conditioning, work hardening and functional restoration for workers with back and neck pain. *Cochrane Database of Systematic Reviews*, **3**, John Wiley & Sons, Ltd, Chichester, UK, DOI: 10.1002/14651858.CD001822.

28 Ostelo, R.W.J.G., van Tulder, M.W., Vlaeyen, J.W.S., Linton, S.J., Morley, S.J. and Assendelft, W.J.J. (2005) Behavioural treatment for chronic low-back pain. *Cochrane Database of Systematic Reviews*, **1**, John Wiley & Sons, Ltd, Chichester,

UK, DOI: 10.1002/14651858.CD002014. pub2.

29 Kabat-Zinn, J., Lipworth, L., Burney, R. and Sellers, W. (1986) Four year follow-up of a meditation based program for the self regulation of chronic pain; treatment outcomes and compliance. *The Clinical Journal of Pain*, **2**, 159–73.

30 Robinson, V.A., Brosseau, L., Casimiro, L., Judd, M.G., Shea, B.J., Tugwell, P., Wells, G. (2002) Thermotherapy for treating rheumatoid arthritis. *Cochrane Database of Systematic Reviews*, **2**, John Wiley & Sons, Ltd, Chichester, UK, DOI: 10.1002/14651858.CD002826.

31 Verhagen, A.P., Bierma-Zeinstra, S.M.A., Cardoso, J.R., de Bie, R.A., Boers, M. and de Vet, H.C.W. (2004) Balneotherapy for rheumatoid arthritis. *Cochrane Database of Systematic Reviews*, **1**, John Wiley & Sons, Ltd, Chichester, UK, DOI: 10.1002/14651858.CD000518.

32 Brosseau, L., Yonge, K.A., Robinson, V., Marchand, S., Judd, M., Wells, G. and Tugwell, P. (2003) Thermotherapy for treatment of osteoarthritis. *Cochrane Database of Systematic Reviews*, **4**, John Wiley & Sons, Ltd, Chichester, UK, DOI: 10.1002/14651858.CD004522.

33 French, S.D., Cameron, M., Walker, B.F., Reggars, J.W. and Esterman, A.J. (2006) Superficial heat or cold for low back pain. *Cochrane Database of Systematic Reviews*, **1**, John Wiley & Sons, Ltd, Chichester, UK, DOI: 10.1002/14651858. CD004750.pub2.

34 Price, C.I.M. and Pandyan, A.D. (2000) Electrical stimulation for preventing and treating post-stroke shoulder pain. *Cochrane Database of Systematic Reviews*, **4**, John Wiley & Sons, Ltd, Chichester, UK, DOI: 10.1002/14651858.CD001698.

35 Osiri, M., Welch, V., Brosseau, L., Shea, B., McGowan, J., Tugwell, P. and Wells, G. (2000) Transcutaneous electrical nerve stimulation for knee osteoarthritis. *Cochrane Database of Systematic Reviews*, **4**, John Wiley & Sons, Ltd, Chichester, UK, DOI: 10.1002/14651858.CD002823.

36 Gadsby, J.G. and Flowerdew, M.W. (2006) Transcutaneous electrical nerve stimulation and acupuncture-like transcutaneous electrical nerve stimulation for chronic low back pain. *Cochrane*

Database of Systematic Reviews, **1**, John Wiley & Sons, Ltd, Chichester, UK, DOI: 10.1002/14651858.CD000210.pub2.

37 Brosseau, L., Yonge, K.A., Robinson, V., Marchand, S., Judd, M., Wells, G. and Tugwell, P. (2003) Transcutaneous electrical nerve stimulation (TENS) for the treatment of rheumatoid arthritis in the hand. *Cochrane Database of Systematic Reviews*, **2**, John Wiley & Sons, Ltd, Chichester, UK, DOI: 10.1002/14651858. CD004377.

38 Carroll, D., Moore, R.A., McQuay, H.J., Fairman, F., TramËr, M. and Leijon, G. (2000) Transcutaneous electrical nerve stimulation (TENS) for chronic pain. *Cochrane Database of Systematic Reviews*, **4**, John Wiley & Sons, Ltd, Chichester, UK, DOI: 10.1002/14651858.CD003222.

39 Brosseau, L., Casimiro, L., Robinson, V., Milne, S., Shea, B., Judd, M., Wells, G. and Tugwell, P. (2001) Therapeutic ultrasound for treating patellofemoral pain syndrome. *Cochrane Database of Systematic Reviews*, **4**, John Wiley & Sons, Ltd, Chichester, UK, DOI: 10.1002/14651858.CD003375.

40 Robinson, V.A., Brosseau, L., Peterson, J., Shea, B.J., Tugwell, P. and Wells, G. (2001) Therapeutic ultrasound for osteoarthritis of the knee. *Cochrane Database of Systematic Reviews*, **3**, John Wiley & Sons, Ltd, Chichester, UK, DOI: 10.1002/14651858.CD003132.

41 Casimiro, L., Brosseau, L., Robinson, V., Milne, S., Judd, M., Wells, G., Tugwell, P. and Shea, B. (2002) Therapeutic ultrasound for the treatment of rheumatoid arthritis. *Cochrane Database of Systematic Reviews*, **3**, John Wiley & Sons, Ltd, Chichester, UK, DOI: 10.1002/14651858.CD003787.

42 Green, S., Buchbinder, R. and Hetrick, S. (2003) Physiotherapy interventions for shoulder pain. *Cochrane Database of Systematic Reviews*, **2**, John Wiley & Sons, Ltd, Chichester, UK, DOI: 10.1002/14651858.CD004258.

11
Interventional Treatments for Chronic Pain

Alexander J. Clark and Christopher C. Spanswick

> • Although injection therapies are widely used in the treatment of chronic pain there is very little research published on their efficacy.
>
> • There are however anecdotal reports in the literature indicating that although adverse events are rare they can be catastrophic.
>
> • The efficacy of spinal cord stimulation is supported by research evidence for a variety of intractable chronic pain conditions but technical and surgical problems are still common.

Interventional treatments for chronic pain include the injection of local anesthetics and other medications at the site of pain, or along the course of the nerve path leading from it. In addition, devices may be surgically implanted that deliver electrical current or medication directly to the spinal cord. Interventional treatments have been used for many years and especially for the past 30 years. In spite of this lengthy period, few studies have been undertaken that demonstrate that these treatments actually benefit the patient more than placebo. The literature is comprised mostly of numerous case reports and small series attesting to benefit but not proving benefit.

The potential morbidity and mortality associated with these treatments [1, 2] is of concern; however, the prevalence of adverse events is unknown. Cases of meningitis and deep muscle abscesses have been linked to the injection of paravertebral muscle (half of them "trigger point" injections), facet joint injection, epidural injection and intrathecal injection.

This chapter will review the existing evidence for commonly used interventional approaches in the treatment of chronic pain so that the reader can weigh the evidence for these approaches.

Chronic Pain: A Health Policy Perspective
Edited by S. Rashiq, D. Schopflocher, P. Taenzer, and E. Jonsson
Copyright © 2008 WILEY-VCH Verlag GmbH & Co. KGaA, Weinheim
ISBN: 978-3-527-32382-1

11.1
Cervical and Lumbar Epidural Injections

Epidural injection involves the placement of a therapeutic substance immediately outside the outermost and thickest protective covering of the spinal cord and the fluid in which it resides. It is a standard part of the practice of anesthesia and has, for instance, revolutionized the provision of pain relief in childbirth. The use of epidural steroid injections for low back and leg pain dates back many decades. There are a number of anecdotal reports of efficacy and a number of uncontrolled studies, which at best provide conflicting evidence.

In a literature review, completed in 1995, of randomized trials of epidural steroid injections in the treatment of low back pain and/or sciatica [3] twelve trials were identified, all with flaws in their study design. Six studies showed benefit and six showed either no benefit or worse outcomes after epidural steroid injection. The best quality studies showed inconsistent results and any benefits appeared to be only short term. Therefore the efficacy of epidural steroids was not established. A significant number of side effects and complications, including headache, backache, water retention, fever, bacterial meningitis and epidural abscess, were noted.

In 2003 a randomized double blind study concluded that there was no difference between epidural steroid or saline injections for sciatica [4] and in 2005 a review of various treatments [5] concluded that the use of epidural steroid injections could be an effective treatment modality but this statement was qualified noting the lack of current evidence.

An editorial in the *British Medical Journal* [6] concluded that in spite of the lack of good evidence from clinical trials, clinical experience suggests that there was still a place for epidural steroid injections. However, another review suggested that the benefits of epidural steroid injections for sciatica are transient and not cost effective [7].

Yet another review noted there was a trend towards a positive benefit [8] but no clear evidence. The authors outlined the possible reasons for the lack of clear evidence including small sample size, variations in patient groups, and variation in procedure.

The evidence for cervical epidural steroid injections is even less clear than for lumbar epidural injections. There are few published data. In two studies the data presented were on small numbers of patients (17 and 27 per group) and neither study included a control group. The initial outcomes seemed promising, particularly for radicular symptoms (attributable to a nerve root) rather than central spinal pain, but the potential adverse consequences can be catastrophic and even fatal [9, 10].

A further issue is that successful placement of a needle into the epidural space can be challenging, even with the use of X-ray guidance [5, 11, 12] and there are a number of serious published complications [13], some of which are life threatening. Overall side effects and complication rates are high (0.5% to 2.5%). Most of these are minor and self-limiting, such as, headache but some are life threatening, such as meningitis. Excessive or frequent steroid administration, by any route, can cause many side effects although the long-term effects of frequent repeated epidural steroid injections are unknown.

In summary, the weight of evidence does not support the use of epidural steroid injections for chronic pain. However, our clinical experience suggests that selected patients with primarily radicular pain can benefit from epidural injections as a part of a multidisciplinary multimodal treatment plan. Patients should be made aware of the very rare but potentially catastrophic complications that may occur.

11.2
Trigger Point Injections

Trigger point injections (TPIs) are frequently used when specific trigger points or tender areas are noted in muscles, as seen in widespread or regional myofascial pain syndromes [14]. Local anesthetics are used, sometimes in combination with steroids.

There is no evidence that TPIs are any better than therapeutic ultrasound when both are done in conjunction with neck stretching exercises [15].

There are no data to suggest that TPIs with either steroid or local anesthetics alone provide long-lasting benefit for patients with chronic low back pain [16]. Furthermore, there is conflicting evidence about the effectiveness of TPIs for the short-term relief of low back pain.

Recently a clinical trial comparing TPIs using botulinum toxin A vs. bupivacaine showed that both were comparable in reducing pain of short duration (four weeks), but there was no placebo control in this study [17].

A recent systematic review of invasive procedures for low back pain concluded that TPIs have not clearly been shown to be effective and cannot be recommended [18].

TPIs should only be considered where short-term reduction in pain may allow other modalities of care to be established.

11.3
Intravenous Lidocaine Infusion

Intravenous local anesthetic infusions have primarily been used in the treatment of complex regional pain syndrome, neuropathic pain (including peripheral nerve injury, postherpetic neuralgia, and diabetic neuralgia) and whiplash related disorders.

Lidocaine doses up to 6 mg/kg over 30–60 minutes have been utilized. There is limited evidence for short-term benefit but patients rarely experience any long-term benefit [19–21]. The best evidence is in the treatment of neuropathic pain [22]. Full resuscitative equipment and ECG monitoring is recommended because of the rare but serious risk of toxicity including seizure and cardiac arrhythmia.

11.4
Intravenous Regional Sympathetic Blockade

Intraveneous regional sympathetic blockade (IRSB), the injection of agents into a limb isolated by tourniquet that block sympathetic nerve transmission, has been used for many years in the treatment of Complex Regional Pain Syndrome (CRPS).

There is no evidence that IRSB with guanethidine reduces the pain associated with CRPS [23, 24].

Various other medications (ketanserin, calcium channel blockers, biphosphonates) have also been used but lack any support from randomized controlled trials.

11.5
Paravertebral/Paraspinal Blocks

Paravertebral injections, both into the musculature and close to the spine, are frequently used to treat low back pain and sciatica.

There is a lack of evidence for efficacy for these injections [18]. Two case reports concerning significant complications of these techniques have appeared recently documenting extensive abscess formation involving the entire paravertebral musculature descending to the level of the mid thighs after repeated paravertebral injection of local anesthetics, corticosteroid and botulinum toxin [25] and spinal abscesses and meningitis [1]. Reports of such adverse outcomes are rare; however, there are no available systematic data to show how often such adverse effects occur.

11.6
Cervical/Lumbar Facet Joint Injection and Radiofrequency Denervation

Injection into and radio frequency destruction of the facet joints of the spine (the paired hinge-like joints that allow individual vertebrae to articulate with adjacent vertebae) are performed for low back and neck pain that is thought to arise from these joints, often due to osteoarthritis or injury.

There is inadequate evidence to support the use of facet joint injections in the treatment of patients with chronic low-back pain [16, 18]. There is some evidence that facet joint injections can be used to predict the outcome of radiofrequency denervation in selected patients with low-back pain [16]. However, there is conflicting evidence on the short-term effect of radiofrequency denervation on pain and disability in chronic low back pain [26].

There is limited evidence that cervical facet joint radiofrequency denervation provides short-term relief of chronic neck pain and cervicobrachial pain [26].

Cervical facet injection/denervation should only be undertaken with X-ray guidance and there are the risks of inadvertent spinal or arterial injection, which can result in spinal cord damage [27].

There is some evidence that facet joint injections can predict the outcome of facet joint denervation in some patients with low back pain, but little evidence of benefit as a sole treatment for low back or neck pain. There is limited evidence that cervical facet joint denervation is beneficial. Appropriate skill sets and facilities are required to provide these interventions.

11.7
Sympathetic Blockade

Temporary interruption of the sympathetic nervous system to a painful limb or viscus can be useful in the diagnosis of sympathetically mediated pain, but there is considerable variation in the amount of sympathetic transmission of pain in such conditions [28]. In addition, the response to sympathetic blockade may be confounded by both a placebo response and variations in the accuracy of the block. The accuracy of the blockade may be confirmed by the use of diagnostic imaging and contrast but monitoring of the limb temperature can also give a good guide to whether sympathetic blockade has occurred [29, 30]. The immediate results of temporary sympathetic blockade should be interpreted with care. It should be evaluated after 24 hours as an immediate response can be produced which is a placebo effect [30].

Sympathetic blocks occasionally have complications that are very significant. For example, the delayed intraspinal passage of local anesthetic following a stellate ganglion block has been reported [31]. Sympathetic blockade can be performed at several levels – cervical, thoracic, lumbar and sacral:

11.7.1
Stellate Ganglion Block

Stellate ganglion blocks have been used for the treatment of many conditions, from hay fever, angina, headache, deafness and vasospasm in peripheral vessels to persistent pain in the upper limb. There is no evidence to support the long-term use of stellate ganglion block with local anesthetics in upper limb pain In skilled hands it is a safe procedure, but the side effects can be significant and the patient should be warned regarding the potential for rare but serious or catastrophic consequences [32].

The clinical effects of sympathetic blockade (in terms of analgesia) seem to outlast the expected length of action of the local anesthetic used [33]. A recent Cochrane review revealed the scarcity of good evidence and was unable to draw a conclusion with regard to efficacy of sympathetic blocks in the treatment of CRPS [24].

11.7.2
Lumbar Sympathetic Block

Lumbar sympathetic blocks with local anesthetics may be used to determine whether a painful lower limb has a sympathetically mediated pain component.

Lumbar sympathetic plexus blocks should always be performed with diagnostic imaging to ensure accurate placement.

There is possibly a place for lumbar sympathetic blockade in patients with severe peripheral ischemia, rest pain and ulcers. The use of neurolytic agents in such cases is justified as the patients are often not fit enough to undergo surgical sympathectomy [32]. Neurolytic sympathectomy cannot otherwise be recommended in non-cancer neuropathic pain.

The evidence for neurolytic sympathectomy is of poor quality and comes mostly from uncontrolled trials and cannot be recommended in most situations. Complications of the procedure may be important and include both worsening pain and/or production of a new pain syndrome [34].

11.7.3
Coeliac and Splanchnic Nerve Blocks

Both coeliac and splanchnic nerve blocks have been used for pain control in chronic non-cancer abdominal pain and in pain from cancer involving abdominal organs. There is limited evidence to support their use in cancer pain [35, 36]. There have been no systematic reviews in chronic non-cancer abdominal pain.

These procedures should be performed with the use of imaging. CT guidance may offer the advantage of better visualization of important structures and thereby reduce the risk of serious complications including pneumothorax, paraplegia and aortic dissection [37]. The incidence of catastrophic consequences may be as high as 2% [38].

There is no evidence that blocks of the sympathetic nervous system to intra-abdominal organs with local anesthetics or neurolytic agents offer any long-term benefit in chronic non-cancer pain. These sympathetic blocks can have serious complications.

11.8
Spinal Cord Stimulation

Spinal cord stimulation involves placement of an electrode or electrodes in the epidural space overlying the spinal cord, usually in the lower thoracic or cervical regions. It is used in a number of conditions including failed back surgery syndrome, CRPS, angina, lower limb ischemia and peripheral neuropathic pain syndromes.

There is limited evidence from systematic reviews in favor of spinal cord stimulation (SCS) for failed back surgery syndrome (FBSS) and complex regional pain syndrome (CRPS) in follow-up of 6–12 months. There is insufficient evidence to assess the benefits and harms of SCS for the relief of other types of chronic pain [39, 40]. A recent RCT of FBSS versus conventional medical management demonstrated better pain relief and improved health-related quality of life and functional capacity for SCS [41].

There is some evidence to suggest that SCS is better than standard conservative treatment to improve limb salvage in the clinical situation in patients with inoperable chronic critical leg ischemia [42]. Retrospective studies have suggested SCS is effective in reducing the frequency of angina attacks and medication use, however no prospective, randomized, blinded, placebo-controlled trials have been undertaken [43].

Complications appear to be frequent, with re-intervention required in about 15 to 32% of patients, infection in 3% and implantation problems in 9% [41, 42].

11.9
Intrathecal Medication Pumps

The use of intrathecal (IT) medications delivered from an implanted pump has been advocated in clinical scenarios where other routes of medication administration and other modalities of treatment have failed. This has been regarded as the "last chance" or "last hope" care. There is some evidence from case series that IT medication pumps can be helpful but no RCT evidence. These are generally of short-term use (less than 1 year) [44]. Most existing evidence is for patients who have cancer pain. There is no evidence of efficacy in long-term use in patients with non-cancer pain. Complications and device issues appear to be frequent.

There are insufficient data currently to recommend this modality in chronic non-cancer pain.

11.10
Conclusion

Interventional therapies are widely performed for chronic pain, but the evidence for their efficacy is poor, and many of them can cause devastating complications. They appeal most strongly to the patient and clinician who believe that chronic pain is principally attributable to a physical defect in the painful body part but, as we learn more about these conditions, this model becomes increasingly difficult to uphold. These therapies are also often performed as first-line specialist therapy by well-meaning clinicians who cannot offer anything else. There is a strong argument for abandoning some of them altogether until properly conducted studies demonstrate their utility. In the meantime, careful consideration of the potential benefits and risks is required along with documented informed consent from the patient. Repeated procedures can be progressively demoralizing to the patient and foster therapeutic dependence on passive instead of active treatments. Interventional treatments should be performed in the context of multidisciplinary programs. When such programs are not available, the onus is on the clinician to demonstrate that he/she is doing more good than harm and that the benefit makes a significant difference in the patient's quality of life and function.

An honest assessment of morbidity and mortality related to injection procedures for pain is needed. It is unknown at this time how often these adverse outcomes occur.

More research is required into whether injection/interventional modalities are beneficial or harmful to our patients. Health care providers and policy makers need to be encouraged to support and participate in more studies to answer these questions.

References

1 Gaul, C., Neundörfer, B. and Winter-holler, M. (2005) Iatrogenic (para-) spinal abscesses and meningitis following injection therapy for low back pain. *Pain*, **116**, 407–10.

2 Butler, S.H. (2005) Primum non nocere – first do no harm. *Pain*, **116**, 175–6.

3 Koes, B.W., Scholten, R.J.P.M., Jan, M.A. *et al.* (1995) Efficacy of epidural steroid injections for low-back pain and sciatica: a systematic review of randomized clinical trials. *Pain*, **63**, 279–88.

4 Valat, J.-P., Giraudeau, B., Rozenberg, S. *et al.* (2003) Epidural corticosteroid injections for sciatica: a randomized, double blind, controlled clinical trial. *Annals of the Rheumatic Diseases*, **62**, 639–43.

5 Grabois, M. (2005) Management of chronic low back pain. *American Journal of Physical Medicine and Rehabilitation*, **84**, S29–41.

6 Samanta, A. and Samanta, J. (2004) Is epidural injection of steroids effective for low back pain? The evidence is equivocal, but clinical experience favours its use in some patients. *British Medical Journal*, **328**, 1509–10.

7 Price, C., Arden, N., Colgan, L. and Rogers, P. (2005) Cost-effectiveness and safety of epidural steroids in the management of sciatica. *Health Technology Assessment*, **9**, 1–58.

8 McLain, R.F., Kapural, L. and Mekhail, N.A. (2005) Epidural steroid therapy for back and leg pain: mechanisms of action and efficacy. *The Spine Journal*, **5**, 191–201.

9 Rozin, L., Rozin, R., Koehler, S.A. *et al.* (2003) Death during transforaminal epidural steroid nerve root block (C7) due to perforation of the left vertebral artery. *The American Journal of Forensic Medicine and Pathology*, **24**, 351–5.

10 Tripathi, M., Nath, S.S. and Gupta, R.K. (2005) Paraplegia after intracord injection during attempted epidural steroid injection in an awake patient. *Anesthesia and Analgesia*, **101**, 1209–11.

11 Vad, V.B., Bhat, A.L., Lutz, G.E. and Canmmisa, F. (2002) Transforaminal epidural steroid injections in lumbosacral radiculopathy: a prospective randomized study. *Spine*, **27**, 11–16.

12 Rathmell, J.P. and Benzon, H.T. (2004) Editorial transforaminal injection of steroids: should we continue? *Regional Anesthesia and Pain Medicine*, **29**, 397–9.

13 Stoll, A. and Sanchez, M. (2002) Epidural hematoma after epidural block: implications for its use in pain manage-ment. *Surgical Neurology*, **57**, 235–40.

14 Simons, D.C., Travell, J.C. and Simons, L.S. (1999) *Myofascial Pain and Dysfunction: the Trigger Point Manual*, 2nd edn, Lippincott, Williams and Wilkins.

15 Esenyel, M., Caglar, N. and Aldemir, T. (2000) Treatment of myofascial pain. *American Journal of Physical Medicine and Rehabilitation*, **79**, 48–52.

16 Resnick, D.K., Choudhri, T.F., Dailey, A.T. *et al.* (2005) Guidelines for the performance of fusion procedures for degenerative disease of the lumbar spine. Part 13: injection therapies, low-back pain, and lumbar fusion. *Journal of Neurosurgery. Spine*, **2**, 707–15.

17 Graboski, C.L., Gray, D.S. and Burnham, R.S. (2005) Botulinum toxin A versus

bupivacaine trigger point injections for the treatment of myofascial pain syndrome: A randomized double blind crossover study. *Pain*, **118**, 170–5.

18 van Tulder, M.W., Koes, B., Seitsalo, S. and Malmivaara, A. (2006) Outcome of invasive treatment modalities on back pain and sciatica: an evidence based review. *Eur Spine Journal*, **15** (Suppl 1), S82–92.

19 O'Gorman, D.A. and Raja, S.N. (2000) Infusion tests and therapies in the management of chronic pain, in *Practical Management of Pain* (ed. P.P. Raj), Mosby, St Louis, pp. 723–31.

20 Attal, N., Gaude, V., Brasseur, L., Dupuy, M., Guirimand, F., Parker, F. and Bouhassira, D. (2000) Intravenous lidocaine in central pain: a double-blind, placebo-controlled, psychophysical study. *Neurology*, **54**, 564–74.

21 Kalso, E., Tramer, M.R., McQuay, H.J. and Moore, R.A. (1998) Systemic local anaesthetic-type drugs in chronic pain: a systematic review. *European Journal of Pain*, **2**, 3–14.

22 Challapalli, V., Tremont-Lukats, I.W., McNicol, E.D., Lau, J. and Carr, D.B. (2005) Systemic administration of local anesthetic agents to relieve neuropathic pain. *The Cochrane Database of Systematic Reviews*, **4**. CD003345. DOI: 10.1002/14651858.CD003345.

23 Jadad, A.R., Carroll, D., Glynn, C.J. and McQuay, H.J. (1995) Intravenous regional sympathetic blockade for pain relief in reflex sympathetic dystrophy: a systematic review and a randomized, double-blind crossover study. *Journal of Pain and Symptom Management*, **10**, 13–20.

24 Cepeda, M.S., Carr, D.B. and Lau, J. (2005) Local anesthetic sympathetic blockade for complex regional pain syndrome. *The Cochrane Database of Systematic Reviews*, **4**. CD004598. DOI: 10.1002/14651858.CD004598.

25 Puehler, W., Brack, A. and Kopf, A. (2005) Extensive abscess formation after repeated paravertebral injections for the treatment of chronic back pain. *Pain*, **113**, 427–9.

26 Niemisto, L., Kalso, E., Malmivaara, A., Seitsalo, S. and Hurri, H. (2003) Radiofrequency denervation for neck and back pain. *The Cochrane Database of Systematic Reviews*, **1**. CD004058. DOI: 10.1002/14651858.CD004058.

27 Heckmann, J.G., Maihofner, C., Lanz, S., Rauch, C. and Neundorfer, B. (2006) Transient tetraplegia after cervical facet joint injection for chronic neck pain administered without imaging guidance. *Clinical Neurology and Neurosurgery*, **108**, 709–11.

28 Treede, R.D., Davis, K.D., Campbell, J.N. and Raja, S.N. (1992) The plasticity of cutaneous hyperalgesia during sympathetic ganglion blockade in patients with neuropathic pain. *Brain*, **115**, 607–21.

29 Tran, K.M., Frank, S.M., Raja, S.N. *et al.* (2000) Lumbar sympathetic block for sympathetically maintained pain: changes in cutaneous temperatures and pain perception. *Anesthesia and Analgesia*, **90**, 1396–401.

30 Price, D.D., Long, S.M., Wilsey, B. and Rafii, A. (1998) Analysis of peak magnitude and duration of analgesia produced by local anesthetics injected into sympathetic ganglia of Complex Regional Pain Syndrome patients. *The Clinical Journal of Pain*, **14**, 216–26.

31 Balaban, B., Baklaci, K., Taskaynatan, M.A. and Mohur, H. (2005) Delayed subdural block as an unusual complication following stellate ganglion blockade. *The Pain Clinic*, **17**, 407–9.

32 Breivik, H. (2003) Sympathetic blocks, in *Clinical Pain Management; Practical Applications and Procedures* (eds H. Bretivik, W. Campbell and C. Eccleston), Arnold Publishers, London, pp. 233–46.

33 Sayson, S.C. and Ramamurthy, S. (2004) Sympathetic blocks, in *Principles and Practice of Pain Medicine* (eds C.A. Warfield and Z.H. Bajwa), 2nd edn, McGraw Hill, New York, pp. 650–4.

34 Mailis-Gagnon, A. and Furlan, A. (2002) Sympathectomy for neuropathic pain. *The Cochrane Database of Systematic Reviews*, **1**. CD002918. DOI: 10.1002/14651858. CD002918.

35 Polati, E., Finco, G., Gottin, L., Bassi, C., Pederzoli, P. and Ischia, S. (1998) Prospective randomized double-blind trial of neurolytic coeliac plexus block in patients with pancreatic cancer. *The British Journal of Surgery*, **85**, 199–201.

36 Kinoshita, H., Denda, S., Shimoji, K., Ohtake, M. and Shirai, Y. (1996) Paraplegia following coeliac plexus block by anterior approach under direct vision. *Masui–Japanese Journal of Anesthesiology*, **45**, 1244–6.

37 Perello, A., Ashford, N.S. and Dolin, J. (1999) Coeliac plexus block using computed tomography guidance. *Palliative Medicine*, **13**, 419–25.

38 Kinoshita, H., Denda, S., Shimoji, K., Ohtake, M. and Shirai, Y. (2005) Percutaneous splanchnic nerve radiofrequency ablation for chronic abdominal pain. *ANZ Journal of Surgery*, **75**, 640–4.

39 Mailis-Gagnon, A., Furlan, A.D., Sandoval, J.A. and Taylor, R. (2004) Spinal cord stimulation for chronic pain. *The Cochrane Database of Systematic Reviews*, **3**. CD003783. DOI: 10.1002/14651858.CD003783.

40 Taylor, R.S. (2006) Spinal cord stimulation in complex regional pain syndrome and refractory neuropathic back and leg pain/failed back surgery syndrome: results of a systematic review and meta-analysis. *Journal of Pain and Symptom Management*, **31**, (Suppl. 4), S13–19.

41 Kumar, K., Taylor, R.S., Jacques, L. *et al.* (2007) Spinal cord stimulation versus conventional medical management for neuropathic pain: a multicentre randomized controlled trial in patients with failed back surgery syndrome. *Pain*, **132**, 179–88.

42 Ubbink, D.T. and Vermeulen, H. (2005) Spinal cord stimulation for non-reconstructable chronic critical leg ischaemia. *The Cochrane Database of Systematic Reviews*, **3**. CD004001. DOI: 10.1002/14651858.CD004001.

43 Yu, W., Maru, F., Edner, M., Hellstrom, K., Kahan, T. and Persson, H. (2004) Spinal cord stimulation for refractory angina pectoris: a retrospective analysis of efficacy and cost-benefit. *Coronary Artery Disease*, **15**, 31–7.

44 Ghafoor, V.L., Epshteyn, M., Carlson, G.H., Terhaar, D.M., Charry, O. and Phelps, P.K. (2007) Intrathecal drug therapy for long-term pain management. *American Journal of Health-System Pharmacy*, **64**, 2447–61.

12
Multidisciplinary Pain Clinic Treatment

Eldon Tunks

- Multidisciplinary pain clinics involve collaboration among professionals with complementary training and skills to address the multifaceted problems associated with chronic pain.

- Treatment approaches very widely across settings but typically involve psychological and rehabilitation models with a wide variety of treatment modalities. Most care models incorporate principles of support, learning how to cope, promoting function, and sustaining function in the community.

- There is evidence that the individual treatment approaches used by multidisciplinary pain clinics are effective as stand-alone interventions.

- There is strong evidence that multidisciplinary pain clinics are clinically and economically effective.
  - This suggests that access to multidisciplinary pain clinics is a valid evidence supported option in the spectrum of care required to adequately manage chronic pain.
  - Effective chronic pain care requires collaboration between healthcare providers in primary, secondary and tertiary care as well as other critical stakeholders including employers, and disability carriers.

The first trial of multidisciplinary and behavioral treatment for chronic pain came out of the Seattle Pain Clinic, headed at that time by Dr John Bonica [1]. Since then, clinical trials concerning the efficacy of behavioral treatment for chronic pain have been numerous and have involved cognitive, conditioning, and biofeedback techniques. Psychoeducational and multidisciplinary pain management program approaches often, or usually, involve cognitive therapy principles, often directed by psychologists or mental health professionals.

Chronic Pain: A Health Policy Perspective
Edited by S. Rashiq, D. Schopflocher, P. Taenzer, and E. Jonsson
Copyright © 2008 WILEY-VCH Verlag GmbH & Co. KGaA, Weinheim
ISBN: 978-3-527-32382-1

12.1
What Constitutes "Multidisciplinary" Care for Chronic Pain?

There is no one formula for "multidisciplinary" or "multimodal" therapy for chronic pain but in general it involves collaboration between clinical professionals who have complementary training and skills to address a multifaceted problem. It is more than a clinical involvement with multiple professionals. In most cases it involves a psychological model (behavioral, cognitive behavioral, psychoeducation or coping skills training) combined with active exercise. It may also involve a rehabilitation model with ergonomics assessment, work hardening, and work re-entry management.

In this chapter the focus is on multidisciplinary treatment for chronic pain management (i.e. comprehensive pain programs). Because multidisciplinary programs usually include behavioral treatments, active exercise, and patient education, we will also consider briefly what is known about these individual modalities that are usually included in multidisciplinary treatment, but we will give main attention to the efficacy of multimodal programs into which these treatments are combined.

Multidisciplinary treatment is typically aimed at patients with chronic pain lasting greater than three months where the patient is unemployed or work-disabled. The activities will typically include:

- Assess for risk factors that may prolong or complicate recovery.
- Improve coping skills, fitness, and appropriate role function.
- Treat medical and psychological comorbidity that stand in the way of recovery.
- Reduce barriers to functioning (e.g. orthopedic aids, environmental modifications).
- Plan for after-care, including vocational rehabilitation as appropriate.

12.2
Chronic Low Back Pain

12.2.1
Behavior Therapy

There is some evidence that behavioral treatment is better than no treatment for chronic low back pain intensity, and there is some evidence that behavioral treatment results in greater functional improvement. At least short-term benefit in chronic back pain relief was noted with cognitive-respondent therapies and relaxation therapy. (Cognitive therapies are based on identification of negative thoughts and catastrophizing and teaching the patient to replace them with coping thoughts. Respondent therapies are usually based on "desensitizing" a patient to fear of harm during activity by setting quotas to gradually increase exercise and activity, in an environment in which the therapist gives support, advice and reassurance. Relaxation therapies, including hypnosis and progressive relaxation or applied relaxation are techniques for reducing tension that is contributing to pain and anxiety).

Behavioral treatments, combined respondent and cognitive therapy, and progressive relaxation therapy, are more effective for relief of chronic back pain, at least in the short term, compared to waiting list control or no treatment, and there is no evidence that any one type of behavioral treatment is significantly more effective than another [2, 3].

12.2.2
Exercise and Other Physical Modalities

There is inconsistent evidence from multiple RCTs for the efficacy of exercise on measures of pain and function in chronic back pain [4, 5].

12.2.3
Patient Education and Back Schools

For both short- and long-term findings, results tended to favor patient education when compared to placebo or no treatment [6]. Tulder *et al.* [4] found evidence that an intensive back school program within an occupational setting was more effective than no treatment.

However, there is limited evidence for greater efficacy on follow-up when group education is compared to other active treatment, on measures of pain recurrence, pain intensity, pain duration, and duration of sick leave [4, 6].

12.2.4
Multidisciplinary Treatment (but Without Workplace Intervention)

A systematic review [7] of 65 studies of multidisciplinary treatment for chronic back pain found proportionally greater improvement for multidisciplinary programs on outcomes of work, medication use, healthcare use, activity, and pain behavior. Patients treated in a multidisciplinary clinic were almost twice as likely to return to work compared to the untreated or single modality treated patients (68% versus 36%). The overall benefit was seen both in the short-term (less than six months) and long-term (more than six months).

Another review [8] of multiple randomized comparisons of multidisciplinary versus control treatment for chronic low back pain found moderate evidence that intensive multidisciplinary rehabilitation reduced pain when compared with outpatient "usual care" or non-multidisciplinary rehabilitation. There was contradictory evidence regarding vocational outcomes. More intensive programs had a significant beneficial outcome for pain and functional status that was not evident in less intensive or twice-weekly outpatient programs.

Jensen *et al.* [9] reported a three year follow-up of the multidisciplinary program for chronic occupational back and neck pain, with prolonged work absence. The experimental treatment groups were cognitive behavioral therapy alone, behaviorally oriented physiotherapy alone, and multimodal treatment combining both physiotherapy and behavior therapy, and a "treatment as usual" control group. The full-time multimodal program was superior to the other three conditions for female patients, reducing the sick leave by 201 days over the follow-up period. There was a non-significant difference in outcomes comparing "treatment as usual", cognitive behavioral or physiotherapy treatment groups.

12.3
Neck Pain

12.3.1
Patient Education

The systematic review by Gross *et al.* [10] concerning controlled trials of patient education for mechanical neck pain concluded that patient education, either alone or in combination with exercise, psychological counseling or drug therapies, was not more effective in reducing neck pain.

12.3.2
Active versus Passive Treatment

For acute whiplash pain there is one large high quality study [11] that suggests that recommending "activity as tolerated", and avoidance of premature physical therapy or immobilization during the first month reduces time to recovery.

A systematic review [12] identified 23 studies of whiplash patients choosing a nonsurgical and nonpharmacological treatment. Only one study included multimodal treatment but ten studies compared passive to a more active treatment. In the single study of multimodal therapy compared to passive (TENS and ultrasound) therapy for acute whiplash patients, the multimodal condition resulted in fewer days off, and nonsignificant improvement in pain. In five out of nine of the other studies, active treatment condition was better than passive treatment such as soft collar and work absence in the outcomes of improving pain and stiffness.

This literature is consistent in the message that passive therapies are inadequate for pain management, and that multimodal (multidisciplinary) treatment is effective active treatment. But if inactivity is ineffective for pain management, what about relaxation therapies or hypnosis? As we will see below, the "activity" is in the patient's active application of these techniques to his/her pain and stress problems, and that practice in the clinic and not at home leads to poor results.

12.4
Headache

A review of migraine treatment [13] found that the headache improvement rates for relaxation with biofeedback were similar to the improvement rates for propranolol (55%), and much superior to improvement rates for placebo (12.2%) or "no treatment" conditions (1.1%).

A review [14] of behavioral treatments for recurrent tension headache found that behavioral treatments resulted in a greater degree of headache improvement when compared to control conditions such as placebo biofeedback or headache self-monitoring. No significant difference in efficacy emerged in comparing one type of behavioral intervention with another (EMG biofeedback, relaxation training, or combined behavioral treatments).

In 2002 the U.S. Headache Consortium published a series of meta-analyses concerning best evidence regarding medical and non-pharmacological treatment of acute and chronic headache. Campbell *et al.* [15] reviewed relaxation, feedback, cognitive behavioral treatment and various combinations. Relaxation techniques were found to be modestly, but significantly effective. There appeared to be no specific advantage for hypnosis over other psychological techniques. Thermal biofeedback alone, or in combination with other techniques, showed no convincing advantage over other behavioral techniques. Several studies of EMG biofeedback suggested somewhat greater efficacy for this type of biofeedback. Cognitive behavioral therapy (CBT) appeared to be efficacious, but combining CBT with thermal biofeedback did not increase effectiveness. The conclusions of the review were that all behavioral treatments may be considered for prevention of migraine, and behavioral treatments may be combined with medication for additional clinical migraine relief, but there is insufficient evidence regarding hypnosis for headache.

A review [16] of pre-post trials and randomized controlled trials dealing with various kinds of biofeedback therapy for chronic migraine headache with at least six-month follow-up, found that in comparison to waiting list control, biofeedback yielded a significant but small to medium effect size. However, no significant advantage emerged comparing biofeedback to false feedback, relaxation, or ergotamine medication. Biofeedback in combination with home training yielded nearly 20% greater effect sizes than outpatient therapy alone. A review [17] of RCTs concerning psychological treatment, mostly for chronic headache in children and adolescents, revealed strong evidence that behavioral treatments, particularly relaxation and cognitive behavioral therapy, are effective in reducing the severity and frequency of chronic headache pain. There was insufficient evidence to judge the effectiveness of behavioral therapies for other outcomes (mood, function, or disability).

12.5
Temporomandibular Joint Pain and Dysfunction

Turner *et al.* [18] reported a detailed analysis of their work on mediators and predictors of therapeutic change in cognitive behavioral therapy for chronic temporomandibular joint pain. (A mediator is a variable that is responsible for part of the effect of treatment on an outcome, and is treated and improves during treatment).

Mediators significantly associated with outcome on measures of "one-year activity interference" and "pain intensity" included patient-reported perceived control, disability, self-efficacy, fear of harm, and catastrophizing. Those with greater baseline depression had greater disability scores at one year. They concluded that "change in perceived pain control" was the mediator that most explained total treatment effect on each outcome. Patients who at the baseline (beginning treatment) had more pain sites, more depression, somatization, rumination, catastrophizing and stress before treatment had higher activity interference at one year. However, patients could respond favorably to CBT despite various baseline characteristics.

12.6
Fibromyalgia/Chronic Fatigue Syndrome

A review [19] of randomized controlled trials of multidisciplinary treatment for fibromyalgia suffered from studies of low quality, and conclusions regarding efficacy of multidisciplinary treatment for this problem could not be drawn.

12.7
Unspecified or Mixed Pain Disorders

Many older studies, mostly of poor quality, that concerned mostly chronic back pain but that did not clearly specify the diagnoses or chronicity, yielded the following results.

12.7.1
Various Behavioral and Cognitive-Behavioral Treatments

A review [20] of clinical studies published from 1960 to 1985, concerning relief of pain using relaxation, biofeedback, hypnosis, placebo, TENS, and no treatment found that only biofeedback and relaxation achieved improvements greater than what was reported with "no treatment".

A review [21] of back pain in primary care found that the CBT improved depression, anxiety, and somatic symptoms at six months and those who continued relaxation practice improved more than those who did not.

A review [22] of controlled trials concerning cognitive behavioral therapy and behavior therapy for chronic pain in adults, excluding headache, found that CBT resulted in significantly greater improvements in pain experience, positive cognitive coping and appraisal, and reduced behavioral pain expression. There was no significant change in measures of mood, catastrophizing, and social functioning.

From these older studies, despite their lower quality, the consistent message is that patients who continue to practise what they learn are more likely to benefit from CBT, and that different behavioral techniques – cognitive behavioral, relaxation, and biofeedback-assisted relaxation, may all be effective for pain relief, improvement in mood, and coping skills.

12.7.2
Patient Education and Back Schools

A meta-analysis [23] was conducted of controlled studies dealing with back schools (a short term educational and exercise program focusing on optimal back care) published from 1977 to 1993, and the control groups were placebo, waiting lists, or minimal treatment. Within six months of treatment, the back school had the strongest effect on improving back posture and movement and on acquiring content information, and medium effects on spinal mobility, recurrent back pain and health care utilization. For pain intensity, functional status and analgesic intake, positive effect sizes were not identified. Evidence was not found for efficacy beyond six months.

12.7.3
Multidisciplinary Treatment

Cutler *et al.* [24] reviewed 22 studies published from 1977 to 1991. Inclusion criteria were identification of patient work status before intervention and on follow-up, and the proportions of patients employed at follow-up. Outcome measured within groups indicated a 32 to 41% increase in employment among those not working initially, and the proportions of those returning to work were substantially higher compared to outcomes in the control groups who were either rejected for treatment due to lack of insurance or who dropped out of treatment. (However, these control groups are likely to be biased toward individuals with risk factors for poor prognosis.)

Another meta-analysis [25] found that magnitude of benefit was greater for comprehensive rehabilitation and back school compared to back school alone, but there was insufficient evidence to conclude that back schools yielded greater benefit than other control conditions.

A study [26] comparing the outcomes of randomized and nonrandomized patients in an eight week outpatient multimodal program, or a four-week intensive inpatient program, or waiting list control, found that at one year the intensive inpatient program resulted in greater likelihood of patients maintaining their treatment gains compared to once weekly multimodal outpatient treatment.

The overall message from these older and lower quality studies is that comprehensive treatment is more likely to be efficacious than single modalities, and more intensive daily programs are more likely to be beneficial than are programs conducted once weekly.

12.8
Cost-Effectiveness of Multidisciplinary Pain Management Programs

There is now consistent agreement from several different sources that multidisciplinary programs are cost effective. Jensen *et al.* [9] reported that in their program there were statistically significant cost reductions in terms of reduced sick leave and some savings in disability pension for the female group receiving comprehensive treatment, compared to control conditions (cognitive behavioral therapy alone, behaviorally oriented physiotherapy alone, and a "treatment as usual" control group).

Gatchel and Okifuji [27] conducted a narrative review of studies reporting treatment outcomes using comprehensive programs for chronic pain. They estimated work return rates for comprehensive programs compared to outcomes of control conditions, noting that, on average, 66% of patients in the comprehensive pain programs versus 27% in the control programs returned to work. They estimated significant saving on disability and health care costs as a result of these comprehensive programs.

Weir *et al.* [28] identified an historical cohort and conducted a telephone interview-based survey and mailed questionnaire, with economic analysis of 571 patients referred to a specialty pain clinic. The measures included demographic, economic and health care utilization variables, and validated psychosocial adjustment and coping measures. Using these measures, 81% were "fairly adjusted or poorly adjusted psychosocially" before treatment. Poor adjustment was correlated with lack of intimate caring social relationships, pain having high impact on their lives, low perceived resilience, unemployment, constant pain, and perceiving pain as affecting all aspects of their lives. There was a significant economic benefit for patients attending the multidisciplinary pain service, compared with those who had not attended, with substantial improvement in cost of health care utilization. Even one consultation in the multidisciplinary pain clinic reduced the subsequent costs associated with the chronic pain. It is notable that in patients with chronic pain, functional impairment, and depression represent a high cost for the healthcare system, and that such patients can benefit from multidisciplinary specialty pain management intervention, and that this multidisciplinary treatment is shown to be cost-effective.

A comparative study [29] was conducted with preassessment of patients into prognosis (good, medium, and poor) for return to work based on psychological and physiotherapy findings. Patients were then randomly assigned to ordinary general practitioner (GP) treatment, light behaviorally oriented physiotherapy treatment with follow-up, and extensive multidisciplinary treatment. After 14 months of further follow-up, rates of work return were determined. The cost of lost productivity was used as a measure of cost-benefit. Good and medium prognosis patients benefited from all types of treatment, whereas poor prognosis patients responded best to intensive multidisciplinary treatment. A calculation was made based on the estimate of cost-benefits if good prognosis patients were to be managed by GPs, medium prognosis patients by behaviorally oriented physiotherapy, and poor prognosis patients by interdisciplinary treatment. The result of the estimate was that the overall cost saving per patient would be $800 US if treatment were to be stratified in this fashion.

It is notable that all of these studies were conducted with different methodologies and in different clinics, and that all have shown cost-effectiveness of multidisciplinary programs for chronic pain patients, who are usually very costly for the healthcare system.

12.9
Recommendations

Considering particularly the evidence of high efficacy from the Sherbrooke group that combined evidence-based elements in rehabilitation [30–32], involved stakeholders, and involved injured workers in the rehabilitation process in a timely fashion (see Chapter 15), and the evidence on cost-benefit for multidisciplinary programs [9, 27–29], the recommendation would be to develop similar programs to deliver timely active rehabilitation, especially for injured workers, in collaboration with stakeholders, with associated clinical trials to document efficacy and identify areas for further improvement. This would require collaboration between academic centers, work sites, workers compensation services, primary care providers, and groups representing workers.

This does not address all chronic pain sufferers. An analogous process should be considered for those who are eligible for accident benefits under no-fault insurance. The challenge would be bringing together the insurance, primary healthcare, rehabilitation, and attorney stakeholders, with common agreement on a common model. The present accident-benefits model is still inclined toward the tort system which undermines the comprehensiveness and timeliness demonstrated in the Sherbrooke model.

Chronic pain patients who are not employed and/or who are disabled at the time of first attendance at a specialty clinic represent another challenge. These may yet benefit from clinic-based tertiary care multidisciplinary chronic pain services. Both for community-based clinics and for university-based clinics that identify themselves as offering pain management, there is a lack of coherence in terms of the treatment model, comprehensiveness, pain interventionist versus rehabilitation focus, waiting list, and intake procedures. It is universally the experience that all pain clinics are severely oversubscribed, with waiting lists of months to years, because of the scale of demand and under-funding problems in Canada's health-care. There are also discrepancies between funding through accident benefits and Workers' Compensation versus much less availability of such services through public funding. These are challenges that must be addressed.

With regard to the ideal philosophy of a multidisciplinary program, the focus should be early restoration of function as well as relief of distress [see Figure 12.1]. The three pivotal questions to be addressed for each patient are

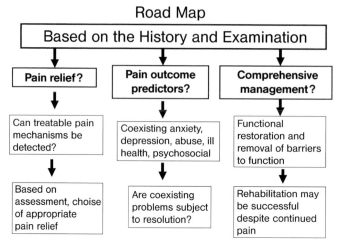

Figure 12.1 Principles of multidisciplinary chronic pain rehabilitation.

1. What is the mechanism of the pain and the most appropriate mechanism(s) for *pain relief*?

2. What is the comorbidity (psychological or physical or psychosocial factors that affect health outcomes) that must be addressed to *remove barriers for change*?

3. What learning and changes and adaptations are necessary on the part of the patient and family and workplace in order to restore adaptive functioning? (i.e. rehabilitation)

Rehabilitation must be based on the principles of support, learning better ways to cope, promoting function, sustaining the support in the transition to the community, and reintegration into normal life. Most multidisciplinary programs lack the resources for that scope of aftercare and outreach.

References

1 Fordyce, W., Fowler, R., Lehmann, J. and de Lateur, B. (1968) Some implications of learning in problems of chronic pain. *Journal of Chronic Diseases*, **21**, 179–90.

2 Tulder, M.W., Ostelo, R.W.J.G., Vlaeyen, J.W.S., Linton, S.J., Morley, S.J. and Assendelft, W.J.J. (2000) Behavioral treatment for chronic low back pain. *The Cochrane Database of Systematic Reviews*, 2000, Issue 2, Art. No.: CD00214, accessed June 30, 2008 (http://www.cochranelibary.com).

3 Ostelo, R.W.J.G., Tulder, M.W., Vlaeyen, J.W.S., Linton, S.J., Morley, S.J. and Assendelft, W.W.J. (2007) Behavioral treatment for chronic low back pain. *The Cochrane Database of Systematic Reviews*, **2**, 2008, accession number 0075320-100000000-D2103, accessed June 30, 2008 (http://www.cochranelibary.com).

4 Tulder, M.W., Koes, B.W. and Bouter, L.M. (1997) Conservative treatment of acute and chronic nonspecific low back pain: a systematic review of randomized

controlled trials of the most common interventions. *Spine*, **22**, 2128–56.

5 Panel, P. (2001) Philadelphia panel evidence-based clinical practice guidelines on selected rehabilitation interventions for low back pain. *Physical Therapy*, **81**, 1641–74.

6 Cohen, J.E., Goel, V., Frank, J.W., Bombardier, C., Peloso, P. and Guillemin, F. (1994) Group education interventions for people with low back pain. An overview of the literature. *Spine*, **19**, 1214–22.

7 Flor, H., Fydrich, T. and Turk, D.C. (1992) Efficacy of multidisciplinary pain treatment centers: a meta-analytic review. *Pain*, **49**, 221–30.

8 Guzman, J., Esmail, R., Karjaainen, K., Malmivaara, A., Irvin, E. and Bombardier, C. (2001) Multidisciplinary rehabilitation for chronic low back pain: systematic review. *BMJ*, **322**, 1511–16.

9 Jensen, I.B., Bergstrom, G., Ljungquist, T. and Bodin, L. (2005) A 3-year follow-up of a multidisciplinary rehabilitation program for back and neck pain. *Pain*, **115**, 273–83.

10 Gross, A.R., Aker, P.D., Goldsmith, C.H. and Peloso, P. (1998) Patient education for mechanical neck disorders. *The Cochrane Database of Systematic Reviews*, **2**, 2008, accession number 0075320-100000000-00330, accessed June 30, 2008 (http://www.cochranelibary.com).

11 Malmivaara, A., Hakkinen, U., Timo, A., Heinrichs, M-L, Koskenniemi, L., Kuosma, E., Lappi, S., Paloheimo, R. *et al.* (1995) The treatment of acute low back pain – bed rest, exercises, or ordinary activity? *The New England Journal of Medicine*, **332**, 351–5.

12 Verhagen, A.P., Scholten-Peeters, G.G.G.M., van Wijngaarden, S., de Bie, R.A., Bierma-Zeinstra, S.M.A. (2008) Conservative treatments for whiplash. *The Cochrane Database of Systematic Reviews*, **2**, 2008, accession number 00075320-100000000-02365, accessed July 9, 2008 (http://www.cochranelibary.com).

13 Holroyd, K.A. and Penzien, D.B. (1990) Pharmacological versus non-pharmacological prophylaxis of recurrent migraine headache: a meta-analytic review of clinical trials. *Pain*, **42**, 1–13.

14 Holroyd, K.A. and Penzien, D.B. (1986) Client variables and the behavioral treatment of recurrent tension headache: a meta-analytic review. *Journal of Behavioral Medicine*, **9**, 515–36.

15 Campbell, J.K., Penzien, D.B. and Wall, E.M. (2002) Evidence-based guidelines for migraine headache: behavioral and physical treatments. *The US Headache Consortium*, 2002, accessed June 30, 2008 (http://www.aan.com/professionals/practice/pdfs/g10089.pdf).

16 Nestoriuc, Y. and Martin, A. (2007) Efficacy of biofeedback for migraine: a meta-analysis. *Pain*, **128**, 111–27.

17 Eccleston, C., Morley, S., Williams, A., Yorke, L. and Mastroyannopoulou, K. (2002) Systematic review of randomized controlled trials of psychological therapy for chronic pain in children and adolescents, with a subset meta-analysis of pain relief. *Pain*, **99**, 157–65.

18 Turner, J., Holtzman, S. and Mancl, L. (2007) Mediators, moderators, and predictors of therapeutic change in cognitive-behavioral therapy for chronic pain. *Pain*, **127**, 276–86.

19 Karjalainen, K., Malmivaara, A., van Tulder, M., Roine, R., Jauhiainen, M., Hurri, H. and Koes, B. (1999) Multidisciplinary rehabilitation for fibromyalgia and musculoskeletal pain in working age adults. *The Cochrane Database of Systematic Reviews*, **2**, 2008, accession number 00075320-100000000-01349, accessed June 30, 2008 (http://www.cochranelibary.com).

20 Malone, M.D. and Strube, M.J. (1988) Meta-analysis of non-medical treatments for chronic pain. *Pain*, **34**, 231–44.

21 Turner, J.A. (1996) Educational and behavioral interventions for back pain in primary care. *Spine*, **21**, 2851–9.

22 Morley, S., Eccleston, C. and Williams, A. (1999) Systematic review and meta-analysis of randomized controlled trials of cognitive behavior and behavior therapy for chronic pain in adults, excluding headache. *Pain*, **80**, 1–13.

23 Maier-Riehle, B. and Harter, M. (2001) The effects of back schools – a meta-analysis. *International Journal of Rehabilitation Research*, **24**, 199–206.

24 Cutler, R.B., Fishbain, D.A., Rosomoff, H.L., Abdel-Moty, E., Khalil, T.M. and Rosomoff, R.S. (2004) Does non-surgical pain center treatment of chronic pain return patients to work? A review and meta-analysis of the literature. *Spine*, **19**, 643–52.

25 di Fabio, R.P. (1995) Efficacy of comprehensive rehabilitation programs and back school for patients with low back pain: a meta-analysis. *Physical Therapy*, **75**, 865–78.

26 Williams, A.C., Nicholas, M.K., Richardson, P.H., Pither, C.E. and Fernandes, J. (1999) Generalizing from a controlled trial: the effects of patient preference versus randomization on the outcome of inpatient versus outpatient chronic pain management. *Pain*, **83**, 57–65.

27 Gatchel, R.J., Noe, C.E., Pulliam, C., Robbins, H., Deschner, M., Gajraj, N.M. and Vakharia, A.S. (2002) A preliminary study of multidimensional pain inventory profile differences in predicting treatment outcome in a heterogeneous cohort of patients with chronic pain. *The Clinical Journal of Pain*, **18**, 139–43.

28 Weir, R., Browne, G.B., Tunks, E., Gafni, A. and Roberts, J. (1992) A profile of users of specialty pain clinic services: predictors of use and cost estimates.

Journal of Clinical Epidemiology, **45**, 1399–415.

29 Haldorsen, E.M.H., Grasdal, A.S., Skouen, J.S., Risa, A.E., Kronholm, K. and Ursin, H. (2002) Is there a right treatment for a particular patient group? Comparison of ordinary treatment, light multidisciplinary treatment, and extensive multidisciplinary treatment for long-term sick-listed employees with musculoskeletal pain. *Pain*, **95**, 49–63.

30 Loisel, P., Durand, P., Abenhaim, L., Gosselin, L., Simard, R., Turcotte, J. and Esdaile, J.M. (1994) Management of occupational back pain: the Sherbrooke model. Results of a pilot and feasibility study. *Occupational and Environmental Medicine*, **51**, 597–602.

31 Loisel, P., Abenhaim, L., Durand, P., Esdaile, J.M., Suissa, S., Gosselin, L., Simard, R., Turcotte, J. *et al.* (1997) Lemaire J A population-based, randomized clinical trial on back pain management. *Spine*, **22**, 2911–8.

32 Loisel, P.J., Poitras, S., Durand, M.J., Champagne, F., Stock, S., Diallo, B. and Tremblay, C. (2002) Cost-benefit and cost-effectiveness analysis of a disability prevention model for back pain management: a six year follow up study. *Occupational and Environmental Medicine*, **59**, 807–15.

13
Complementary and Alternative Medicine Approaches to Chronic Pain

Mark A. Ware

- Pain is a common reason for using complementary and alternative medicine (CAM).

- Patients who use CAM may have different beliefs and expectations of conventional medicine than those who do not.

- Clinicians should understand CAM because it has implications for the effectiveness and safety of conventional care.

- While several good quality trials and reviews have been conducted, many CAM therapies have not been rigorously evaluated.

- Existing CAM policy and regulatory frameworks in Canada could facilitate the evaluation of CAM utilization patterns and health care costs, and pilot projects of integrative pain programs in academic centers could provide valuable feasibility and effectiveness assessments.

13.1
Introduction

Complementary and alternative medicine (CAM) is a term used to describe a group of diverse medical and healthcare systems, practices, and products that are not presently considered to be part of conventional medicine [1]. The annual prevalence of use of CAM in industrialized countries ranges from 20% to 60% [2] and appears to have stabilized in recent years [3]. Data for Canada is emerging; in 2003 an estimated 20% of Canadians aged 12 or older visited alternative healthcare providers [4]. In 2005, 71% of Canadians had tried a natural health product at one time or another, with 38% using such products on a daily basis, 11% weekly, and 37% only during certain seasons [5]. Since these products and practices are invariably paid for out of patients' pockets, the extent of CAM use reflects not only a perceived need for more healthcare options, but also reflects an economic cost that is not represented in provincial healthcare budgets.

Chronic Pain: A Health Policy Perspective
Edited by S. Rashiq, D. Schopflocher, P. Taenzer, and E. Jonsson
Copyright © 2008 WILEY-VCH Verlag GmbH & Co. KGaA, Weinheim
ISBN: 978-3-527-32382-1

The reasons why patients use CAM may include *positive* factors such as perceived safety and effectiveness, holistic approaches to health, self-control, more time and physical contact with therapist in a relaxing environment, and accessibility; and *negative* factors such as perceived failings of conventional medicine (adverse events, long waiting lists, focus on interventional approaches, lack of efficacy) and mistrust of the medical establishment and Western science [6].

Chronic pain is one of the most common clinical syndromes for which patients seek out and use CAM [7]. Prior to the mid-nineteenth century, pain was treated with herbal remedies as crude extracts or dry herbs. Several modern pharmacological approaches to pain have their origins in herbal remedies including morphine from *Papaverum somniferum* (poppy), aspirin from *Silax alba* (white willow), and more recently cannabinoids from *Cannabis sativa* (marijuana). Surveys of patients with chronic pain syndromes suggest that CAM use is prevalent with estimates of 30–50% (see below). This chapter will review the reasons why physicians treating patients with chronic pain need to know about CAM, explore the current evidence behind selected CAM treatments, and consider the economic and policy implications of integrating CAM into modern chronic pain management.

13.2
Why Do Pain Physicians Need to Know About CAM?

Physicians managing patients with chronic pain should be aware of CAM use by their patients for several reasons. These include the possibility that patients may:

- be spending resources and time on CAM practices;
- be at risk of drug–herb interactions;
- be unaware of safety aspects of the CAM therapies being used;
- be seeking treatment of symptoms which should be known to their treating physician;
- have attitudes towards conventional treatments which may influence compliance.

Awareness of a patient's CAM use and the underlying motivations for such use will inform the physician's understanding of the patient behind the pain. A reasoned and balanced approach to discussing CAM use may help improve the therapeutic relationship between doctor and patient. Physicians may even face malpractice suits if they fail to identify safety issues concerning CAM use or if they recommend treatments which are unsafe [8].

13.3
Epidemiology of CAM in Chronic Pain

13.3.1
Classification of CAM

The US-based National Center for Complementary and Alternative Medicine has classified CAM into five major categories (see Table 13.1) [1]. This classification of CAM into whole systems and four domains is a convenient approach to what can be a bewildering array of treatment approaches. However by breaking down CAM into biological, body-based, energy-based and mind–body approaches, such a schema obscures the fact that many of these therapies have at their core a holistic approach to wellness which makes many CAM therapies more similar than discrete. Defining and classifying CAM is also transitory; several therapeutic approaches that were once considered CAM are now considered part of mainstream pain medicine; these approaches include biofeedback, massage therapy and even, to some extent, needling techniques.

Chronic pain is a common reason for the use of CAM. Table 13.2 shows a summary of the prevalence of, and reasons for, CAM use, as well as the main modalities used, in several pain populations. Several broad trends emerge from these studies. CAM use is positively associated with younger age, being married, a higher level of education, and a higher household income. In addition, CAM use

Table 13.1 NCCAM classification of CAM.

Domain	Description	Examples
Whole medical systems[a]	Built upon complete system of theory and practice	Traditional Chinese medicine, Ayurvedic medicine, homoeopathy, naturopathic medicine
Mind–body medicine	Techniques designed to enhance the mind's capacity to affect bodily function and symptoms	Meditation, prayer, music therapy (previously cognitive behavior, biofeedback)
Biologically-based practices	Use substances found in nature	Herbs, foods, vitamins, essential oils
Manipulative and body-based practices	Based on manipulation and/or movement of one or more parts of the body	Chiropractic, osteopathy, massage therapy
Energy medicine	Involves the use of energy fields	Qi Gong, Reiki, therapeutic touch, magnetic fields

a Note this group of systems may include therapeutic approaches from other domains.

Table 13.2 Summary of selected CAM use surveys among patients with chronic pain.

Disease	Prevalence of CAM use	Examples of common types of CAM used	Reference
TMJ pain	22%	Relaxation, chiropractic	[9]
Chronic back pain	39%	Chiropractic, massage, acupuncture	[10]
Osteoarthritis	40%	Vitamins, celery, fish oil	[11]
Spinal cord injury	40%	Acupuncture, massage, chiropractic	[12]
Fibromyalgia	56%		[13]
Peripheral neuropathy	43%	Megavitamins, magnets, acupuncture	[14]

is negatively associated with pain intensity and positively associated with greater analgesic use, depression, and co-morbidity. Most patients appear to use conventional medical services in addition to CAM [10].

Among patients with peripheral neuropathy, CAM users report using magnets and acupuncture (30% each), herbal remedies (22%) and chiropractic (22%) [14]. In this study, CAM use was associated with the presence of diabetic neuropathy and burning pain. Chiropractic and massage are frequently used by patients with chronic low back pain [15]. In a questionnaire study of 170 chronic pain patients attending a pain clinic in Montreal, 96 (56%) had used a CAM modality in the past month, of which the most common were nutritional supplements (34%), massage (11%) and meditation (7%) (Figure 13.1) [16].

13.3.2
First Nations Healing Traditions and Pain

Little is known of the prevalence of the use of herbal medicine for pain among Canadian First Nations peoples. The compounding of herbal preparations is not regulated under the Natural Health Products regulations. The use of healing rituals such as sweat lodges, smudging, pipe smoking, the use of song, sun dances and prayer, the concept of the "medicine wheel", and herbs such as sage, sweet-grass and cedar have all been described. It is interesting to note that the use of herbs such as echinacea, goldenseal, St John's Wort, and evening primrose by First Nations healers is also increasingly receiving interest from a wider population. It is clear that within First Nations cultures and, increasingly, within Western society, indigenous healers are given a credibility that goes beyond our current

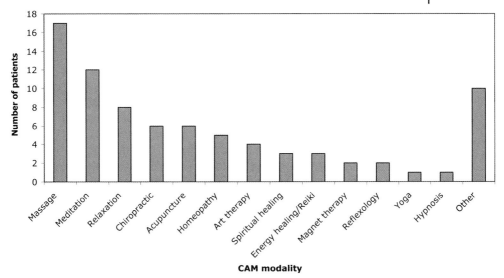

Figure 13.1 CAM modalities used among 170 chronic noncancer pain patients. (Ware *et al.*, submitted).

paradigm of evidence-based medicine and points to a fundamental role of the healer in society. This role is crucial in addressing the suffering that is associated with chronic pain.

13.4
Clinical Evidence of Safety and Efficacy

Systematic reviews carried out under the Cochrane Collaboration are widely regarded as authoritative and highly respected evidence for treatment effects. Reviews have recently been conducted looking at several CAM modalities. A search of the Cochrane Library group for reviews on ("complementary medicine" and "chronic pain" (search date 3 Dec 2007) found 78 published reviews. Reviews for conditions such as irritable bowel syndrome, Crohn's disease, venous or arterial insufficiency and complications, dysmenorrhea and for therapies described as "behavioral interventions" were excluded. Eleven reviews were selected, including three on low back pain, two on osteoarthritis, and two on headache (including migraine); herbal remedies were the focus of four reviews, two were on acupuncture and two were on massage. The results are summarized in Table 13.3.

A recent review of omega-3 fatty acid supplementation for inflammatory joint pain found these supplements to be "attractive adjunctive treatment for joint pain associated with rheumatoid arthritis, inflammatory bowel disease and dysmennorhea" [28].

Table 13.3 Cochrane Library systematic reviews of CAM treatments
for chronic pain conditions, by year of publication.

Disease	Intervention	Year	Number of trials	Results	Reference
Low back pain	Herbal medicine	2006	10	Harpagophytum procumbens, Salix alba and Capsicum frutescens seem to reduce pain more than placebo	[17]
Mechanical neck disorders	Massage	2006	19	No practice recommendations made	[18]
Osteoarthritis	Glucosamine	2005	20	Improvement in pain was noted for one (Rotta) preparation but not for others; no effect on WOMAC pain and functional outcomes	[19]
Low back pain	Acupuncture and dry-needling	2005	35	For chronic low-back pain, acupuncture more effective for pain relief and functional improvement than no treatment or sham treatment immediately after treatment and in the short-term only	[20]
Chronic/ recurrent headache	Noninvasive physical treatments	2004	22	Spinal manipulation may be effective for migraine and chronic tension-type headache. Both spinal manipulation and neck exercises may be effective for cervicogenic headache	[21]
Migraine prevention	Feverfew	2004	5	No convincing evidence of efficacy	[22]
Low back pain	Massage	2002	8	Massage may be beneficial for subacute and chronic low back pain	[23]
Idiopathic headache	Acupuncture	2001	26	Evidence supportive of acupuncture	[24]
Chronic pelvic pain	All interventions	2000	14	Writing therapy and static magnetic field therapy show some evidence of short-term efficacy	[25]
Osteoarthritis	Herbal medicine	2000	5	Convincing evidence for avocado-soybean unsaponifiables	[26]
Rheumatoid arthritis	Herbal medicine	2000	11	Some potential benefit for the use of GLA	[27]

Music therapy has been evaluated for pain relief in 51 studies involving a variety of pain conditions including procedural pain (28 studies), post-operative pain (14 studies), chronic noncancer pain (3 studies), cancer (2 studies), experimental pain (2 studies), and labor pain (2 studies) [29]. Listening to music reduces pain by an average of 0.5 on a 10 point scale (95% CI 0.2–0.9), and increases likelihood of 50% pain relief by 70% (NNT = 5). Post-operative opioid requirements were also reduced.

13.4.1
Other Trials of CAM for Chronic Pain

The following is a list of studies in which CAM therapies have been studied for pain disorders organized (alphabetically) by CAM modality. This list is not meant to be exhaustive but is intended to show how extensive efforts have been made to study CAM for chronic pain. The evidence base is gradually accumulating.

13.4.1.1 Acupuncture
Acupuncture has been systematically reviewed for chronic neck pain with moderate evidence of effectiveness [30]. Evidence for effectiveness in low back pain has not been convincing [31]. For shoulder pain and lateral elbow pain data were also insufficient to draw conclusions [32, 33]. Positive trials have also been reported for chronic tension-type headache [34, 35], osteoarthritis of the knee [36], chronic neck and shoulder pain [37, 38], and low back pain [39, 40]. Acupuncture has not been found to be useful for fibromyalgia [41]. Numerous systematic reviews have been conducted but methodological issues are important considerations [42].

13.4.1.2 Chiropractic
Evidence of clinical benefit from chiropractic manipulation in low back pain is modest but patient satisfaction is high [43–45].

13.4.1.3 Energy Therapy
Therapeutic touch was found to be inferior to progressive muscle relaxation for pain and distress in adults with degenerative arthritis [46] but was associated with improved self-efficacy compared to cognitive behavior therapy in adults with chronic pain [47].

13.4.1.4 Meditation/Relaxation
Mindfulness meditation has not been shown to be better than standard therapy for fibromyalgia [48]. Breath therapy similarly was not better than physical therapy for chronic low back pain [49].

13.4.1.5 Movement Therapies
Iyengar yoga has been shown to have positive effects in chronic low back pain [50]. T'ai Chi has been shown to improve range of motion in rheumatoid arthritis [51]. Qigong has been shown to be effective in reducing pain and distress in fibromyalgia [52].

13.4.1.6 Osteopathy

Osteopathic treatment of low back pain was compared to standard care and found to have similar effects on pain and function but the osteopathic group used less medication and physiotherapy [53].

13.5
Safety Concerns

The Cochrane reviews cited above focus on evidence of efficacy; the reporting of adverse events is not included in review summaries and does not form part of standard clinical trial quality scores. For many physicians and for patients, the possible harms that may arise from using CAM treatments are a major concern, and so the standard evidence base is sorely lacking in this information. As discussed above, harms from CAM may be direct (a result of the treatment itself) or indirect (a result of patients not using a possibly effective conventional treatment). Indeed, one approach towards consideration of CAM may be driven by safety first: if the direct risk of harm is low or negligible (e.g. homoeopathy), then focus shifts to indirect harms, and if these are acceptable, then the issue of efficacy becomes a balance between specific and non-specific (e.g. placebo) effects. It is up to the clinician to determine the risk–benefit ratio in any given circumstance.

Specific safety issues arise with particular therapies and these are discussed below.

13.5.1
Natural Health Products

The active components of many natural health products (NHPs) are metabolized by similar pathways to modern medicines. It is therefore important to recognize whether NHPs are being taken, and specifically which ones, and to be aware of potential interactions. While many people appear to take NHPs safely, potential interactions have been identified [54]. For example, non-steroidal anti-inflammatory drugs (NSAIDs) have the potential to interact with herbs that have antiplatelet activity (e.g. gingko biloba, garlic, ginger, bilberry, dong quai, feverfew, meadowsweet, turmeric and willow) and with those containing coumadin (chamomile, motherworth, horse chestnut, fenugreek and red clover). The risk of hepatotoxicity may be increased when acetaminophen is taken in conjunction with herbs such as kava. The use of sedative herbs such as chamomile, valerian, and kava may increase CNS depression due to opioid medications, and opioid effectiveness may be reduced by ginseng.

13.5.2
Manipulation Therapies

Chiropractic has been shown to have a very small risk of stroke due to vertebral artery dissection (cervical manipulation; relative risk 1:1 million) and disc herniation (1:3 million) [55].

13.5.3
Acupuncture

Acupuncture risks are minimized by using disposable needles only and by ensuring that treatment is only undertaken by licensed acupuncturists. Local risks include pain, tiredness and bleeding; faintness and pneumothorax are rare [56].

13.6
Research on CAM: A Matter of Control?

It is clear from most systematic reviews that the methodological weaknesses of CAM trials hamper the ability to draw solid conclusions from such work. The gold standard RCT relies on the existence of a credible control group (active or inert/placebo) and the fact that the subject, and ideally the investigator, are blinded to the allocation. Such an approach is unquestionably difficult with many CAM treatments, and while methodological issues must not be used as an excuse not to do CAM research, interpreting the results requires a level of acceptance that the perfect study is not likely to be possible. However, if good quality observational studies and pragmatic trials are carried out, the data may still be useful in determining overall evidence of effectiveness. Certainly, safety and health economic data can be collected systematically and these factors may well be the major issues in policy regarding CAM in the future.

13.7
Policy Implications

Since most CAM treatments are paid for out of the patients' pocket, the costs of such practices are not included in the costs of socialized medicine. Private insurers cover some CAM practices to a limited extent; for example, some policies may allow a limited number of massage therapy sessions per year if prescribed by a physician and performed by a registered massage therapist. Similar approaches may be taken for acupuncture and chiropractic care.

13.7.1
Cost Effectiveness

Few cost-effectiveness studies have been done in CAM practices. One European study of spa treatment for fibromyalgia (FM) found improved quality of life at an incremental cost of 885 Euro per patient per year [57]. In a study of CAM use among FM patients with insurance, annual health care costs for CAM users were the same as those who did not use CAM, even though the disease burden was greater for CAM users [13].

13.7.2
Regulating CAM in Canada

Healthcare *practice* falls under provincial jurisdiction in Canada, which for CAM includes licensing of practitioners and reimbursement policies under provincial insurance plans. Therefore the offering of CAM services, either by medical practitioners or by other practitioners, is regulated differently in each province. Healthcare *products* (e.g. natural health products such as vitamins, herbal supplements) are regulated federally by the Natural Health Products Directorate [58] in a manner similar to the regulation of drugs by the Therapeutic Products Directorate. Patients and practitioners should be aware of how such products are licensed and the conditions around which they are marketed and sold.

13.7.3
Teaching CAM in Medical Education

Little is known of the teaching of CAM in pain education. The IASP curriculum for pain education includes a section on CAM [59] but the extent to which this material is included in most standardized curricula is not known. A recent project based on the IASP curriculum did not contain CAM education [60], despite the fact that chronic pain is one of the most common reasons for CAM use [7].

13.7.4
Towards an Integrative Pain Medicine?

Should safe and effective CAM therapies be integrated into pain clinic services? Are they cost effective? A 2005 survey of 39 US academic health centers found that while 23 offered CAM services (particularly acupuncture, massage, dietary supplements, mind-body therapies, and music therapy), none had written policies concerning malpractice liability or credentialing practices [61]. Nothing is currently known of the policy issues regarding integration of such practices in chronic pain centers in Canada. The mechanics of providing integrated (CAM plus conventional) care in Canada deserve to be further explored.

13.8
Conclusions

This chapter has addressed some of the main issues involved in the consideration of CAM practices and pain management. There is a clear discrepancy between the public perception of the role of CAM in pain self-care and the availability of such care in multidisciplinary healthcare programs (with respect to both access and cost). Too little is known about the efficacy of many CAM modalities to justify major initiatives in widespread and publicly funded CAM integration in pain management. However, existing CAM policy and regulatory frameworks in Canada could facilitate the evaluation of CAM utilization patterns and healthcare costs, and pilot projects of integrative pain programs in academic centers could provide valuable feasibility and effectiveness assessments. Such data would enlighten the debate surrounding CAM and chronic pain and would allow health policy decisions to be based on hard economic facts.

References

1 NCCAM (2007) What is CAM? http://nccam.nih.gov/health/whatiscam/ (17 June, 2008).

2 Ernst, E. (2000) The role of complementary and alternative medicine. *British Medical Journal*, 321, 1133–5.

3 Tindle, H.A., Davis, R.B., Phillips, R.S. and Eisenberg, D.M. (2005) Trends in use of complementary and alternative medicine by US adults: 1997–2002. *Alternative Therapies in Health and Medicine*, 11, 42–9.

4 Baseline Natural Health Products Survey among Consumers, March 2005. http://www.hc-sc.gc.ca/dhp-mps/pubs/natur/eng_cons_survey_eng.php (17 June, 2007).

5 Park, J. (2005) Use of alternative health care. *Health Reports*, 16, 39–41.

6 Ernst, E. Complementary medicine for pain. http://www.wellcome.ac.uk/en/pain/microsite/medicine1.html (23 August, 2007).

7 Barnes, P., Powell-Griner, E., McFann, K. and Nahin, R. (2004) Complementary and Alternative Medicine Use among Adults: United States, 2002.

8 Cohen, M.H. (2007) Malpractice and CAM (Complementary and Alternative Medicine) Update. http://www.camlawblog.com/malpractice-and-risk-management-264-malpractice-and-cam-complementary-and-alternative-medicine-update.html (24 August, 2007).

9 Raphael, K.G., Klausner, J.J., Nayak, S. and Marbach, J.J. (2003) Complementary and alternative therapy use by patients with myofascial temporomandibular disorders. *Journal of Orofacial Pain*, 17, 36–41.

10 Foltz, V., St Pierre, Y., Rozenberg, S. *et al.* (2005) Use of complementary and alternative therapies by patients with self-reported chronic back pain: a nationwide survey in Canada. *Joint Bone Spine*, 72, 571–7.

11 Zochling, J., March, L., Lapsley, H., Cross, M., Tribe, K. and Brooks, P. (2004) Use of complementary medicines for osteoarthritis – a prospective study. *Annals of the Rheumatic Diseases*, 63, 549–54.

12 Nayak, S., Matheis, R.J., Agostinelli, S. and Shifleft, S.C. (2001) The use of complementary and alternative therapies for chronic pain following spinal cord injury: a pilot survey. *The Journal of Spinal Cord Medicine*, 24, 54–62.

13 Lind, B.K., Lafferty, W.E., Tyree, P.T., Diehr, P.K. and Grembowski, D.E. (2007) Use of complementary and alternative medicine providers by fibromyalgia patients under insurance coverage. *Arthritis and Rheumatism*, 57, 71–6.

14 Brunelli, B. and Gorson, K.C. (2004) The use of complementary and alternative medicines by patients with peripheral neuropathy. *Journal of the Neurological Sciences*, 218, 59–66.

15 Sherman, K.J., Cherkin, D.C., Connelly, M.T. *et al.* (2004) Complementary and alternative medical therapies for chronic low back pain: What treatments are patients willing to try? *BMC Complementary and Alternative Medicine*, 4, 9.

16 Yang, J., Lau, N.M., Ware, M.A., Shir, Y. and Collet, J.P. (2004) Patterns and prevalence of complementary and alternative medicine use among patients with chronic non-cancer pain. *Journal of Complementary and Integrative Medicine*, 2, 57.

17 Gagnier, J.J., van Tulder, M., Berman, B. and Bombardier, C. (2006) Herbal medicine for low back pain. *Cochrane Database of Systematic Reviews*, CD004504.

18 Haraldsson, B.G., Gross, A.R., Myers, C.D. *et al.* (2006) Massage for mechanical neck

disorders. *Cochrane Database of Systematic Reviews*, **3**, CD004871.

19 Towheed, T.E., Maxwell, L., Anastassiades, T.P. *et al.* (2005) Glucosamine therapy for treating osteo-arthritis. *Cochrane Database of Systematic Reviews*, CD002946.

20 Furlan, A.D., van Tulder, M.W., Cherkin, D.C. *et al.* (2005) Acupuncture and dry-needling for low back pain. *Cochrane Database of Systematic Reviews*, CD001351.

21 Bronfort, G., Nilsson, N., Haas, M. *et al.* (2004) Non-invasive physical treatments for chronic/recurrent headache. *Cochrane Database of Systematic Reviews*, CD001878.

22 Pittler, M.H. and Ernst, E. (2004) Feverfew for preventing migraine. *Cochrane Database of Systematic Reviews*, CD002286.

23 Furlan, A.D., Brosseau, L., Imamura, M. and Irvin, E. (2002) Massage for low back pain. *Cochrane Database of Systematic Reviews*, CD001929.

24 Melchart, D., Linde, K., Fischer, P. *et al.* (2001) Acupuncture for idiopathic headache. *Cochrane Database of Systematic Reviews*, CD001218.

25 Stones, R.W. and Mountfield, J. (2000) Interventions for treating chronic pelvic pain in women. *Cochrane Database of Systematic Reviews*, CD000387.

26 Little, C.V. and Parsons, T. (2001) Herbal therapy for treating osteoarthritis. *Cochrane Database of Systematic Reviews*, CD002947.

27 Little, C. and Parsons, T. (2001) Herbal therapy for treating rheumatoid arthritis. *Cochrane Database of Systematic Reviews*, CD002948.

28 Goldberg, R.J. and Katz, J. (2007) A meta-analysis of the analgesic effects of omega-3 polyunsaturated fatty acid supplementation for inflammatory joint pain. *Pain*, **129**, 210–23.

29 Cepeda, M.S., Carr, D.B., Lau, J. and Alvarez, H. (2006) Music for pain relief. *Cochrane Database of Systematic Reviews*, CD004843.

30 Trinh, K.V., Graham, N., Gross, A.R. *et al.* (2006) Acupuncture for neck disorders. *Cochrane Database of Systematic Reviews*, **3**, CD004870.

31 Furlan, A.D., van Tulder, M., Cherkin, D. *et al.* (2005) Acupuncture and dry-needling for low back pain: an updated systematic review within the framework of the cochrane collaboration. *Spine*, **30**, 944–63.

32 Green, S., Buchbinder, R., Barnsley, L. *et al.* (2002) Acupuncture for lateral elbow pain. *Cochrane Database of Systematic Reviews*, CD003527.

33 Green, S., Buchbinder, R. and Hetrick, S. (2005) Acupuncture for shoulder pain. *Cochrane Database of Systematic Reviews*, CD005319.

34 Melchart, D., Streng, A., Hoppe, A. *et al.* (2005) The acupuncture randomised trial (ART) for tension-type headache – details of the treatment. *Acupuncture in Medicine*, **23**, 157–65.

35 Ebneshahidi, N.S., Heshmatipour, M., Moghaddami, A. and Eghtesadi-Araghi, P. (2005) The effects of laser acupuncture on chronic tension headache – a randomised controlled trial. *Acupuncture in Medicine*, **23**, 13–18.

36 Witt, C., Brinkhaus, B., Jena, S. *et al.* (2005) Acupuncture in patients with osteoarthritis of the knee: a randomised trial. *Lancet*, **366**, 136–43.

37 He, D., Veiersted, K.B., Hostmark, A.T. and Medbo, J.I. (2004) Effect of acupuncture treatment on chronic neck and shoulder pain in sedentary female workers: a 6-month and 3-year follow-up study. *Pain*, **109**, 299–307.

38 White, P., Lewith, G., Prescott, P. and Conway, J. (2004) Acupuncture versus placebo for the treatment of chronic mechanical neck pain: a randomized, controlled trial. *Annals of Internal Medicine*, **141**, 911–19.

39 Kerr, D.P., Walsh, D.M. and Baxter, D. (2003) Acupuncture in the management of chronic low back pain: a blinded randomized controlled trial. *The Clinical Journal of Pain*, **19**, 364–70.

40 Meng, C.F., Wang, D., Ngeow, J., Lao, L., Peterson, M. and Paget, S. (2003) Acupuncture for chronic low back pain in older patients: a randomized, controlled trial. *Rheumatology (Oxford)*, **42**, 1508–17.

41 Assefi, N.P., Sherman, K.J., Jacobsen, C., Goldberg, J., Smith, W.R. and Buchwald, D. (2005) A randomized clinical trial of

acupuncture compared with sham acupuncture in fibromyalgia. *Annals of Internal Medicine*, **143**, 10–19.

42 Sood, A., Sood, R., Bauer, B.A. and Ebbert, J.O. (2005) Cochrane systematic reviews in acupuncture: methodological diversity in database searching. *Journal of Alternative and Complementary Medicine*, **11**, 719–22.

43 Haas, M., Groupp, E. and Kraemer, D.F. (2004) Dose-response for chiropractic care of chronic low back pain. *The Spine Journal*, **4**, 574–83.

44 Hawk, C., Long, C.R., Rowell, R.M., Gudavalli, M.R. and Jedlicka, J. (2005) A randomized trial investigating a chiropractic manual placebo: a novel design using standardized forces in the delivery of active and control treatments. *Journal of Alternative and Complementary Medicine*, **11**, 109–17.

45 Hoiriis, K.T., Pfleger, B., McDuffie, F.C. *et al.* (2004) A randomized clinical trial comparing chiropractic adjustments to muscle relaxants for subacute low back pain. *Journal of Manipulative and Physiological Therapeutics*, **27**, 388–98.

46 Eckes Peck, S.D. (1997) The effectiveness of therapeutic touch for decreasing pain in elders with degenerative arthritis. *Journal of Holistic Nursing*, **15**, 176–98.

47 Smith, D.W., Arnstein, P., Rosa, K.C. and Wells-Federman, C. (2002) Effects of integrating therapeutic touch into a cognitive behavioral pain treatment program. Report of a pilot clinical trial. *Journal of Holistic Nursing*, **20**, 367–87.

48 Astin, J.A., Berman, B.M., Bausell, B., Lee, W.L., Hochberg, M. and Forys, K.L. (2003) The efficacy of mindfulness meditation plus Qigong movement therapy in the treatment of fibromyalgia: a randomized controlled trial. *The Journal of Rheumatology*, **30**, 2257–62.

49 Mehling, W.E., Hamel, K.A., Acree, M., Byl, N. and Hecht, F.M. (2005) Randomized, controlled trial of breath therapy for patients with chronic low-back pain. *Alternative Therapies in Health and Medicine*, **11**, 44–52.

50 Williams, K.A., Petronis, J., Smith, D. *et al.* (2005) Effect of Iyengar yoga therapy for chronic low back pain. *Pain*, **115**, 107–17.

51 Han, A., Robinson, V., Judd, M., Taixiang, W., Wells, G. and Tugwell, P. (2004) Tai chi for treating rheumatoid arthritis. *Cochrane Database of Systematic Reviews*, CD004849.

52 Haak, T. and Scott, B. (2007) The effect of Qigong on Fibromyalgia (FMS): a controlled randomized study. *Disability and Rehabilitation*, 1–9.

53 Andersson, G.B., Lucente, T., Davis, A.M., Kappler, R.E., Lipton, J.A. and Leurgans, S. (1999) A comparison of osteopathic spinal manipulation with standard care for patients with low back pain. *The New England Journal of Medicine*, **341**, 1426–31.

54 Abebe, W. (2002) Herbal medication: potential for adverse interactions with analgesic drugs. *Journal of Clinical Pharmacy and Therapeutics*, **27**, 391–401.

55 Oliphant, D. (2004) Safety of spinal manipulation in the treatment of lumbar disk herniations: a systematic review and risk assessment. *Journal of Manipulative and Physiological Therapeutics*, **27**, 197–210.

56 Ernst, E. and White, A.R. (2001) Prospective studies of the safety of acupuncture: a systematic review. *The American Journal of Medicine*, **110**, 481–5.

57 Zijlstra, T.R., Braakman-Jansen, L.M., Taal, E. and Rasker, J. (2007) J, and van de Laar, MA. Cost-effectiveness of Spa treatment for fibromyalgia: general health improvement is not for free. *Rheumatology (Oxford)*, **46**, 1454–9.

58 Natural Health Products Directorate, Health Canada (2007) http://www.hc-sc.gc.ca/dhp-mps/prodnatur/index_eng.php (30 August, 2007).

59 Charlton, J.E. (ed.) (2005) Complementary therapies, in *Core Curriculum for Professional Education in Pain*, IASP Press, Seattle, Chap. 24, pp. 1–4.

60 Watt-Watson, J., Hunter, J., Pennefather, P. *et al.* (2004) An integrated undergraduate pain curriculum, based on IASP curricula, for six Health Science Faculties. *Pain*, **110**, 140–48.

61 Cohen, M.H., Sandler, L., Hrbek, A., Davis, R.B. and Eisenberg, D.M. (2005) Policies pertaining to complementary and alternative medical therapies in a random sample of 39 academic health centers. *Alternative Therapies in Health and Medicine*, **11**, 36–40.

14
Chronic Pain Self-Management

Michael Hugh McGillion, Sandra LeFort, and Jennifer Stinson

- Evidence is clear that self-management interventions are highly effective, cost-effective adjuncts to the usual care for reducing chronic pain and related disability.

- Self-management education interventions emphasize the formation of a collaborative partnership between patients and healthcare professionals.

- The most successful self-management education programs bolster individuals' confidence to achieve optimal functioning, promote acceptance of illness-induced limitations, and promote positive ways of thinking, feeling and behaving.

Despite the high prevalence and deleterious impact of chronic pain, access to timely and appropriate care remains problematic. Many patient pain-related education programs are tertiary-based and lack sufficient scope and complexity to address the multidimensional experience of chronic pain. There is a paucity of accessible community-based approaches to chronic pain management across countries. Many with chronic pain are overwhelmed by the impact of their pain problem and lack adequate knowledge and skills to manage their pain at home, on a day-to-day basis. Alternate models of care that emphasize the patient's critical role in pain management need to be made readily available in communities where people live and work. As an adjunct to the usual care, pain self-management education has received increased attention as an effective means of providing an accessible, community-based approach that enables people to better deal with everyday problems that result from chronic pain, thus improving health related quality of life (HRQL).

This chapter provides a brief history and overview of the concept of self-management, a discussion of seminal pain self-management work in the United States argues that self-efficacy is a key mechanism behind successful pain self-management, and presents detailed examples of current self-management research programs for specific adult and child chronic pain populations in Canada. The cost-effectiveness of self-management, as well as policy implications for future

Chronic Pain: A Health Policy Perspective
Edited by S. Rashiq, D. Schopflocher, P. Taenzer, and E. Jonsson
Copyright © 2008 WILEY-VCH Verlag GmbH & Co. KGaA, Weinheim
ISBN: 978-3-527-32382-1

research, knowledge translation and dissemination of pain self-management programs in Canada are also discussed.

14.1
Self-Management: Brief History and Overview

Following a swell of epidemiological research on the prevalence of chronic illness in the 1960s and 70s [1], self-management emerged as a priority for health science researchers in the 1980s and 90s as a major class of patient education-related interventions. Standard health care delivery models were criticized as narrow in scope, and designed solely to address acute illnesses when it was realized that chronic illnesses without cure were prevalent throughout the world [1, 2]. Similarly, traditional patient education models were criticized as lacking adequate scope and complexity to address an aging population, increases in co-morbidities, and the diverse needs of individuals who must manage the day-to-day impact of chronic illness at home.

Traditional patient education models are focused on technical self-care skills and specific disease-related information. The healthcare professional is understood to be the authority on the patient's self-care priorities and information is imparted to the patient who acts as a passive recipient of information. Patient compliance is assumed to lead to improved health-related behaviors and outcomes [1–3].

In contrast, self-management education interventions emphasize the formation of a collaborative partnership between patients and healthcare professionals. At the onset of this partnership, the patient's every-day problems in living with chronic illness are carefully described (typically through iterative discussions) and prioritized. These priorities–which can often differ from those of the healthcare professionals [2, 3]–dictate the direction and scope of intervention for each patient [4, 5–9].

Barlow and colleagues conducted a large-scale systematic review of self-management interventions in order to determine the major tenets of self-management education programs and self-management designs that are the most effective [9]. Their review of 145 studies from the world literature found that when compared to usual care, self-management programs consistently improve patients' (i) knowledge, (ii) performance of self-management behaviors such as exercise, relaxation, energy conservation, and stress reduction, and (iii) various aspects of physical and emotional functioning such as exercise capacity and mood status. They also found that the majority of self-management programs to date are designed for adult patients. They can be delivered in a variety of formats including (i) individual counseling, (ii) small group sessions, or (iii) a combined individual and group-based approach. Each mode of delivery is equally effective. Regardless of format, most programs utilize a multi-component approach, offering an array of self-management techniques that participants can rehearse and incorporate into their day-to-day lives. The choice of technique(s) often depends on the individual's personality, lifestyle and confidence level. Clinical and community center-based settings are the most typical locations for program delivery that are led by a variety of tutors or facilitators, including healthcare professionals, peers, or fellow patients.

A critical feature of most successful self-management education programs, including those reviewed by Barlow and colleagues, is a firm grounding in social, cognitive and/or behavioural theories, targeting individuals' (i) confidence to achieve optimal functioning, (ii) acceptance of their chronic illness-induced limitations, and (iii) positive ways of thinking, feeling and behaving [1–3, 5–9]. One such theory commonly applied to more successful programs is self-efficacy theory. Preeminent psychologist Albert Bandura [10–12] defined the concept of self-efficacy as "The exercise of human agency through people's beliefs in their capabilities to produce desired effects by their actions" [12]. He argued that inherent in human nature is the need for control in everyday circumstances. Not simply a matter of knowing what to do, self-efficacy reflects one's perceived capacity to organize and integrate cognitive, social and behaviral skills in order to meet a variety of purposes in managing illness day-to-day [10–12].

In putting self-efficacy theory into practice, self-management programs improve patients' confidence to achieve optimal health by providing opportunities for (i) skills mastery – practising self-management techniques in a supportive environment, (ii) modelling – learning positive health behaviors from facilitators and peers, (iii) reinterpretation of symptoms – careful examination of illness-related beliefs that may lead to maladaptive behaviors (such as being overly sedentary as a means to avoid continued pain), and (iv) social persuasion – the support and encouragement of one's friends, family and healthcare professionals [13]. Barlow and colleagues fashioned the following well-accepted definition of contemporary self-management education, espousing these self-efficacy enhancing principles [9]:

> Self-management refers to the individual's ability to manage the symptoms, treatment, physical and psychological consequences and lifestyle changes inherent in living with a chronic condition. Efficacious self-management encompasses the ability to monitor one's condition and the effect of the cognitive, behavioral, and emotional responses necessary to maintain a satisfactory quality of life, thus a dynamic and continuous process of self-regulation is established (p. 178).

Depending on the nature of the chronic illness being addressed, pain relief has been emphasized as a major goal for those enrolled in self-management programs. Designed for those living with arthritis, the Arthritis Self-Management Program (ASMP), designed by Lorig and colleagues at the Stanford University Patient Education Research Center, has been a seminal model for effective pain self-management, worldwide [14].

14.2
The Arthritis Self-Management Program

The ASMP is a standardized, six-week, self-management program, delivered in two-hour weekly sessions to small groups of 8 to 10 patients [14]. It is based on the adult education principles of relevance, reinforcement, feedback, individualization and facilitation. Didactic presentation is minimized. It is also process-oriented, facilitating individual exploration of a range of self-management approaches including (i) self-help/self-management principles, (ii) discussion of key factors for optimal living with chronic disease such as diet, fitness, nutrition and medication regimes, (iii) cognitive-behavioral approaches to pain management and emotional responses, and (iv) behavioral modification strategies. Nearly 20 years after its original inception, the ASMP continues to be delivered globally by both generalist health care providers and by trained lay leaders at a cost ranging from $0 to $600 US per course [14, 15].

Although not marketed as a pain self-management program, the ASMP has been a major contributor to the global pain self-management effort. It is now widely disseminated through national arthritis societies on three continents [14–19]. Multiple large-scale randomized controlled trials of the ASMP have shown that this program significantly improves levels of self-efficacy for those with chronic arthritis pain. These enhancements in self-efficacy have consistently mediated clinically significant improvements in pain, HRQL, knowledge, exercise and relaxation behaviors, and depression and disability for up to four years post-intervention, without any formal reinforcement of program content [19]. Sustained reductions in healthcare costs have also been reported as a result of the program, including 40% fewer physician visits and average savings of $648 US and $189 US for persons with rheumatoid arthritis and osteoarthritis, respectively [19].

Marks *et al.* [20] found that results of the ASMP evaluation studies have consistently supported the following major tenets of Bandura's Self-Efficacy Theory [21] (i) the strength of a person's belief in his/her capability to do a specific task or achieve a certain result is a good predictor of motivation and behavior, (ii) self efficacy belief can be enhanced through performance mastery, modeling, reinterpretation of symptoms, and social persuasion, (iii) enhanced self-efficacy leads to improved behavior, motivation, thinking patterns and emotional well-being.

As a highly successful and cost-effective self-management model for improving pain and related functional status outcomes, the ASMP, as well as a later generic version of this program, were adapted and tested in Canada for two different adult chronic pain conditions chronic pain, and cardiac pain arising from chronic stable angina.

14.3
The Chronic Pain Self-Management Program

LeFort [22] developed the *Chronic Pain Self-Management Program* (CPSMP) in 1995 by adapting the ASMP in order to make it more directly applicable to people with chronic pain. Modifications were made with respect to (i) myths and information about chronic pain, (ii) understanding acute and chronic pain, (iii) pacing activity and rest, (iv) exercise, (v) communicating about chronic pain, (vi) breathing and body awareness, and (viii) medications.

Topics	Week					
	1	2	3	4	5	6
Self-help principles	✓					
Debunking myths	✓					
What is chronic pain?	✓					
Balancing rest/activity	✓			✓		
Exercise/ROM Dance	✓	✓	✓	✓	✓	✓
Pain management/ relaxation		✓	✓	✓	✓	✓
Depression			✓			
Nutrition				✓		
Evaluating non-traditional treatments					✓	
Problem-solving	✓	✓	✓	✓	✓	✓
Communication skills		✓			✓	
Medications						✓
Fatigue						✓
Feedback/contracting	✓	✓	✓	✓	✓	✓

Figure 14.1 Chronic pain self-management program overview.

Like the ASMP, the CPSMP was designed to enhance self-efficacy via a standardized small-group intervention format (7–9 patients per group), delivered by a nurse facilitator, in two-hour weekly sessions, over six weeks. Figure 14.1 provides a detailed overview of the program content and format. During the first week's session, there is detailed discussion of self-help principles and the meaning or impact of chronic pain for each individual participant. Each session then builds upon the last, introducing various cognitive and behavioral self-management techniques through mini-lectures, information sharing, and brainstorming. As with the ASMP, didactic presentation is kept to a minimum. With an emphasis on facilitating optimism and mutual support, participants are encouraged to set weekly, incremental self-management goals that address the impact of chronic pain in their lives. For example, a participant with severe chronic low back pain may set a goal to be able to walk short distances once or twice weekly, and work toward optimizing his/her walking ability over time, and at a realistic pace. The completion of weekly action plans is a critical component of the goal setting process. In order to complete an action plan, participants identify a specific short-term goal they want to achieve and then plan in detail how these goals will be reached (i.e. what needs to be done and when, as well as how much or how often). To maximize motivation and confidence, participants are encouraged to share their action plans with their fellow group members and report on their progress at the beginning of each session from week to week.

LeFort *et al.* [23] conducted a single-site randomized controlled trial, involving 110 participants, in order to determine the efficacy of the CPSMP for improving pain and functional status-related outcomes. This trial found that the CPSMP resulted in significant improvements in pain, dependency on others, sense of vitality, various aspects of role functioning, life satisfaction and self-efficacy and resourcefulness to manage pain three months post intervention. Like the ASMP trials [19], these improvements were mediated by participants' enhanced perceptions of their personal self-efficacy. Overall, the results of the CPSMP study pointed to a successful standardized self-management intervention based on a sound study design (i.e. random allocation, blind assessors, use of a standard protocol and patient education materials, and intention-to-treat analysis).

Because the CPSMP efficacy trial was delivered by a single registered nurse facilitator at one site, LeFort *et al.* [22] conducted a subsequent larger-scale, multi-site effectiveness trial of the CPSMP across three Canadian provinces. This 2003–2004 trial (*n* = 279) demonstrated that the positive effects on mental health and resourcefulness of the initial CPSMP trial were retained when the program was delivered by generic healthcare providers in the context of their regular workload. These benefits appear to have been maintained in a sub-sample of participants who were followed for 12 months post-intervention (S. LeFort, personal communication, Nov. 6, 2007). Non-statistically, a trend toward reductions in indirect pain-related costs for male participants of the program (e.g. forgone income and leisure time of spouses and others engaged in informal care giving) was also noted.

Additional analyses of the participants' anecdotal remarks found that at the beginning of the CPSMP, the participants felt isolated, overwhelmed, frustrated by their pain-related limitations and generally powerless. They felt in constant pain and described their experience in terms of losses. At the end of the CPSMP, the participants indicated that, while they still had pain, they felt they had more control over the pain, were less alone, more hopeful and were able to identify concrete ways that they could help themselves [23]. These positive changes reflected the strength of the program to enhance perceived self-efficacy and feelings of being validated or understood.

The CPSMP evaluation trials provided strong evidence for the effectiveness of this standardized and adapted program to improve a variety of pain and functional-status related outcomes for those with chronic pain.

14.4
The Chronic Angina Self-Management Program

Cardiac pain arising from chronic stable angina (CSA) pectoris is a ubiquitous and cardinal symptom of ischemic heart disease, characterized by pain or discomfort in the chest, upper abdomen, back, arm(s), shoulders, neck, jaw and/or teeth [24]. This condition is distressing, often resulting in recurrent pain episodes, poor general health, impaired role functioning, activity restriction, and reduced capacity for self-care [25–36]. Recent prevalence data suggest that more than 6 500 000 Americans may be living with CSA [37]. National Population Health Survey data suggest that 36% of Canadians with ischemic heart disease (IHD) live with angina symptoms and related daily activity restriction [38]. The Canadian Laboratory Center for Disease Control also found that in 1995, 16% of all physician visits related to heart disease in Canada (29.6 million visits) involved a complaint of angina [39].

CSA also imposes numerous direct and indirect societal costs. A recent economic pilot study by McGillion *et al.* [40] found that the total median annualized societal cost of CSA in Ontario was $12 615 (Canadian) per person (2003–2005). In the United Kingdom, the direct cost of CSA in 2000, including prescriptions, admissions, outpatient referrals, and procedures, was estimated at £669 000 000, accounting for 1.3% of the total National Health Service expenditure [41].

Akin to the situation for those living with chronic pain, there was a critical need to develop a pain self-management intervention. In 2004, McGillion *et al.* interviewed CSA patients and clinicians to determine the specific pain self-management learning needs of CSA patients [42]. Based on these findings, the Chronic Angina Self-Management Program (CASMP) was formed by adapting the Chronic Disease Self-Management Program [21] which was itself an adaptation of the original ASMP. Like the CPSMP program, the CASMP is a standardized self-management program delivered by a registered nurse facilitator in a small group (8 to 10 patients) format including two-hour sessions weekly over a six-week period. Figure 14.2 outlines the content of the CASMP in detail.

In a recent RCT, McGillion *et al.* [43] found that the CASMP was effective for improving CSA patients' self-efficacy, anginal pain, physical functioning and general health status in the short term. As with the 2004 CPSMP trial, this study found positive shifts in the meaning of pain following self-management training [44]. During the first program session, CSA was described by the participants as a major negative life change characterized by fear, frustration, limitations and anger. At the end of the program, CSA was more positively understood as a broad and ongoing pain problem requiring continual self-management in order to retain desired life goals and optimal functioning [44].

Until recently, little to no high-quality data syntheses had been conducted in order to determine the overall effectiveness of self-management interventions as an adjunct to usual care, across various community-based settings and countries. Therefore, in 2007, McGillion *et al.* [40] used meta-analytic techniques to determine the effectiveness of self-management interventions for improving symptoms, HRQL and psychological well-being in CSA [45]. Seven RCTs, involving 949 participants across seven countries were reviewed. While intervention format and processes varied, self-efficacy enhancement was a major focus for each program. Overall, self-management training was found to reduce angina and the need for nitroglycerine by approximately 3 and 4 times per week, respectively, in the short-term. Major improvements in physical functioning and general health status were also found. The overall ability of self-management to improve psychological well-being for CSA patients remains unknown; the investigators were unable to pool these data due to major differences in the way psychological well-being was measured in each trial.

CASMP Program Overview						
	Week 1	Week 2	Week 3	Week 4	Week 5	Week 6
Overview of Self-management and Chronic Angina	✓					
Making an Action Plan	✓	✓	✓	✓	✓	✓
Relaxation/Cognitive Symptom Management	✓		✓	✓	✓	✓
Feedback/Problem-solving		✓	✓	✓	✓	✓
Common Emotional Responses to Cardiac Pain: Anger/Fear/Frustration		✓				
Staying Active/Fitness		✓	✓			
Better Breathing			✓			
Fatigue/Sleep Management			✓			
Energy Conservation				✓		
Eating for a Healthy Heart				✓		
Monitoring Angina Symptoms and Deciding When to Seek Emergency Help				✓		
Communication				✓		
Angina and Other Common Heart Medications					✓	
Evaluating New/Alternative Treatments					✓	
Cardiac Pain and Depression					✓	
Monitoring Angina Pain Symptoms and Informing the Health Care Team						✓
Communicating with Health Care Professionals About Your Cardiac Pain						✓
Future Self-Management Plans						✓

Figure 14.2 Chronic angina self-management program (CASMP) overview.

14.5
Canadian-Based Pain Self-Management Research in Children

Self-management interventions targeting children with chronic pain remains understudied. A new Canadian-based program of pain self-management research for children involves the development and testing of *Teens Taking Charge: Managing Arthritis On-line*, web-based intervention for adolescents living with Juvenile Idiopathic Arthritis (JIA). JIA is the most common chronic pediatric rheumatic disease in Western Europe and North America [46]. Its course is unpredictable and children commonly experience pain, joint swelling, stiffness and fatigue that may restrict physical and social interactions [47]. Management of JIA is complex, involving multiple therapies, with constant monitoring. As children with JIA age, they are expected to assume more responsibility for their disease management [48]. However, adolescent adherence to JIA management activities is sub-optimal. Poor adherence and inappropriate self-management behaviors may reduce the potential benefits of treatment [49]. The vast majority of adolescents do not receive comprehensive education or self-management training due to (i) difficulty accessing these services (e.g. long wait times); (ii) limited availability of trained professionals (e.g. psychologists), especially in rural areas; (iii) costs, such as lost paid work time; and (iv) acceptability of these programs [50, 51]. Enhanced awareness and greater self-management early in the JIA disease trajectory may be critical in reducing the impact of JIA on youth.

E-health technologies can offer an innovative approach to improving the health service delivery and acceptability of self-management interventions. Web-based self-management programs have several advantages including (i) a secure server to deliver, customize and record homework sessions, (ii) opportunities for ongoing communication with healthcare team member and peers, (iii) private access to information at any time, (iv) increased treatment convenience, (v) personalized treatment via selection of relevant learning modules, and (vi) high appeal for computer savvy adolescents [52]. While the internet has emerged as a main health information source for youth [53], few self-administered multi-media programs for children and adolescents with chronic illnesses have been developed and validated [54]. Some preliminary evidence suggests that computer-based, self-administered treatments for children and families are efficacious in children with asthma [55, 56], diabetes [57, 58], and headaches [59]. For example, Gerber *et al.* [58] demonstrated the feasibility of an internet diabetes self-management program for adolescents and young adults, transitioning to adult medical care centers.

Murray *et al.* [52] conducted a systematic review of interactive health communication applications or internet-based interventions that included health information plus at least one of the following: (i) behavioral change support, (ii) decision-making support, and (iii) social support for adults and children. These interventions are meant to augment the information and support provided by the patient's healthcare professionals. Internet-based interventions can: (i) provide health information, (ii) enable informed decision-making, (iii) promote healthy behaviors, (iv) promote peer exchange (sharing information and emotional support), promote self-care, and (v) manage the demand for health services. The review found that internet-based interventions were effective for improving knowledge and helping patients to feel more socially supported. These interventions may also improve behavioral and clinical outcomes as well as enhance self-efficacy; their ability to improve emotional and economic-related outcomes remains unknown.

In Step 1 of the *Teens Taking Charge* project, Stinson *et al.* [60] conducted a qualitative study using individual and focus group interviews to explore the self-management needs of adolescents with JIA, and determine the acceptability of a web-based self-management approach for this patient group. The participants articulated how they developed effective self-management strategies through a process of "letting go" from others who had managed their illness (healthcare professionals, parents) and "gaining control" over managing their illness on their own. Two strategies that assisted in this process were gaining knowledge and skills to manage the disease, and enhancing understanding through social support. Five additional sub-themes emerged including: (i) possessing knowledge and awareness about JIA, (ii) listening to and challenging care-providers, (iii) communicating with the doctor, (iv) managing pain, and (v) managing emotions.

The next steps of this research program include the development and testing of the intervention. These steps will lay the foundation for a multi-center RCT, to test the effectiveness of the program for reducing pain and improving HRQL outcomes for adolescents with JIA. The end goal of the *Teens Taking Charge* research program is to disseminate an efficient and effective web-based self-management intervention as a means of knowledge exchange for adolescents with JIA. The program's potential may also be farther reaching. It is foreseeable that this program could ultimately be integrated into the pre-licensure curricula of healthcare professional programs across Canada, in order to better prepare them to promote effective self-management strategies in young people with arthritis. It is envisioned that this program will be adapted to other chronic pain conditions in children such as fibromyalgia and those with pain-related disability syndromes.

14.6
Summary and Policy Implications

It is clear that self-management interventions are highly effective adjuncts to the usual care for reducing chronic pain and improving HRQL, particularly for the adult arthritis and chronic pain populations. These interventions are also cost-effective, across divergent chronic disease populations. For painful chronic conditions, in particular, the savings associated with participation in Lorig *et al.*'s ASMP make a strong case [19]. Major direct and system-related savings have also been found for those with other painful conditions. For example, at one year post-intervention, significantly fewer hospital stays (approximately 1 per person) and 60% fewer primary care physician visits have been reported for those with inflammatory bowel disease (IBD) and irritable bowel syndrome (IBS), respectively [61, 62].

Until recently, chronic pain self-management in Canada has not been a major priority. A continued emphasis on tertiary-based, pharmacological, and invasive interventions has historically overshadowed the value of pain self-management training. In order to successfully integrate pain self-management as a mainstay of cost-effective treatment, funding support for continued research, development and dissemination of these programs is critical. Some key questions that remain unanswered could have major implications for successful implementation of pain self-management training nation-wide. The question of one's "preparedness" or "readiness" to engage in pain self-management training, for example, is an emerging area of research. Recent Canadian work by Hadjistavropoulos and Shymkiw [63] suggests that one's beliefs about perceptions about control may be key factors in one's intention to engage in pain self-management practices. Other key questions to be examined include the potential impact of pain self-management interventions on adverse health outcomes and long-term Canadian health service utilization patterns. The relative effectiveness of various modes of delivery of pain self-management interventions (e.g. group based vs. internet-based), across adult and child pain populations, is also largely unexplored territory. Key competencies, practice guidelines, and training programs must also be developed in order to adequately prepare healthcare professionals for educating and supporting patients to self-manage their pain.

Support for the dissemination and integration of existing pain self-management models within the Canadian health care system is also a major priority. Most self-management interventions are developed and tested within academic centers or research institutes, formally and/or informally linked with a variety of hospital and community-based settings [64]. Therefore, successful dissemination of self-management programs depends on strong partnerships between researchers and key stakeholders working together to navigate the complexities of "transitioning" these programs into existing and diverse system infrastructures [64, 65]. Ideally, stakeholder partners from government, health service agencies, and patient groups should be encouraged to be involved at the onset of pain self-management research programs and/or implementation strategies [64]. Also critical to garnering the support needed for the proliferation of pain self-management in Canada is effective knowledge translation. Policy makers and the public require timely, user-friendly products of pain self-management research, in non-scientific language. The current federal government emphasis on knowledge transfer and translation, as an expert field of inquiry, is a testament to the complexity of translating research into useful, publicly accessible information.

The CPSMP research program provides a current example of putting dissemination and knowledge transfer partnership principles into practice. In 2004, the Institute of Musculoskeletal Health and Arthritis (IMHA) of the Canadian Institutes of Health Research (CIHR) created the Knowledge Exchange Task Force (KETF) (http://www.cihr-irsc.gc.ca/e/27297.html); a strategic initiative to begin an active dialogue between researchers and end-users/stakeholders in the area of chronic pain and its management [66]. The KETF includes representatives from national and provincial organizations (e.g. Arthritis Societies, Osteoporosis Canada, the Canadian Paraplegic Association, seniors groups, and others), and representatives of health professional organizations. The first research topic chosen for presentation and active dialogue was chronic pain self-management. LeFort engaged in dialogue with KETF members about (i) the research process, (ii) trial design, (iii) how to interpret the study results, and (iv) ways to best communicate these results to key stakeholders. Since that time, a number of Canadian pain researchers have been invited to present their findings at each KETF meeting, usually held twice a year [66].

As the KETF has matured, its members have evolved in their roles to become "research ambassadors". They have translated the results of pain research into key messages in order to raise public awareness. Short user-friendly written materials, incorporating these messages, have been distributed to various national organizations that represent key stakeholder interests in pain self-management research. The KETF research ambassadors now plan to invite policy makers to their next round of meetings, a vital next step in furthering their knowledge transfer efforts.

Acknowledgments

The authors gratefully acknowledge the participation of the participants living with chronic pain for participating in the respective programs of pain self-management research. Portions of the Chronic Angina Self-Management Program first appeared in or are derived from the Chronic Disease Self-Management Program Leader's Master Trainer's Guide (1999). Those portions are Copyright 1999, Stanford University. Portions of the Chronic Pain Self-Management Program first appeared in or are derived from the Arthritis Self-Help Program Leader's Guide (1995). These portions are Copyright 1995, Stanford University.

References

1 Holman, H. and Lorig, K. (2004) Patient self-management: A key to effectiveness and efficiency in care of chronic disease. *Public Health Reports*, **119**, 239–43.

2 Bodenheimer, T., Lorig, K., Holman, H. and Grumbach, K. (2002) Patient self-management of chronic disease in primary care. *Journal of the American Medical Association*, **288**, 2469–75.

3 McGillion, M.H., Watt-Watson, J.H., Kim, J. and Graham, A. (2004) Learning by heart: A focused groups study to determine the psychoeducational needs of chronic stable angina patients. *The Canadian Journal of Cardiovascular Nursing*, **14**, 12–22.

4 McGillion, M., Watt-Watson, J., LeFort, S. and Stevens, B. (2007) Positive shifts in the meaning of cardiac pain following a psychoeducation program for chronic stable angina. *The Canadian Journal of Nursing Research*, **39**, 48–65.

5 Holman, H. and Lorig, K. (2000) Patients as partners in managing chronic disease (2000). *British Medical Journal*, **320**, 526–7.

6 Barlow, J.H., Shaw, K.L. and Harrison, K. (1999) Consulting the "experts": children and parents' perceptions of psycho-educational interventions in the context of juvenile chronic arthritis. *Health Education Research*, **14**, 597–610.

7 McGillion, M., Watt-Watson, J., Stevens, B., LeFort, S. and Coyte, P. (2008) Randomized controlled trial of a psychoeducation program for the self-management of chronic cardiac pain. *Journal of Pain and Symptom Management* (April 3, 2008, Epub ahead of print).

8 LeFort, S., Gray-Donald, K., Rowat, K.M. and Jeans, M.E. (1998) Randomised controlled trial of a community based psychoeducation program for the self-management of chronic pain. *Pain*, **74**, 297–306.

9 Barlow, J., Wright, C., Sheasby, J., Turner, A. and Hainsworth, J. (2002) Self-management approaches for people with chronic conditions: a review. *Patient Education and Counseling*, **48**, 177–87.

10 Bandura, A. (1986) *Social Foundations of Thought and Action: A Social Cognitive Theory*, Prentice Hall, Englewood Cliffs, New Jersey.

11 Bandura, A. (1977) *Social Learning Theory*, Prentice-Hall, Englewood Cliffs, New Jersey.

12 Bandura, A. (1997) *Self-Efficacy: The Exercise of Control*, W.H. Freeman, New York.

13 Lorig, K. (2003) Self-management education: More than nice extra. *Medical Care*, **41**, 699–702.

14 Lorig, K. (1986) Development and dissemination of an arthritis patient education course. *Family and Community Health*, **9**, 23–32.

15 Lorig, K., Lubeck, D., Kraines, R.G., Selenznick, M. and Holman, H.R. (1985) Outcomes of self-help education for patients with arthritis. *Arthritis and Rheumatism*, **28**, 680–5.

16 Lorig, K., Lubeck, D., Selenznick, M., Brown, B.W., Ung, E. and Holman, R. (1989) The beneficial outcomes of the arthritis self-management course are inadequately explained by behavior change. *Arthritis and Rheumatism*, **31**, 91–5.

17 Lorig, K. and Holman, H.R. (1989) Long-term outcomes of an arthritis study: effects of reinforcement efforts. *Social Science and Medicine*, **20**, 221–4.

18 Lorig, K., Mazonson, P. and Holman, H.R. (1993) Evidence suggesting that health education for self-management in patients with chronic arthritis has maintained health benefits while reducing health care costs. *Arthritis and Rheumatism*, **36**, 439–46.

19 Lorig, K. and Holman, H.R. (1993) Arthritis self-management studies: a twelve year review. *Health Education Quarterly*, **20**, 17–28.

20 Marks, R., Allegrante, J.P. and Lorig, K. (2005) A review and synthesis of research evidence for self-efficacy enhancing interventions for reducing chronic disability: Implications for health education practice (Part II). *Health Promotion Practice*, **6**, 148–56.

21 Lorig, K., Stewart, A., Ritter, P., González, V.M., Laurent, D. and Lynch, J. (1996) Conceptual basis for the chronic disease self-management study, in *Outcome Measures for Health Education and Other Health Care Interventions*, Sage Publications, Inc, Thousand Oaks, pp. 1–7.

22 LeFort, S.M. (1998) Randomized controlled trial of a community-based nursing intervention for those experiencing chronic non-malignant pain. (Doctoral dissertation, McGill University, 1998). *Dissertation Abstracts International, DAI-B*, **60** (12), 6023.

23 LeFort, S., Gray-Donald, K., Rowat, K.M. and Jeans, M.E. (1998) Randomised controlled trial of a community based psychoeducation program for the self-management of chronic pain. *Pain*, **74**, 297–306.

24 LeFort, S., Watt-Watson, J. and Webber, K. (2003) Results of a randomized trial of the chronic pain self-management program in three Canadian provinces. *Pain Research and Management*, **8** (SupplB), 73.

25 LeFort, S., Webber, K. and Watt-Watson, J. (2003) Before and after the chronic pain self-management program: "What chronic pain means to me". *Pain Research and Management*, **8** (SupplB), 74.

26 Stewart, S., Inglis, S. and Hawkes, A. (2006) *Chronic Cardiac Care: A Practical Guide to Specialist Nurse Management*, Blackwell Publishing Ltd, Massachusetts.

27 Brorsson, B., Bernstein, S.J., Brook, R.H. and Werko, L. (2001) Quality of life of chronic stable angina patients four years after coronary angioplasty or coronary artery bypass surgery. *Journal of Internal Medicine*, **249**, 47–57.

28 Brorsson, B., Bernstein, S.J., Brook, R.H. and Werko, L. (2002) Quality of life of patients with chronic stable angina before and 4 years after coronary artery revascularization compared with a normal population. *Heart*, **87**, 140–5.

29 Caine, N., Sharples, L.D. and Wallwork, J. (1999) Prospective study of health related quality of life before and after coronary artery bypass grafting: outcome at 5 years. *Heart*, **81**, 347–51.

30 Erixson, G., Jerlock, M. and Dahlberg, K. (1997) Experiences of living with angina pectoris. *Nursing Science and Research in the Nordic Countries*, **17**, 34–8.

31 Gardner, K. and Chapple, A. (1999) Barriers to referral in patients with angina:

Qualitative study. *British Medical Journal*, **319**, 418–21.

32 Lyons, R.A., Lo, S.V. and Littlepage, B.N.C. (1994) Comparative health status of patients with 11 common illnesses in Wales. *Journal of Epidemiology and Community Health*, **48**, 388–90.

33 MacDermott, A.F.N. (2002) Living with angina pectoris: a phenomenological study. *European Journal of Cardio-vascular Nursing*, **1**, 265–72.

34 Miklaucich, M. (1998) Limitations on life: women's lived experiences of angina. *Journal of Advanced Nursing*, **28**, 1207–15.

35 Pocock, S.J., Henderson, R.A., Seed, P., Treasure, T. and Hampton, J. (1996) Quality of life, employment status, and anginal symptoms after coronary artery bypass surgery: 3-year follow-up in the randomized intervention treatment of angina (RITA) trial. *Circulation*, **94**, 135–42.

36 Spertus, J.A., Jones, P., McDonell, M., Fan, V. and Fihn, S.D. (2002) Health status predicts long-term outcome in outpatients with coronary disease. *Circulation*, **106**, 43–9.

37 Spertus, J.A., Salisbury, A.C., Jones, P.G., Conaway, D.G. and Thompson, R.C. (2004) Predictors of quality of life benefit after percutaneous coronary intervention. *Circulation 2004*, **110**, 3789–94.

38 Wandell, P.E., Brorsson, B. and Aberg, H. (2000) Functioning and well-being of patients with type 2 diabetes or angina pectoris, compared with the general population. *Diabetes and Metabolism 2000*, **26**, 465–71.

39 American Heart Association (2006) *Heart Disease and Stroke Statistics-2006 Update*, American Heart Association, Dallas, Texas.

40 McGillion, M., Croxford, R., Coyte, P., Watt-Watson, J. and Stevens, B. (2007) Cost of illness for chronic stable angina patients. *Pain Research and Management*, **12**, 137.

41 Stewart, S., Murphy, N., Walker, A., McGuire, A. and McMurray, J.J.V. (2003) The current cost of angina pectoris to the National Health Service in the UK. *Heart*, **89**, 848–53.

42 McGillion, M.H., Watt-Watson, J.H., Kim, J. and Graham, A. (2004) Learning by heart: A focused groups study to determine the psychoeducational needs of chronic stable angina patients. *Canadian Journal of Cardiovascular Nursing*, **14**, 12–22.

43 McGillion, M., Watt-Watson, J., Stevens, B., LeFort, S., Coyte, P. and Graham, A. Randomized controlled trial of a psychoeducation program for the self-management of chronic cardiac pain. *Journal of Pain and Symptom Management* (in press).

44 McGillion, M., Watt-Watson, J., LeFort, S. and Stevens, B. (2007) Positive shifts in the meaning of cardiac pain following a psychoeducation program for chronic stable angina. *CJNR*, **39**, 48–65.

45 McGillion, M., Arthur, H., Victor, C., Watt-Watson, J. and Cosman, T. (2008) Effectiveness of psychoeducational interventions for improving symptoms, health-related quality of life, and psychological well being in patients with stable angina. *Current Cardiology Reviews*, **4**, 1–11.

46 Saurenman, R.K., Rose, J.B., Tyrrell, P., Feldman, B.M., Laxer, R.M., Schneider, R. and Silverman, E.D. (2007) Epidemiology of juvenile idiopathic arthritis in a multiethnic cohort. *Arthritis and Rheumatism*, **56**, 1974–84.

47 Shaw, K.L., Southwood, T.R., Duffy, C.M. and McDonagh, J.E. (2006) Health related quality of life in adolescents with juvenile idiopathic arthritis. *Arthritis and Rheumatism*, **55**, 199–207.

48 Harrington, J., Kirk, A. and Newman, S. (1999) Developmental issues in adolescence and the impact of rheumatic disease, in *Adolescent Rheumatology* (eds D.A. Isenberg and J.J. Miller), Martin Duntiz Ltd, London, pp. 21–33.

49 Rapoff, M.A. (2002) Assessing and enhancing adherence to medical regimens for juvenile rheumatoid arthritis. *Pediatric Annals*, **31**, 373–9.

50 Barlow, J.H., Shaw, K.L. and Harrison, K. (1999) Consulting the "experts": children's and parents' perceptions of psycho-educational interventions in the context of juvenile chronic arthritis. *Health Education Research*, **14**, 597–610.

51 Elgar, F.J. and McGrath, P.J. (2003) Self-administered psychosocial treatments for children and families. *The Journal of Clinical Psychiatry*, **59**, 321–39.

52 Murray, E., Burns, J., See, T.S. and Lai, R. (2005) Interactive Health Communication Applications for people with chronic disease. *Cochrane Database of Systematic Reviews*, **4**, 1–25.

53 Gray, N.J., Klein, J.D., Noyce, P.R. and Tracy, S. (2005) Health information-seeking behavior in adolescents: the place of the Internet. *Social Science and Medicine*, **20**, 1467–78.

54 Barlow, J.H. and Ellard, D.R. (2004) Psycho-educational interventions for children with chronic disease, parents and siblings: an overview of the research evidence base. *Child: Care, Health and Develop*, **30**, 637–45.

55 Shegog, R., Bartholomew, K., Parcel, G.S., Sockrider, M.M., Masse, L. and Abramson, S.L. (2001) Impact of a computer-assisted educational program on asthma self-management behavior. *Journal of the American Medical Informatics Association*, **8**, 49–61.

56 Krishna, S., Francisco, B.D., Balas, E.A., Koning, P., Graff, G.R. and Madsen, R.W. (2003) Internet-enabled interactive multimedia asthma education program: a randomized trial. *Pediatrics*, **111**, 503–10.

57 Brown, S.J., Lieberman, D.A., Gemeny, B.A., Fan, Y.C., Wilson, D.M. and Pasta, D.J. (1997) Educational video game for juvenile diabetes: results of a controlled trial. *Medical Informatics*, **22**, 77–89.

58 Gerber, B.S., Solomon, M.C. and Shaffer, T.L. (2007) Evaluation of an Internet diabetes self-management training program for adolescents and young adults. *Diabetes Technology and Therapeutics*, **9**, 60–7.

59 Hicks, C.L., von Baeyer, C.L. and McGrath, P.J. (2006) Online psychological treatment for pediatric recurrent pain: A randomized evaluation. *Journal of Pediatric Psychology*, 1–13.

60 Stinson, J., Toomey, P.C., Stevens, B.J., Kagan, S., Duffy, C.M., Huber, A., Malleson, P., McGrath, P. *et al.* (2008) Asking the experts: exploring the self-management needs of adolescents with arthritis. *Arthritis Care and Research*, **59**, 65–72.

61 Kennedy, A.P., Nelson, E., Reeeves, D., Richardson, G., Roberts, C., Robinson, A., Rogers, A.E., Schulper, M. *et al.* (2004) A randomized controlled trial to assess the effectiveness and cost of a patient oriented self-management approach to chronic inflammatory bowel disease. *Gut*, **53**, 1639–45.

62 Robinson, A., Lee, V., Kennedy, A., Middleton, L., Rogers, A., Thompson, D.G. and Reeves, D. (2006) A randomised controlled trial of self-help interventions in patients with a primary care diagnosis of irritable bowel syndrome. *Gut*, **55**, 643–8.

63 Hadjistavropoulos, H. and Shymkiw, J. (2007) Predicting readiness to self-manage pain. *The Clinical Journal of Pain*, **23**, 259–66.

64 Lorig, K., Hurwicz, M.L., Sobel, D., Hobbs, M. and Ritter, P.L. (2005) A national dissemination of an evidence-based self-management program: a process evaluation study. *Patient Education and Counseling*, **59**, 69–79.

65 Bruce, B., Lorig, K. and Laurent, D. (2007) Participation in patient self-management programs. *Arthritis and Rheumatism*, **57**, 851–4.

66 Brachaniec, M., Tillier, W. and Dell, F. (2006) The Institute of Musculoskeletal Health and Arthritis (IMHA) Knowledge Exchange Task Force: an innovative approach to knowledge translation. *The Journal of the Canadian Chiropractic Association*, **50**, 8–13.

Part Three
Special Challenges in Chronic Pain Management

15
Chronic Pain in Persons with Mental Health Issues
Charl Els

- Chronic pain is associated with a wide range of emotional responses and psychiatric disorders including depression which is present in 30 to 50% of cases.

- Available data indicate that depression is a consequence rather than a cause of chronic pain.

- Patients with co-morbid chronic pain and psychiatric disorders are poorly served by the current chronic pain and psychiatric care systems as both are under resourced and have little experience or expertise related to these comorbidities.

- A number of new pharmacological therapies and treatment strategies have particular promise in serving this population. Research targeted to exploring clinical outcomes of novel treatment approaches for this population is needed.

- New models of care that encourage the collaboration of mental health and chronic pain professionals to develop a strong evidence-based approach to managing patients with these comorbidities are required.

15.1
Background

Chronic pain frequently coexists with a range of emotional symptoms, varying degrees of psychopathology and with mental disorders, and may be an associated symptom of a number of psychiatric illnesses. Pain is widely recognized [1, 2] to have both psychological and physical components, and one of the early pioneers of psychiatry, Sigmund Freud, elaborated on the existence of a mind–body connection at a time when it was not widely endorsed [3].

Chronic pain patients with pronounced psychopathology tend to be complicated and difficult to manage [4], and furthermore, mentally ill persons are often dis-

Chronic Pain: A Health Policy Perspective
Edited by S. Rashiq, D. Schopflocher, P. Taenzer, and E. Jonsson
Copyright © 2008 WILEY-VCH Verlag GmbH & Co. KGaA, Weinheim
ISBN: 978-3-527-32382-1

enfranchised from medically necessary services [5]. Evidence [6] suggests that collaboration between pain specialists and mental health professionals is not the norm. It is also unusual to encounter case-finding, accurate screening for, and management of pain symptoms and syndromes in contemporary psychiatric practice.

The goal of this section is to provide an overview of findings and emerging themes that may influence health policy. Three key areas that are explored are:

1. The prevalence and scope of psychiatric illness in chronic pain patients.
2. The systemic barriers to mentally ill persons accessing medically necessary services.
3. Specific trends in the pharmacotherapy of comorbidity.

15.2
The Prevalence of Psychopathology with Chronic Pain

Pain is both an emotional *and* a physical experience, and a range of expected emotional responses to chronic pain may arise, including anxiety, depressed mood, anger, bereavement, frustration, irritation, and isolation. These symptoms may or may not reach the threshold for the diagnosis of a mental disorder. There are a variety of characteristics of patients with chronic pain [7]:

1. Preoccupation with pain
2. Strong and ambivalent dependency needs
3. Feelings of isolation and loneliness
4. Characterological masochism
5. Inability to take care of self needs
6. Passivity
7. Lack of insight into patterns of self-defeating behavior
8. Inability to deal appropriately with anger and hostility
9. Use of pain as a symbolic means of communication

The psychological comorbidities found in chronic pain patients are [8]:

1. Organic diagnosis elusive
2. Search for cure
3. Hostility/anger at physician for failure to diagnose or cure or for perceived incorrect input into legal situation
4. Anger at employer negligence in injury or for forced job loss
5. Financial issues causing stress
6. Dissatisfaction with medical care either for failure to diagnose/cure or for inability to get perceived necessary medical care
7. Anger at carer for above or for other issues, such as late payment of benefits
8. Litigation stress, such as dissatisfaction with lawyer
9. Confusion over conflicting diagnosis and recommendations
10. Anger with spouse over solicitousness or under solicitousness

11. Spousal depression
12. Poor coping strategies
13. Spousal problems/stress such as perceived nonsupport or blaming or not believing the patient's pain
14. Fear of pain
15. Poor self-esteem
16. Pre-injury job stress
17. Childhood victimization
18. Unrealistic expectations about treatment

Emotional disorders associated with chronic pain syndrome [7] are:

- Somatoform disorders
  - Somatization disorder
  - Undifferentiated somatoform disorder
  - Conversion disorder
  - Pain disorder
  - Hypochondriases
  - Body dysmorphic disorder
  - Somatoform disorder Not otherwise specified
- Mood disorders
- Anxiety disorders
- Personality disorders
- Psychological factors affecting a medical condition
  - Mental disorder affecting medical condition
  - Psychological symptoms affecting medical condition
  - Personality traits or coping style affecting medical condition
  - Maladaptive health behaviors affecting medical condition
  - Stress-related physiological response affecting medical condition
  - Other or unspecified psychological factors affecting medical condition
- Malingering
- Factitious disorder
- Schizophrenia and other psychotic disorders
- Substance-related disorders

Suffering from a condition for which the medical profession may not find a suitable solution may be associated with a sense of anger, hostility, confusion, fear, despair, hopelessness, and other symptoms of significant subjective distress. If no objectively demonstrable tissue damage is apparent to explain the chronic pain, care providers may erroneously attribute it to a mental condition, indirectly contributing to a sense of frustration and subjective distress.

The prevalence rates of psychiatric comorbidity in chronic pain patients have been studied. Comorbidities have been found to range widely [9]. Methodological flaws may in part explain the wide ranges. A review [10] summarizes the expected high levels of comorbidity, and describes the most prevalent conditions in each of the categories. Depression appears to be the psychiatric disorder that predominates. Patten *et al.* [11] report that the lifetime prevalence rates of a major depressive disorder are estimated at 12.2%. In a population with chronic pain, about one third will suffer from major depression, and 30–55% will have depressive symptoms [12].

The co-occurrence of depression and pain in epidemiological samples does not, however, provide any evidence of potential causality. In another review [13] the question is posed whether chronic pain -associated depression is an antecedent or a consequence of chronic pain; 191 empirical studies related to the association were reviewed. The principal finding that depression is more common in chronic pain patients than in healthy controls frames depression as a consequence of the presence of chronic pain. One hypothesis explaining the co-occurrence is that of vulnerability or predisposition to the development of depression. There also appears to be a growing body of evidence suggesting that depression lowers pain tolerance, increases analgesic requirements and exacerbates the debilitating effects associated with chronic pain [14]. It is also noteworthy [15] that chronic pain is also associated with post-traumatic stress disorder, panic disorder, agoraphobia, and a number of other conditions. It appears that comorbidity is the expectation and not the exception.

Associated with the major psychiatric disorders is the potential for risk of harm to self or others. The risk of suicide in those suffering from chronic pain is associated with a positive family history of suicide [16], but there has been no definitive study demonstrating the association between suicidal ideation and pain duration or pain severity.

15.3
The Systemic Barriers to Access

Like other disenfranchised and marginalized populations, persons with mental illness appear to have inequitable access to medically necessary services [5], and specialized pain management options are no exception. Conversely, persons with chronic pain infrequently have optimal access to specialized mental health services and such referrals are often reserved for those with the most marked psychopathology. Individuals with severe and persistent mental illness may be especially prone to experiencing these disparities in services, in part related to the very nature of the illness. Cognitive deficits and psychotic features in particular may result in suboptimal skills in advocating for equity in their own management. Putative safety concerns may rank higher on the hierarchy of priorities than chronic pain.

Allocation of sufficient resources to mental health has been a historical and ongoing management concern, and pain management in the mentally ill has not been a priority. Access to sufficient resources at the point of delivery of integrated mental health and addiction services, to include pain management screening and management protocols, as well as psychosocial screening and management for those with chronic pain, may help to remove the existing disparities in access to integrated pain and mental health care.

15.4
Pharmacotherapeutic Considerations

The development of pragmatic solutions and effective protocols for managing chronic pain in the mentally ill is imperative. An integral part of mental health care and pain management is a longitudinal integrated psychosocial focus combined with pharmacotherapy. Considerations for medication management of pain in persons with mental illness include: (i) Risk of abuse, (ii) addictive potential of a drug, (iii) the potential of the drug to exacerbate existing psychopathology and (iv) the potential of the drug to improve existing psychopathology. The risk of addiction in persons with mental illness is elevated, and the lifetime prevalence of addiction in those with mental illness has been estimated as almost one in two [17].

The evidence for pharmacotherapy for pain management in chronic pain patients with concurrent mental illness is considerably weaker than that for the non-mentally ill population because research protocols typically exclude mentally ill persons.

The existing range of pharmacotherapeutic options for pain management is not contra-indicated *per se* in persons with mental illness, and similarly, the psychotropic drugs used to treat mental illness are not *per se* contra-indicated in those with chronic pain. However, such combinations may cumulatively contribute to specific side-effects. The use of opiates and psychotropics may both contribute to exacerbate a common side-effect of both classes, namely decreased gastrointestinal motility and subsequent constipation. Further, the use of opiates in combination with sedating psychotropic drugs may contribute to further sedation and psychomotor retardation, as well as possible cognitive impairment [6].

Psychotropics may also impact on the efficacy of certain opiates used in the treatment of pain. Two commonly used antidepressants, fluoxetine and paroxetine, may significantly inhibit the efficacy of codeine.

A recent addition to the market: a combination of buprenorphine plus naloxone (combination drug: Suboxone) suppresses the symptoms of opioid withdrawal, decreases cravings for opioids, reduces illicit opioid use, blocks the effects of other opioids if used concurrently, enhances treatment retention of opioid addicted persons, and displays inherent and pronounced analgesic qualities. The combination of buprenorphine (partial mu-opioid agonist) and naloxone (mu-opioid antagonist) in a single formulation (sublingual) is available in Canada, and buprenorphine (Subutex) is available as a single formulation in the United States, but not Canada.

For the management of certain subtypes of pain, Health Canada [18] has approved access to medical marijuana for sufferers of a number of specific subtypes of chronic pain. The use of cannabinoids, synthetic or naturally occurring, may be associated with the exacerbation of psychotic symptoms in persons with preexisting psychotic illness or with a vulnerability to developing psychosis, and may further be associated with the relapse into psychosis in persons with schizophrenia. The use of cannabinoids should be conducted with great caution in individuals with a history of mental illness.

15.5
Proposed Systems' Level Changes

In a universal health care system that is free at the point of delivery, like that of Canada, disparities in management of comorbid chronic pain and psychopathology may contribute to excess morbidity and healthcare spending. Further research into the association between pain and psychopathology may eventually contribute to remedying such disparity. Advocating for increased efforts in psychopathology screening and management in pain clinic settings, and for optimal pain management in mental health settings appears prudent.

The greatest challenge remains provision of equitable access in a universal healthcare system, but easy solutions are evasive. These challenges are at least as intractable as the problem of co-occurring pain and psychopathology itself.

References

1 Al-Anon Family Group Inc. (1992) *Courage to Change: One Day at a Time*, Al-Anon Family Group Headquarters Inc., Virginia Beach, VA.

2 International Association for the Study of Pain, http://www.iasp-pain.org (URL accessed July 2007).

3 Breuer, J. and Freud, S. (1895) Studies on hysteria, in *The Standard Edition of the Complete Psychological Works of Sigmund Freud* (ed. J. Strachey), Hogarth.

4 Dolin, S.J. and Stephens, J.P. (1998) Pain clinics and liaison psychiatry. *Anaesthesia*, **53**, 317–19.

5 Kisely, S., Smith, M., Lawrence, D., Cox, M., Campbell, L.A. and Maaten, S. (2007) Inequitable access for mentally ill patients to some medically necessary procedures. *Canadian Medical Association Journal*, **176**, 779–84.

6 Marazziti, D., Mungai, F., Vivarelli, L., Presta, S. and Dell'Osso, B. (2006) Pain and psychiatry: a critical analysis and pharmacological review. *Clinical Practice and Epidemiology in Mental Health 2,31*.

http://www.cpementalhealth.com/content/2/1/31 (accessed 1 July 2007).

7 Aronoff, G.M. (1992) *Evaluation and the Treatment of Chronic Pain*, Williams & Wilkins, Baltimore.

8 Fishbain, D.A. (2005) Polypharmacy treatment approaches to the psychiatric and somatic comorbidities found in patients with chronic pain. *American Journal of Physical Medicine and Rehabilitation*, **84**, S56–63.

9 Blair, M.J., Robinson, R.L., Katon, W. and Kroenke, K. (2003) Depression and pain comorbidity a literature review. *Archives of Internal Medicine*, **163**, 2433–45.

10 Fishbain, D.A. (1999) Approaches to treatment decisions for psychiatric comorbidity in the management of the chronic pain patient. *The Medical Clinics of North America*, **83**, 737–60.

11 Patten, S.B., Want, J.L., Williams, J.V., Currie, S., Beck, C.A., Maxwell, C.J. and el-Guebaly, N. (2006) Descriptive epidemiology of major depression in Canada. *Canadian Journal of Psychiatry*, **51**, 84–90.

12 Birket-Smith, M. (2001) Somatization and chronic pain. *Acta Anaesthesiologica Scandinavica*, **45**, 1114–20.

13 Fishbain, D.A., Cutler, R. and Rosomoff, H.L. (1997) Chronic pain-associated depression: antecedent or consequence of chronic pain? A review. *The Clinical Journal of Pain*, **13**, 116–37.

14 Merskey, H. (1965) The effect of chronic pain upon the response to noxious stimuli by psychiatric patients. *Journal of Psychosomatic Research*, **8**, 405–19.

15 Sharp, J. and Keefe, B. (2005) Psychiatry in chronic pain: a review and update. *Current Psychiatry Reports*, **7**, 213–19.

16 Smith, M.T., Perlis, M.L. and Haythornwaite, J.A. (2004) Suicidal ideation in outpatients with chronic musculoskeletal pain: an exploratory study of the role of sleep onset insomnia and pain intensity. *The Clinical Journal of Pain*, **20**, 111–18.

17 Regier, D.A., Farmer, M.E., Rae, D.S., Locke, B.Z., Keith, S.J., Judd, L.L. and Goodwin, F.K. (1990) Comorbidity of mental disorders with alcohol and other drug abuse. Results from the epidemio-logic catchment area (ECA) study. *Journal of the American Medical Association*, **264**, 2511–18.

18 Health Canada: Marijuana medical access regulations, http://www.hc-sc.gc.ca/dhp-mps/marihuana/index_e.html (URL accessed July 2007).

16
Chronic Pain in Injured Workers

Douglas P. Gross, Martin Mrazik, and Iain Muir

- Pain is prevalent in workers and is an ongoing reason for work disability.

- Successful management strategies include the implementation of contin-uum of care models aimed at providing timely, effective health care in the most appropriate setting.

- Continuum of care models for workers in pain include early assessment, identification of recovery barriers, and targeted interventions for those workers most at risk of prolonged work loss.

- Coordinated management approaches are very different from the unidisci-plinary primary care currently provided in most jurisdictions and would require major funding and policy changes.

- Other potential solutions include alterations to current care provision struc-tures to facilitate more effective disability management through closer com-munication and collaboration between all stakeholders (workers in pain, employers, insurance agents and healthcare providers).

16.1
Introduction

Pain is a leading reason for lost productive work time. Pain is reported by workers in all types of jobs and across all categories of industry. While many workers con-tinue to function at work despite pain, others seek assistance in the form of healthcare or indemnity benefits. This section will review the epidemiology of pain at work, factors associated with pain-related work disability, as well as potential interventions that have been successful for reducing the large burden associated with pain in the injured worker.

Chronic Pain: A Health Policy Perspective
Edited by S. Rashiq, D. Schopflocher, P. Taenzer, and E. Jonsson
Copyright © 2008 WILEY-VCH Verlag GmbH & Co. KGaA, Weinheim
ISBN: 978-3-527-32382-1

16.2
Background

16.2.1
Epidemiology

While workers experience pain in multiple body parts, back pain is the leading cause of work loss. The high prevalence and social and economic impact of back pain has led to its inclusion as one of four musculoskeletal conditions specifically targeted by the Bone and Joint Decade (2000–2010) initiative endorsed by the World Health Organization [1].

A recent systematic review presenting data on the prevalence of back pain yielded point prevalence estimates from 12 to 33%, one-year prevalence estimates from 22 to 65% and lifetime prevalence estimates from 11 to 84% [2]. Another recent review of the literature on back pain prevalence estimated the point prevalence specifically in North America at 5.6% [3]. A recent survey undertaken in Alberta found a lifetime back pain prevalence of 84% and 1-week prevalence of 34% [4]. The wide ranges in the reported estimates are likely due to variations in the questions asked of respondents, however, it is clear that back pain is extremely common and affects nearly everyone at some time in their life. While back injury claim reporting trends have been downward over the past few years [5–7], the high rate of pain reporting and work loss due to the condition still represents a major health problem [8]. In Alberta, back pain is the leading category of workers' compensation injury claims in terms of frequency and cost [9].

16.2.2
Healthcare Seeking Behavior

Despite the high frequency of pain complaints in workers, management strategies vary considerably. It appears that many sufferers cope independently and do not seek care for their pain. In the Swedish working population, only 5% of individuals experiencing a new episode of back pain sought care for the condition [10]. Within the Alberta survey mentioned above, 12% of people reported doing little to treat or otherwise manage their last episode of back pain [4]. They simply continued their regular daily activities as recommended in most international guidelines for the management of back pain. Approximately 40% reported seeking assistance from a healthcare professional such as a physician, physical therapist or chiropractor. This is similar to study results from Australia, Belgium, USA and the UK where a minority of individuals report seeking care [11–14].

Interestingly those who seek care for pain episodes, especially multiple care episodes, may be no better off or even obtain worse outcomes when compared to those who do not seek care [10, 15–17]. This may represent evidence of iatrogenesis in subjects undergoing repeated and prolonged treatment; however, healthcare seeking behavior is likely influenced by such factors as severity or diminished

ability to cope with a painful condition which are also associated with poor outcomes [18]. Despite these potential confounding influences, if patients seeking
care are not expected to have better outcomes than those who do not seek care,
the value of primary clinic-based care in the management of pain conditions
(beyond ruling out serious pathology and providing reassurance) is uncertain.

16.2.3
Work Disability

Most individuals experiencing time loss from work due to pain, recover and return
to work quickly; however, a minority remains off work for prolonged periods of
time [19, 20]. Those remaining off work are responsible for the majority of costs
due to the condition and present a challenge for treating health professionals. The
total cost of economic loss due to pain is high, estimated to be as much as 1–2%
of the Gross National Product in Western Countries [20]. Furthermore, an intriguing paradox has occurred over the past decades as the incidence and prevalence
of pain appears unchanged, yet, there has been a dramatic increase in disability
associated with musculoskeletal pain. Thus the socioeconomic impact associated
with pain appears to be increasing [21]. This has led to attempts to develop clinical
strategies or predictive models for identifying patients at risk of delayed recovery
[22]. Unfortunately these efforts have had only modest success; however, some
lessons have been learned regarding the characteristics of workers in pain who
are at risk of chronic work disability.

16.3
Characteristics of Workers in Pain

16.3.1
Demographic Factors

The bulk of the existing prediction research has identified demographic variables
as modestly contributing to pain and work-related outcomes. Regardless of how
pain and return to work variables are studied, increased age and lower levels of
education are typically associated with delayed return to work [23].

16.3.2
Psychological Factors

Outcomes associated with psychological factors are diverse and difficult to assimi-
late given the various definitions applied to what constitutes a psychological variable. None-the-less, psychological variables continue to be of primary interest to
researchers. The most consistent findings relate to workers' recovery expectations
and fear of re-injury as key determinants of a worker's eventual return to work.

One study found, after adjusting for baseline demographics, that higher self-reports of pain intensity, higher perceptions of disability, fears that work may increase pain, poor recovery expectations, and catastrophizing were associated with poorer outcomes [24]. Other studies have also demonstrated that recovery expectations moderate work disability within one year post-injury [25, 26].

Several systematic literature reviews summarizing multiple studies of pain outcome in workers have found positive relationships between psychological functioning and RTW outcomes. Hoogendoorn *et al.* [27] reviewed 1363 studies from 1966 to 1997 that investigated the impact of back pain on work. The authors conclude that, among these studies, there was strong evidence for low social support and low job satisfaction as risks for back pain. However, insufficient evidence was found for high work pace or high qualitative demands, job content, job control, and decision latitude. Linton summarized 975 studies, 21 of which met inclusion criterion for review in his study [28]. Of the 11 types of psychological risk factors, strong evidence was found for six factors (job satisfaction, monotonous work, work relations, work demands, stress, and perceived ability to work), moderate evidence for four variables (control, work pace, emotional effort at work, and belief that work is dangerous) and one variable (work content) had inconclusive evidence.

Other systematic reviews have cautioned against over-interpreting findings given the limited number of studies that incorporated rigorous research methodologies. In addition, there is a paucity of prospective studies linking psychological variables to poor outcomes. Recently, Hartvigsen *et al.* [29] systematically reviewed the literature and identified major methodological problems in the majority of studies. Only 40 prospective studies published between 1990 and 2002 met rigorous scientific criteria. The authors concluded "we found moderate evidence for no positive association between perception of work, organization aspects of work, and social support and lower back pain." Therefore, further research that incorporates sound methodologies and prospective studies is needed.

16.3.3
Workplace Factors

A recent systematic analysis by Crook *et al.* [23] of variables in the workplace that affect worker outcomes yielded a handful of methodologically sound, prospective studies that provided evidence that the workplace environment moderates return to work outcomes. The findings suggested, in addition to demographic variables, factors such as availability of modified duties and light mobilization were predictors of faster return to work. In addition, referral for treatment within 30 days of injury and pre-morbid job stability (more than two years on the job) also predicted stronger return to work (RTW) outcomes. Problems with colleagues or problems on the job were also significantly related to poorer RTW outcomes. In another Finnish study of over 900 metal industry workers, low job control and poor supervisor support were significantly related to length of hospitalization following back injury [30].

16.3.4
Workers' Broader Contexts

Research has extended its emphasis on individual variables to include much broader factors such as the work environment and the greater social milieu. For instance, Battie and Bigos [31] examined over 56 variables in 3000 Boeing workers for their relation to future industrial back pain reporting. Physical factors (including past medical history, specific outcomes from a physical examination, and functional movements) that were long suspected to be risk indicators of poor outcomes were included in the study. Surprisingly it was the characteristics of the workplace setting, such as job satisfaction and emotional distress on the job that were most highly correlated with initial reports of pain. In fact, the various physical factors were not significantly associated with the report of back pain.

Moreover, the pervading economic climate has been demonstrated to be intricately associated with pain-related behaviors and disability. A classic study by Clemmer and Mohr examined the relationship between the incidence of lost-time low back claims and periods of economic activities [32]. Their findings reveal increased reports of low-back pain during downturns in economic activity and increased possibilities of being laid off. Therefore, the current treatment and management of pain in the injured worker must be viewed in a larger context of demographic, biological, psychological, societal, and economic influences.

16.4
Role of Compensation/Insurance Systems

Besides the influence of personal and workplace factors, broader societal systems also influence pain-related work disability. Individuals experiencing pain that is caused or exacerbated by work activities are often eligible for workers' compensation benefits. Most industrialized nations have some form of workers' compensation or insurance system to assist injured workers. These systems provide wage replacement benefits and payment for healthcare in patients who are unable to perform regular work duties because of work-related injuries or painful conditions.

Within North America, most compensation systems operate under an injury model [33]. This gives rise to some difficulties in the case of regional pain disorders such as back pain in that often no specific injury event precipitates the condition [34, 35]. Multiple studies of individuals with back pain as well as other painful conditions have shown that receipt of compensation benefits is associated with worse outcomes [36–39]. The explanation for this is unclear. Those receiving workers' compensation may have higher occupational physical demands that are confounding the observed relationship, or the fact these patients are being paid for being off work may have an influence. Clearly, broader societal influences and systems have an influence on workers with pain. Often these influences are supportive, but frequently they are detrimental.

16.5
Management Strategies

16.5.1
Continuums of Care

Efforts at reducing long-term disability have focused on identifying patients at risk of developing chronic problems on whom appropriate interventions can be focused [40, 41]. While results have been promising, current methods are of modest accuracy and some at-risk patients remain unidentified [21]. Recognizing that screening tools are underdeveloped, another alternative is to intervene early on all patients. Unfortunately, extensive early intervention leads to treatment being directed to patients who would likely have recovered without and may actually delay recovery [42]. Clearly, appropriate treatment needs to be applied to appropriately selected patients at the appropriate time for healthcare management to be successful. For these reasons, continuums of care have been implemented to guide the healthcare management of workers in pain.

A continuum of care has been defined broadly as a coordinated array of settings, services, providers, and care levels in which health, medical, and supportive services are provided in the appropriate care setting [43]. This may include treatment in an acute hospital, outpatient department, or community setting. Ideally, the patient receives healthcare at the most appropriate time and site according to their stage of recovery and level of need, and strong continuity and linkages exist between services within the system [44]. Continuum of care models have also been tested in management of whiplash-associated disorders and mild traumatic brain injury in Quebec and in Alberta within workers' compensation claimants filing soft tissue claims [45, 46]. Implementation of these models appears to have resulted in quicker recovery and reduced duration of work disability. Within Alberta, the workers' compensation model also resulted in dramatic cost savings without compromising patient satisfaction with care (estimated cost savings ~$20 million CDN over a one year period) [45]. Continuums of care may assist in guiding the provision of appropriate, evidence-based care at the appropriate time and in the appropriate setting. While research has demonstrated their effectiveness within controlled insurance contexts, it is unknown what effect they would have within the broader health system.

16.5.2
Psychologically-Based Interventions

As discussed earlier in this section, psychosocial variables can play a significant role in the initiation and maintenance of chronic pain. Yet, psychosocial variables are complex, embedded in a larger social context, and often highly unique to individual patients. For this reason, most traditional clinic-based treatments aimed at alleviating pain have been found to be of only marginal benefit [47].

Recent studies have evaluated cognitive behavioral therapy, social marketing interventions, and psychologically-based physical therapy treatment. Each of these approaches aim to alter disability behaviors through altering beliefs and attitudes. Authors of two recent systematic reviews of chronic back pain report that rehabilitation programs are effective at reducing duration of work-related disability only when they contain a component of cognitive behavior therapy [48, 49]. Importantly, the articles also highlight that cognitive behavior therapy embedded in a multi-disciplinary treatment program yields significantly better results when compared with unimodal treatments [50]. However, newer approaches in primary care have shown promising results when such cognitive behavioral principles are incorporated into routine physical therapy practice [51, 52]. In this approach, clinicians are trained to identify psychological barriers to recovery early in the treatment process, then provide goal-directed interventions targeting the observed barriers with the goal of changing beliefs, attitudes and subsequent disability behaviors. This approach to care provision has been found more effective than traditional physical therapy approaches. Although such interventions have promise they are not currently widespread in practice and a challenge will be faced in disseminating these interventions to clinical practice.

16.5.3
Disability Management Model

Disability Management is a relatively new approach to managing the impact of work place injury, illness and disability. It has been defined as "an active process of minimizing the impact of an impairment (resulting from injury, illness or disease) on the individual's capacity to participate effectively in the workplace" [53]. The major goal of disability management is returning the ill or injured employee back to safe, productive, and meaningful work in as timely a manner as possible. Return-to-work interventions are typically focused on what can be done at the workplace, and thus go beyond the traditional reliance on health care interventions. Increasingly employers, insurers and workers compensation boards are including disability management practices into their operational mandate. Recent data indicates that 70% of Canadian employers have taken steps to implement some form of disability management practices [54].

Advocates of disability management programs propose that a well structured program can shorten time off work, reduce disability costs and protect productivity [55]. However, only relatively recently have these claims been supported with strong evidence from the scientific literature [56]. Disability management programs have been found to reduce the number of days off work by half [57] and increase the odds of returning to work by a factor of 2.5 [58]. Of the disability management interventions studied to date, an early offer of modified work and good communication between all stakeholders (specifically the health care provider) are most strongly associated with return to work and lower disability costs [56]. Early contact by the employer, the presence of a return to work coordinator, and a worksite visit have also been found to be effective disability management interventions.

To summarise the key disability management intervention strategies are:

1. Early contact with the worker by the workplace or insurer
2. An offer of modified work (work accommodation offer)
3. Communication between health care providers and the workplace
4. A work site visit that facilitates a return-to-work plan
5. Return-to-work or case coordination

These scientific findings are mirrored in the results of a survey of Canadian companies with established disability management practices [54]. Taken collectively, there does appear to be convergence between the scientific literature and employer opinion surveys that disability management practices can significantly reduce disability and the associated costs.

Disability Management practices typically involve multiple stakeholders who must pull together toward a common goal of return-to-work [59–61]. Potential stakeholders include the worker and their family, the employer, union representatives, health care providers and the insurers. Critical to the success of any disability management program is formal, ongoing collaboration and communication between all key stakeholders [56, 60]. While it may be reasonable to expect all stakeholders to agree to principles of return-to-work; ensuring efficient communication between parties has proven to be rather more challenging. This, in part, may be due to competing motivations among the various stakeholders (Table 16.1) and the reality that stakeholders are operating within their own economic, professional, social and legislative contexts [56].

16.6
Conclusions

While pain is extremely prevalent in workers and an ongoing reason for work disability, some successful management strategies exist. These include the implementation of continuum of care models aimed at providing timely, effective health care in the most appropriate setting. Continuum of care models for workers in pain would include early assessment, identification of recovery barriers, and targeted interventions (often multidisciplinary in nature and aimed at altering beliefs and attitudes about pain-related disability) for those workers most at risk of prolonged work loss and the development of chronic pain. Research has demonstrated dramatic effects of model implementation within controlled insurance environments, however it is unknown what effect they would have within the broader health system. This high-level, coordinated management approach is very different from the unidisciplinary primary care currently provided in most jurisdictions and would require major funding and policy changes. Other potential solutions include alterations to current care provision structures to facilitate more effective disability management through closer communication and collaboration between all stakeholders (workers in pain, employers, insurance agents and healthcare providers).

Table 16.1 Possible motivations and interests of stakeholders (adapted from [59]).

Stakeholder	Motivations and interests
Injured worker	Alleviate pain and distress Quick access to health care Job and financial security Safe return to work
Employer	Maintain productivity and profitability Retain experienced employees Reduce disability costs Ensure compliance with local legislative requirements
Union representative	Protect the rights of injured employee Protect the job security of injured employee Protect the rights of coworkers and Union Members
Health care providers	Protect their patient's health and well-being Fair compensation for service Maintain health care utilization Opportunities to augment existing services
Insurers	Ensure return-to-work for the most efficient costs Balance needs of the injured worker against the costs to the system Maintain cost-effective premiums Increase customer base (private insurers)

References

1 Woolf, A.D. and Pfleger, B. (2003) Burden of major musculoskeletal conditions. *Bulletin of the World Health Organization*, **81**, 646–56.

2 Walker, B.F. (2000) The prevalence of low back pain: a systematic review of the literature from 1966 to 1998. *Journal of Spinal Disorders*, **13**, 205–17.

3 Loney, P.L. and Stratford, P.W. (1999) The prevalence of low back pain in adults: a methodological review of the literature. *Physical Therapy*, **79**, 384–96.

4 Gross, D.P., Ferrari, R., Russell, A.S. et al. (2006) A population-based survey of back pain beliefs in Canada. *Spine*, **31**, 2142–5.

5 Department of Health and Human Services, Centers for Disease Control and Prevention (National Institute for Occupational Safety and Health), Publication Number 2002-119: "Worker

Health Chartbook 2000 – Nonfatal Injury", April 2002.

6 Mustard, C., Cole, D., Shannon, H. *et al.* (2003) Declining trends in work-related morbidity and disability, 1993–1998: a comparison of survey estimates and compensation insurance claims. *American Journal of Public Health*, **93**, 1283–6.

7 Hashemi, L., Webster, B.S. and Clancy, E.A. (1998) Trends in disability duration and cost of workers' compensation low back pain claims (1988–1996). *Journal of Occupational and Environmental Medicine*, **40**, 1110–19.

8 Murphy, P.L. and Volinn, E. (1999) Is occupational low back pain on the rise? *Spine*, **24**, 691–7.

9 Alberta, H.R.E. (2005) Occupational Injuries and Diseases in Alberta: 2005 Summary.

10 Vingard, E., Mortimer, M., Wiktorin, C. et al. (2002) Seeking care for low back pain in the general population: a two-year follow-up study: results from the MUSIC-Norrtalje Study. *Spine*, **27**, 2159–65.

11 Walker, B.F., Muller, R. and Grant, W.D. (2004) Low back pain in Australian adults. Health provider utilization and care seeking. *Journal of Manipulative and Physiological Therapeutics*, **27**, 327–35.

12 Szpalski, M., Nordin, M., Skovron, M.L. et al. (1995) Health care utilization for low back pain in Belgium. Influence of sociocultural factors and health beliefs. *Spine*, **20**, 431–42.

13 Carey, T.S., Evans, A., Hadler, N. et al. (1995) Care-seeking among individuals with chronic low back pain. *Spine*, **20**, 312–17.

14 Waxman, R., Tennant, A. and Helliwell, P. (1998) Community survey of factors associated with consultation for low back pain. *British Medical Journal*, **317**, 1564–7.

15 Gross, D.P. and Battie, M.C. (2005) Predicting timely recovery and recurrence following multidisciplinary rehabilitation in patients with compensated low back pain. *Spine*, **30**, 235–40.

16 Jacobs, P., Schopflocher, D., Klarenbach, S. et al. (2004) A health production function for persons with back problems: results from the Canadian Community Health Survey of 2000. *Spine*, **29**, 2304–8.

17 Cote, P., Hogg-Johnson, S., Cassidy, J.D. et al. (2005) Initial patterns of clinical care and recovery from whiplash injuries: a population-based cohort study. *Archives of Internal Medicine*, **165**, 2257–63.

18 van der Weide, W.E., Verbeek, J.H., Salle, H.J. et al. (1999) Prognostic factors for chronic disability from acute low-back pain in occupational health care. *Scandinavian Journal of Work, Environment and Health*, **25**, 50–6.

19 Frank, J.W., Kerr, M.S., Brooker, A.S. et al. (1996) Disability resulting from occupational low back pain. Part I: What do we know about primary prevention? A review of the scientific evidence on prevention before disability begins. *Spine*, **21**, 2908–17.

20 Norlund, A.I. and Waddell, G. (2000) Cost of back pain in some OECD countries, in *Neck and Back Pain: the Scientific Evidences of Causes, Diagnosis, and Treatment* (eds A.L. Nachemson and E. Jonsson), Lippincott, Williams and Wilkins, Philadelphia, pp. 421–5.

21 Waddell, G. (1998) *The Back Pain Revolution*, Churchill Livingstone, Edinburgh.

22 Linton, S.J., Gross, D., Schultz, I.Z. et al. (2005) Prognosis and the identification of workers risking disability: research issues and directions for future research. *Journal of Occupational Rehabilitation*, **15**, 459–74.

23 Crook, J., Milner, R., Schultz, I.Z. and Stringer, B. (2002) Determinants of occupational disability following a low back injury: a critical review of the literature. *Journal of Occupational Rehabilitation*, **12**, 277–95.

24 Turner, J.A., Franklin, G., Fulton-Kehoe, D. et al. (2006) Worker recovery expectations and fear-avoidance predict work disability in a population-based workers' compensation back pain sample. *Spine*, **31**, 682–9.

25 Grotle, M., Vollestad, N.K., and Veierod, M.B. et al. (2004) Fear-avoidance beliefs and distress in relation to disability in acute and chronic low back pain. *Pain*, **112**, 343–52.

26 Dionne, C.E. (2005) Psychological distress confirmed as predictor of long-term back-related functional limitations in primary care settings. *Journal of Clinical Epidemiology*, **58**, 714–18.

27 Hoogendoorn, W.E., van Poppel, M.N., Bongers, P.M. et al. (2000) Systematic review of psychosocial factors at work and private life as risk factors for back pain. *Spine*, **25**, 2114–25.

28 Linton, S.J. (2001) Occupational psychological factors increase the risk for back pain: a systematic review. *Journal of Occupational Rehabilitation*, **11**, 53–66.

29 Hartvigsen, J., Lings, S., Leboeuf-Yde, C. and Bakketeig, L. (2004) Psychosocial factors at work in relation to low back pain and consequences of low back pain; a systematic, critical review of prospective cohort studies. *Occupational and Environmental Medicine*, **61**, e2.

30 Kaila-Kangas, L., Kivimäki, M., Riihimäki, H. Luukkonen, R., Kirjonen, J., Leino-Arjas, P. (2004) Psychosocial factors at work as predictors of hospitalization for back disorders: a 28-year follow-up of industrial employees. *Spine*, **29**, 1823–30.

31 Battie, M.C. and Bigos, S.J. (1991) Industrial back pain complaints. A broader perspective. *The Orthopedic Clinics of North America*, **22**, 273–82.

32 Clemmer, D.I. and Mohr, D.L. (1991) Low-back injuries in a heavy industry. II. Labor market forces. *Spine*, **16**, 831–4.

33 Hadler, N.M. (1999) *Occupational Musculoskeletal Disorders*, Lippincott Williams and Wilkins, Philadelphia.

34 Hall, H., McIntosh, G., Wilson, L. *et al.* (1998) Spontaneous onset of back pain. *The Clinical Journal of Pain*, **14**, 129–33.

35 Carragee, E., Alamin, T., Cheng, I. *et al.* (2006) Does minor trauma cause serious low back illness? *Spine*, **31**, 2942–9.

36 Hadler, N.M., Carey, T.S. and Garrett, J. (1995) The influence of indemnification by workers' compensation insurance on recovery from acute backache. North Carolina Back Pain Project. *Spine*, **20**, 2710–15.

37 Kennedy, C.A., Manno, M., Hogg-Johnson, S. *et al.* (2006) Prognosis in soft tissue disorders of the shoulder: predicting both change in disability and level of disability after treatment. *Physical Therapy*, **86**, 1013–32.

38 Rainville, J., Sobel, J.B., Hartigan, C. *et al.* (1997) The effect of compensation involvement on the reporting of pain and disability by patients referred for rehabilitation of chronic low back pain. *Spine*, **22**, 2016–24.

39 Atlas, S.J., Chang, Y., Kammann, E. *et al.* (2000) Long-term disability and return to work among patients who have a herniated lumbar disc: the effect of disability compensation. *The Journal of Bone and Joint Surgery. American Volume*, **82**, 4–15.

40 Linton, S.J. and Boersma, K. (2003) Early identification of patients at risk of developing a persistent back problem: the predictive validity of the Orebro Musculoskeletal Pain Questionnaire. *The Clinical Journal of Pain*, **19**, 80–6.

41 Gatchel, R.J., Polatin, P.B., Noe, C. *et al.* (2003) Treatment- and cost-effectiveness of early intervention for acute low-back pain patients: a one-year prospective study. *Journal of Occupational Rehabilitation*, **13**, 1–9.

42 Sinclair, S.J., Hogg-Johnson, S.H., Mondloch, M.V. *et al.* (1997) The effectiveness of an early active intervention program for workers with soft-tissue injuries. The early claimant cohort study. *Spine*, **22**, 2919–31.

43 Position of the American Dietetic Association (2000) Nutrition, aging, and the continuum of care. *Journal of the American Dietetic Association*, **100**, 580–95.

44 Patch, C. and Milosavljevic, M. (1999) The application of the continuum of care model in the re-configuration of nutrition and dietetics services. *Journal of Quality in Clinical Practice*, **19**, 183–7.

45 Stephens, B. and Gross, D.P. (2007) The influence of a continuum of care model on the rehabilitation of claimants with soft tissue injury. *Spine*, **32**, 2898–904.

46 Suissa, S., Giroux, M., Gervais, M. *et al.* (2006) Assessing a whiplash management model: a population-based non-randomized intervention study. *The Journal of Rheumatology*, **33**, 581–7.

47 Vlaeyen, J.W., Teeken-Gruben, N.H., Goossens, M.E. *et al.* (1996) Cognitive-educational treatment of fibromyalgia: a randomized clinical trial. *The Journal of Rheumatology*, **23**, 1237–45.

48 Hoffman, B.M., Papas, R.K., Chatkoff, D.K. and Kerns, R.D. (2007) Meta-analysis of psychological interventions for chronic low back pain. *Health Psychology*, **26**, 1–9.

49 Schonstein, E., Kenny, D., Keating, J. *et al.* (2003) Physical conditioning programs for workers with back and neck pain: a cochrane systematic review. *Spine*, **28**, E391–5.

50 Gatchel, R.J., Peng, Y.B., Peters, M.L. *et al.* (2007) The biopsychosocial approach to chronic pain: scientific advances and future directions. *Psychological Bulletin*, **133**, 581–624.

51 Sullivan, M.J., Ward, L.C., Tripp, D. *et al.* (2005) Secondary prevention of work disability: community-based psychosocial intervention for musculoskeletal disorders. *Journal of Occupational Rehabilitation*, **15**, 377–92.

52 Sullivan, M., Adams, H., Rhodenizer, T. and Stanish, W. (2006) A psychosocial risk factor-targeted intervention for the prevention of chronic pain and disability following whiplash injury. *Physical Therapy*, **86**, 8–18.

53 Shrey, D.E. and Lacerte, M. (1995) *Principles and Practice of Disability Management in Industry*, GR Press, Winterpark, FL.

54 Watson, W. (2003) *Staying@Work: Building on Disability Management*. www.watsonwyatt.com.

55 National Institute for Disability Management and Research (NIDMAR) (2005) *Disability Management Success: A Global Corporate Perspective*, The National Institute for Disability Management and Research.

56 Franche, R.L., Cullen, K., Clarke, J. *et al.* (2005) Workplace-based return-to-work interventions: a systematic review of the quantitative literature. *Journal of Occupational Rehabilitation*, **15**, 607–31.

57 Krause, N., Dasinger, L.K. and Neuhauser, F. (1998) Modified work and return to work: a review of the literature. *Journal of Occupational Rehabilitation*, **8**, 113–39.

58 Loisel, P., Lemaire, J., Poitras, S. *et al.* (2002) Cost-benefit and cost-effectiveness analysis of a disability prevention model for back pain management: a six year follow up study. *Occupational and Environmental Medicine*, **59**, 807–15.

59 Franche, R.L., Baril, R., Shaw, W. *et al.* (2005) Workplace-based return-to-work interventions: optimizing the role of stakeholders in implementation and research. *Journal of Occupational Rehabilitation*, **15**, 525–42.

60 Mortelmans, K., Donceel, P. and Lahaye, D. (2006) Disability management through positive intervention in stakeholders' information asymmetry. A pilot study. *Occupational Medicine*, **56**, 129–36.

61 Pransky, G., Shaw, W., Franche, R.L. and Clarke, A. (2004) Disability prevention and communication among workers, physicians, employers, and insurers – current models and opportunities for improvement. *Disability and Rehabilitation*, **26**, 625–34.

17
Chronic Pain and Addictions
Katherine Diskin and Nady el-Guebaly

- The prevalence of patients with remote or current addiction within the chronic pain population is likely to be in the range of 15 to 30%.
- This is largely due to lifestyle risks among those with addictions rather than addictions initiated by medical prescription of opioids.
- There are strong ethical arguments for using opioids to treat chronic pain where it can be done safely.
- There are well developed strategies that can be employed to safely manage opioid treatment for patients with comorbid addictions.
- Education of healthcare providers on issues related to assessment and management of opioids in patients with chronic pain is currently lacking.

The use of opioids to treat chronic pain complicates any discussion of addiction in the chronic pain population, as opioids are both major analgesics and drugs of abuse [1]. While this discussion will deal with some clinical and regulatory issues related to opioids, Gourlay [2] has pointed out that the issue may become more complex in the future with the wider adoption of cannabinoids for the treatment of chronic pain, particularly if it is concluded that inhalation is the most efficacious means of administration.

When considering addiction in chronic pain we are presented with a definitional dilemma, since the DSM IV (TR) criteria for substance dependence include both withdrawal (physical dependence) and tolerance, symptoms experienced by many patients on long term opioid treatment. The American Pain Society, American Academy of Pain Medicine, and the American Society of Addiction Medicine produced a consensus definition to be used in the context of pain treatment – "addiction is a primary, chronic, neurobiologic disease, with genetic, psychosocial, and environmental factors influencing its development and manifestations. It is characterized by behaviors that include one or more of the following: impaired control over drug use, compulsive use, continued use despite harm, and craving" [3]. However, even this clarification poses difficulties, since many seemingly aberrant drug misusing or drug seeking behaviors ,such as unilaterally increasing dosages and

Chronic Pain: A Health Policy Perspective
Edited by S. Rashiq, D. Schopflocher, P. Taenzer, and E. Jonsson
Copyright © 2008 WILEY-VCH Verlag GmbH & Co. KGaA, Weinheim
ISBN: 978-3-527-32382-1

obtaining prescriptions from multiple prescribers, may be due to insufficient analgesia, patients' anxiety about their treatment, fear of withdrawal symptoms, or other psychological or contextual issues. Moreover, continued prescription of opioids may result in hyperalgesia, which may also stimulate drug seeking behaviors. Ultimately a diagnosis of medication misuse or opioid addiction should be based on clinical judgement, informed by an understanding of the physiological and psychological effects of long-term opioid treatment.

Concerns regarding the potential for iatrogenic opioid addiction affect regulators, prescribers and patients. These concerns may result in the choice of less effective drugs or less effective doses of opioids to treat chronic pain. Thirty five percent of Canadian family physicians reported that they would never prescribe opioids for moderate-to severe chronic pain, and 37% identified addiction as a major barrier to prescribing opioids [4], while a survey of Canadian chronic pain patients found 70% were worried about potential addiction [5]. These concerns are even more significant when chronic pain patients have a current or previous substance abuse history. Prescribers' concerns regarding the provision of opioids to these individuals mean that they are even more likely than the general chronic pain population to experience inadequate levels of analgesia [6, 7]. Regulatory requirements regarding prescriptions and scrutiny of prescribing practices may contribute to underprescribing [7].

The misuse of opioids by individuals who are not suffering from chronic pain is a significant concern. Katz *et al.* [8] noted that prescription opioid abuse as a percentage of all drug abuse cases doubled in the US from 1997 to 2002 and referred to Substance and Mental Health Services Administration (SAMHSA) figures indicating that in the United States prescription opioid abuse surpasses that of most street drugs, including heroin. Concerns over diversion of medication at any point along the supply chain, whether through mistaken or fraudulent prescription practices, theft from warehouses and pharmacies, and purchase over the Internet, affect the accessibility of opioid medications for chronic pain patients. The WHO 2000 guidelines on national opioid control are derived from the principle of balance between the prevention of drug abuse and ensuring the availability of opioid analgesics for medical purposes, with the caveat that efforts to prevent abuse and diversion must not interfere with medical availability. It is often the case that the emphasis has been on prevention of abuse and diversion rather than on availability and appropriate use of opioid analgesics [9].

There are many significant gaps in the literature on chronic pain and addiction, the most significant of which is the wide variability in estimated prevalence of opioid addiction in chronic pain patients. This variability is due to the varying definitions used between studies, with reported rates varying from 5% to 50% [3]. Pain and addiction specialists are hampered by this lack of data, with pain specialists tending to understate the problem of addiction and addiction specialists avoiding the use of opioids in treating pain, except in rare circumstances [10].

It has been argued that there is a very low risk of iatrogenic addiction in patients who do not have a personal or family history of substance abuse, are not involved in a substance abusing subculture and who do not have premorbid psychopathology [6, 11]. Indeed, prior mental health disorders may prove to have as much or more relevance than substance abuse disorders in the development of opioid addiction in chronic pain patients. We do know that chronic pain patients commonly present with comorbid psychiatric disorders, including substance abuse. A recent prospective study of 15 160 veterans who were chronic users of opioids other than methadone found that both prior substance abuse and prior mental health disorders were predictors of opioid abuse/dependence in chronic pain patients, with an odds ratio of 2.34 for substance abuse and 1.46 for mental health disorders. Given the higher prevalence of mental health disorders compared to substance abuse in the sample (45.3% vs. 7.6%), the risk for opioid abuse/dependence was 14% for prior mental health disorder and 4% for prior substance abuse [12]. This finding further supports the need to consider other psychiatric comorbidities as well as comorbid substance abuse as potential risk factors for opioid problems in chronic pain patients. A number of screening tools have been created to assess for the potential risk of problems with opioid medication in chronic pain patients. These instruments generally include assessment of both prior and current substance abuse and prior and current mental health issues [3].

Gourlay *et al.* [13] have developed a "universal precautions" risk management approach involving a 10-step process: appropriate diagnosis, psychological assessment including risk of addictive disorders, informed consent, treatment agreement, pre- and post-intervention assessment of pain and functioning, appropriate trial of opioid therapy +/−adjunctive medication, reassessment of pain score and level of functioning, regular assessment of analgesia, activity, adverse effects and aberrant behavior, periodic review of pain and comorbid diagnoses, documentation. Depending on identified risk factors patients can be triaged into three groups based on this assessment : primary care patients; primary care patients who require specialist support; and patients who require specialty pain management. This approach allows for patients who are at risk for addiction to receive adequate pain relief under close supervision.

Suggestions for improving care for chronic pain patients could include removing barriers to appropriate treatment through regulations that address the addictive potential of opioids without impeding access to medications, improving clinical care by educating physicians and patients regarding assessing risk for addiction and using appropriate precautions, and continued research on the effects of long term opioid treatment.

The World Health Organization, the International Narcotics Control Board and national governments report that opioids are not sufficiently available for medical purposes, due to the low priority of pain care, exaggerated fears of addiction, overly restrictive drug control policies and difficulties in procurement and distribution [14]. To improve availability proposed regulations and existing policies could be evaluated using the model proposed by the Wisconsin School of Medicine and Public Health [15] to identify provisions that enhance and impede pain management. Policies that impede pain management include the implication that opioids are to be used as a last resort, and that the use of opioids is outside of legitimate professional practice and requires additional prescription requirements, while policies that enhance pain management include the recognition of pain management as part of general medical practice, the encouragement of medical use of opioids for pain management, and differentiation between dependence, tolerance and addiction.

Changing regulations to increase access to medications suitable for the treatment of chronic pain could involve increasing the availability of methadone for chronic pain – with a more frequent dosing regimen than when used for addiction treatment. It could also involve working toward making buprenorphine, which is currently used off label for pain treatment in the US [3], more available in Canada. Diversion of prescription opioids could be addressed by continuing the development of abuse deterrent formulations and attempts to develop a formulation that encapsulates the opioid within non-soluble particles, preventing their use for injection [8]. Internationally, addressing internet purchase of medications and working to enhance the integrity of the supply chain to prevent drug thefts would address diversion without adversely affecting patient care [8].

Clinically, the universal precautions approach [13] would serve to reassure regulators, physicians and patients that appropriate measures are being taken to deal with the potential for addiction. This would also entail greater collaboration between pain specialists and addiction specialists. It is likely that this would result in more appropriate analgesia and reduce the abuse of over-the-counter medications by chronic pain patients.

To facilitate adoption of the balanced perspective it would be helpful to educate regulators, physicians and patients regarding the appropriate use of opioids to address moderate to severe chronic pain. For regulators this would involve clarification of the role of and need for appropriate access to medical opioids. For physicians this would involve education to deal with the "dated, duped, disabled" physicians who are more likely to be involved in prescription drug diversion and abuse [10] and training in effective risk management for chronic pain patients. For patients, education could include information regarding the risks/benefits of opioids, differentiation between dependence, tolerance and addiction, and education for family members regarding the use of opioids in pain management.

Finally, it is clear that much more research is required in order to develop a clearer picture of addiction in chronic pain patients. Use of a consistent definition of addiction in chronic pain to develop a trustworthy estimate of prevalence would be extremely helpful in increasing our knowledge of this complex issue.

References

1 Portenoy, R., Payne, R. and Passik, S. (2005) Acute and chronic pain, in *Substance Abuse*, 4th edn (eds J.H. Lowinson, P. Ruiz, R. Millman and J.G. Langrod), Lippincott, Williams & Wilkins, pp. 863–904.

2 Gourlay, D. (2005) Addiction and pain medicine. *Pain Research and Management*, **10**, 38A–43A.

3 Ballantyne, J.C. and LaForge, K.S. (2007) Opioid dependence and addiction during opioid treatment of chronic pain. *Pain*, **127**, 235–55.

4 Kahan, M., Srivastava, A., Wilson, L., Gourlay, D. and Midner, D. (2006) Misuse of and dependence on opioids. *Canadian Family Physician*, **52**, 1081–7.

5 Moulin, D.E., Clark, A., Speechley, M. and Morley-Forster, P. (2002) Chronic pain in Canada – Prevalence, treatment, impact and the role of opioid analgesia. *Pain Research and Management*, **7**, 179–84.

6 Kirsh, K.L., Whitcomb, L., Donaghy, K. and Passik, S. (2002) Abuse and addiction issues in medically ill patients with pain: attempts at clarification of terms and empirical study. *The Clinical Journal of Pain*, **18**, S52–60.

7 Aronoff, G. (2000) Opioids in chronic pain management: is there a significant risk of addiction?. *Current Review of Pain*, **4**, 112–21.

8 Katz, N.P., Adams, E.H., Benneyan, J.C., Birnbaum, H.G., Budman, S.H., Buzzeo, R.W., Carr, D.B., Cicero, T.J. *et al.* (2007) Foundations of opioid risk management. *Clinical Journal of Pain*, **23** (2), 103–18.

9 Joranson, D.E. and Ryan, K.M. (2007) Ensuring opioid availability: methods and resources. *Journal of Pain and Symptom Management*, **33**, 527–32.

10 Ling, W., Wesson, D.R. and Smith, D.E. (2005) Prescription opiate abuse, in *Substance Abuse*, 4th edn (eds J.H. Lowinson, P. Ruiz, R. Millman and J.G. Langrod), Lippincott, Williams & Wilkins, pp. 459–68.

11 Passik, S.D. and Kirsh, K.L. (2004) Opiod therapy in patients with a history of substance abuse. *CNS Drugs*, **18** (1), 13–25.

12 Edlund, M., Steffick, D., Hudson, T., Harris, K. and Sullivan, M. (2007) Risk factors for clinically recognized opioid abuse and dependence among veterans using opioids for chronic non-cancer pain. *Pain*, **129**, 355–62.

13 Gourlay, D., Heit, H. and Almahrezi, A. (2005) Universal precautions in pain medicine: a rational approach to the treatment of chronic pain. *Pain Medicine*, **6**, 107–12.

14 World Health Organization (2007) Achieving balance in national opioids control policy guidelines for assessment, http://www.who.int/entity/cancer/publications2/en/ (accessed July 10, 2007).

15 Paul, P. (2006) *Achieving Balance in Federal and State Pain Policy. A Guide to Evaluation*, 3rd edn, University of Wisconsin School of Medicine and Public Health. Carbone Comprehensive Cancer Center, www.painpolicy.wisc.edu (accessed 10 July 2007).

18
Chronic Pain in Children

Bruce Dick

- Chronic pain is a prevalent problem in children and adolescents.

- Adequate management of pediatric chronic pain is an essential component of standard healthcare practice.

- Despite a considerable accumulation of research and clinical knowledge about managing chronic pain in children, it is still often poorly managed.

- The short- and long-term negative consequences of poorly managed chronic pain can have very significant and pervasive consequences on a developing child.

- Biopsychosocial treatment approaches have been validated as standard clinical practice for managing complex chronic pain in children and adolescents.

- Expert opinion and research findings on healthcare policy for managing chronic pain in children and adolescents suggest that education of healthcare providers and regulation of clinical guidelines, public media campaigns, and the development of multidisciplinary pain management programs are effective and warranted.

18.1
Scope and Historical Context

Historically, pain in children has been misunderstood, sometimes ignored, and even denied [1]. Even relatively recently, the critical importance of recognizing the prevalence and impact of pain in children as well as the multifactorial nature of pain in children still had to be specifically targeted in the research literature [2]. There is considerable evidence that significant negative long-term physical and psychological effects on children result from inadequately managed pain [3, 4]. Possibly the most stark example of this comes from research by Anand and colleagues, who found that infants undergoing surgery with insufficient pain control had markedly higher stress responses that were associated with much poorer

Chronic Pain: A Health Policy Perspective
Edited by S. Rashiq, D. Schopflocher, P. Taenzer, and E. Jonsson
Copyright © 2008 WILEY-VCH Verlag GmbH & Co. KGaA, Weinheim
ISBN: 978-3-527-32382-1

clinical outcomes [5]. Fortunately, our knowledge of chronic pain in children has grown enormously over the past 20 years as the result of a concerted effort by researchers and clinicians across the world. Adequate management of pain has now been more widely recognized as an essential component of standard healthcare practice.

18.2
Overview/Review of the Literature

As in many fields of study, as our understanding of chronic pain in children has expanded, so has our realization that complex interactions of many factors underlie the effects of pain. Children present with unique features that make the assessment and treatment of their pain especially challenging. For example, due to the subjective nature of pain, it can be difficult even for adults to describe their pain. Factors such as a child's developmental level and ability to communicate information about pain exacerbates the challenge of adequately assessing pain [6–8]. Sex differences have also been found to exist in children's coping style and reaction to pain [9].

Recurrent pain problems are a common complaint in children [10] and tend to increase in prevalence with age [11]. Estimates of the occurrence of some common chronic pain problems have been found to be as high as 10–19% of children reporting recurrent abdominal pain [12] and as many as 28% of adolescents reporting chronic headaches [13]. Chronic pain in children and adolescents tends to be more common in females than in males [14].

Chronic pain has the potential to significantly negatively impact many areas of a child's life. These effects can be both physical and psychological and can have marked short- and long-term consequences [15]. Chronic pain in children has been found to be associated with increased disability, poorer mental health, distress, reduced school attendance, and decreased quality of life [16–18]. There is also increasing evidence of the importance of examining the interaction between chronic pain and sleep when creating a treatment program. There is a high level of comorbidity between chronic pain and sleep disorders in children [19, 20]. It has been suggested that treating either of these problems will likely decrease the severity of the symptoms of the other disorder [21].

Primary goals that are often targeted when treating chronic pain in children include decreasing distress, pain chronicity, and pain-related disability. A recent study incorporated a Delphi poll to obtain a consensus statement on factors associated with chronic pain in children from a group of international experts. That research found that factors including having a higher tendency to be focused on somatic cues, being more likely to ruminate about pain, reporting magnified pain, feeling more helpless due to pain, and excessive utilization of healthcare services contribute considerably to the chronicity of children's pain [22]. In addition, a child's fear of activity that could increase pain, self-concept of being disabled by pain, being more likely to ruminate about pain, reporting magnified pain, and feeling more helpless due to pain were also related to long-term pain-related disability [22].

Often, there is no clear treatment protocol for managing complex chronic pain in children. Using a biopsychosocial approach, three broad areas of treatment are addressed. First, it is critical that expert medical assessment be completed in order to ensure that serious underlying organic problems that could be causing chronic pain are identified or ruled out. Typical biological/medical treatment protocols incorporate standard medical therapies including the use of medications that target known or suspected biological mechanisms. A critical issue that must be taken into account is that the biological mechanisms associated with many chronic pain problems are difficult, or, given current medical knowledge, impossible to detect. For example, the majority of cases of chronic abdominal pain are "functional" in nature as no objective evidence of organic pathology can be found to explain them [16]. This does not mean that pain should be conceptualized as malingering, activity avoidance, or psychopathology. In fact, there has been a trend that as new medical diagnostic tests have been developed and improved, the organic causes of a number of functional chronic pain conditions are being better understood [23–25]. Non-pharmacological biological approaches may include physiotherapy and graded exercise. Ideally, activity increases are based upon an errorless learning theory approach where low initial exercise targets are selected and slight incremental increases in activity occur over time.

Secondly, as children with higher levels of distress, anxiety, depression, and negative life events are more likely to be at increased risk of long-term pain and disability [26, 27], psychological treatments are also a key component of the biopsychosocial model. These treatments include modalities such as biofeedback, relaxation training, and cognitive therapies and are often incorporated in order to target the effects of chronic pain on an individual [28]. These targets may include a child's mood, fear of pain, fear of activity, acceptance of pain, and any thought process that might contribute to the disabling effects of pain on the child. Cognitive-behavioral treatment programs that focus on pain education, learning pain management strategies that help to reduce problematic behavioral patterns such as activity cycling and activity avoidance, and reducing pain-related distress, fear, anxiety, and depression have been found to be effective [29].

Finally, best practice standards for chronic pain across the lifespan emphasize the importance of conceptualizing this problem within a social framework [30]. For children, the social context of their pain can be especially complex due to the fact that many aspects of their environment can be completely out of their control. The importance of familial and other factors related to a child's environment has been increasingly recognized in pediatric chronic pain research [31, 32].

18.3
What Public (Health) Policies Would Enable Better Care/Outcomes for Patients, Families and Communities?

Schecter and colleagues [1] shared expert perspective on three primary strategies regarding how to improve pediatric pain research and practice. First, they proposed that education during training and continuing education programs for healthcare professionals be active and interactive, citing research suggesting that passive didactic sessions on pain management are often not effective at changing clinical practice [33, 34]. They also suggest that public advertising campaigns using media, including billboards, the lay press, and electronic and visual media, have the potential to inform the public regarding the need and benefits of adequate pain management and encourage public support of these initiatives. Second, as chronic pain is a multifactorial problem, it is unlikely that a professional from a single discipline could adequately address and manage the many challenges that arise for children with chronic pain. Multidisciplinary program management of chronic pain is another key element of adequate health care for this population. Research has shown that these services have the potential to provide valuable clinical services, research initiatives, training for clinical students, and education for primary care providers [35]. Each of these functions has the potential to foster improved treatment and understanding of chronic pain. While multidisciplinary chronic pain services are costly, if they are organized and managed in a way that allows them to provide these key elements, they have enormous potential for benefiting children and thereby, society. Finally, Schecter and colleagues advocate that regulatory bodies such as healthcare institutions, professional organizations, and governments should be active in monitoring pain management policies through practice regulations. The value of these strategies is obvious and researchers and clinicians would do well to diligently adopt and promote their use.

18.4
Conclusions

Chronic pain in children is a pervasive and serious health problem [36]. While a great deal has been learned in recent years about how to most effectively assess and treat it, so much is still unknown. It is clear that the cost to the individual child, the child's family, and to society can be great if chronic pain is not adequately managed. Pharmaceutical research for medications in pediatric populations is seriously under funded. Even modest governmental and industry funding has the potential to address serious gaps in our current knowledge. Only through the continued efforts of individuals across governing bodies, institutions, disciplines, and research areas will we be able to expand what we already know regarding the biological, psychological, and social and environmental factors that are related to the development and maintenance of chronic pain in children. In addition, there is also much that we do not understand regarding the short- and long-term physical and psychological consequences of chronic pain in children. Current trends suggest that we are closer than ever to disentangling the complicated puzzle of chronic pain but much remains to be learned. It is critical that strategic funding be implemented that is aimed at stimulating clinical and pharmaceutical research to address current deficiencies.

References

1 Schechter, N.L., Berde, C.B. and Yaster, M. (2003) Pain in infants, children, and adolescents; an overview, in *Pain in Infants, Children, and Adolescents* (eds N.L. Schechter, C.B. Berde and M. Yaster), Lippincott Williams & Wilkins, Philadelphia, pp. 3–18.

2 McGrath, P.J. (1996) There is more to pain measurement in children than "ouch". *Canadian Psychology*, **37**, 63–75.

3 Porter, F.L., Grunau, R.E. and Anand, K.J. (1999) Long-term effects of pain in infants. *Journal of Developmental and Behavioral Pediatrics*, **20**, 253–61.

4 Dooley, J. and Bagnell, A. (1995) The prognosis and treatment of headaches in children – a ten year follow-up. *The Canadian Journal of Neurological Sciences*, **22**, 47–9.

5 Anand, K.J.S. and Aynsley-Green, A. (1987) Randomized trial of fentanyl anesthesia in preterm babies undergoing surgery: effects on stress response. *Lancet*, **8524**, 62–7.

6 Zeltzer, L.K., Barr, R.G., McGrath, P.A. and Schechter, N.L. (1992) Pediatric pain: interacting behavioral and physical factors. *Pediatrics*, **90**, 816–21.

7 McGrath, P.J. and Frager, G. (1996) Psychological barriers to optimal pain management in infants and children. *The Clinical Journal of Pain*, **12**, 135–41.

8 von Baeyer, C.L. (2006) Children's self-reports of pain intensity: scale selection, limitations and interpretation. *Pain Research and Management*, **11**, 157–62.

9 Lynch, A.M., Kashikar-Zuck, S., Goldschneider, K.R. and Jones, B.A. (2007) Sex and age differences in coping styles among children with chronic pain. *Journal of Pain and Symptom Management*, **33**, 208–16.

10 Frare, M., Axia, G. and Battistella, P.A. (2002) Quality of life, coping strategies, and family routines in children with headache. *Headache*, **42**, 953–62.

11 Rothner, A.D. (2001) Headaches in children and adolescents: update 2001. *Seminars in Pediatric Neurology*, **8**, 2–6.

12 Chitkara, D.K., Rawat, D.J. and Talley, N.J. (2005) The epidemiology of childhood recurrent abdominal pain in Western countries: a systematic review. *The American Journal of Gastroenterology*, **100**, 1868–75.

13 Hershey, A.D. (2005) What is the impact, prevalence, disability, and quality of life of pediatric headache? *Current Pain and Headache Reports*, **9**, 341–4.

14 Moore, A.J. and Shevell, M. (2004) Chronic daily headaches in pediatric neurology practice. *Journal of Child Neurology*, **19**, 925–9.

15 Finley, G.A. and McGrath, P.J. (eds) (1998) Measurement of pain in infants and children, in *Progress in Pain Research and Management*, Vol. **10**, IASP Press, Seattle.

16 Di Lorenzo, C., Colletti, R.B., Lehman, H.P., Boyle, J.T., Gerson, W.T., Hyams, J.S., Squires, R.H. and Walker, L.S. (2005) Chronic abdominal pain in children. *Pediatrics*. **115**, e370–81.

17 Merlijn, V.P., Hunfeld, J.A., van der Wouden, J.C., Hazebroek-Kampschreur, A.A., Passchier, J. and Koes, B.W. (2006) Factors related to the quality of life in adolescents with chronic pain. *The Clinical Journal of Pain*, **22**, 306–15.

18 Vetter, T.R. (2007) A primer on health-related quality of life in chronic pain medicine. *Anesthesia and Analgesia*, **104**, 703–18.

19 Luc, M.E., Gupta, A., Birnberg, J.M., Reddick, D. and Kohrman, M.H. (2006) Characterization of symptoms of sleep disorders in children with headache. *Pediatric Neurology*, **34**, 7–12.

20 Isik, U., Ersu, R.H., Ay, P., Save, D., Arman, A.R., Karakoc, F. and Dagli, E. (2007) Prevalence of headache and its association with sleep disorders in children. *Pediatric Neurology*, **36**, 146–51.

21 Miller, V.A., Palermo, T.M., Powers, S.W., Scher, M.S. and Hershey, A.D. (2003) Migraine headaches and sleep disturbances in children. *Headache*, **43**, 362–8.

22 Miró, J., Huguet, A. and Nieto, R. (2007) Predictive factors of chronic pediatric pain and disability. A delphi poll. *The Journal of Pain*, **8**, 774–92 [Epub. ahead of print].

23 Boey, C.C. and Goh, K.L. (2002) Psychosocial factors and childhood recurrent abdominal pain. *Journal of Gastroenterology and Hepatology*, **17**, 1250–3.

24 Miranda, A. and Sood, M. (2006) Treatment options for chronic abdominal pain in children and adolescents. *Current Treatment Options in Gastroenterology*, **9**, 409–15.

25 Shulman, R.J., Eakin, M.N., Jarrett, M., Czyzewski, D.I. and Zeltzer, L.K. (2007) Characteristics of pain and stooling in children with recurrent abdominal pain. *Journal of Pediatric Gastroenterology and Nutrition*, **44**, 203–8.

26 Walker, L.S., Garber, J., Van Slyke, D.A. and Greene, J.W. (1995) Long-term health outcomes in patients with recurrent abdominal pain. *Journal of Pediatric Psychology*, **20**, 233–45.

27 Mulvaney, S., Lambert, E.W., Garber, J. and Walker, L.S. (2006) Trajectories of symptoms and impairment for pediatric patients with functional abdominal pain: a 5-year longitudinal study. *Journal of the American Academy of Child and Adolescent Psychiatry*, **45**, 737–44.

28 McGrath, P.J., Dick, B.D. and Unruh, A.M. (2003) Psychologic and behavioral treatment of pain in children and adolescents, in *Pain in Infants, Children, and Adolescents* (eds N.L. Schechter, C.B. Berde and M. Yaster), Lippincott Williams & Wilkins, Philadelphia, pp. 303–16.

29 Eccleston, C., Bruce, E. and Carter, B. (2006) Chronic pain in children and adolescents. *Paediatric Nursing*, **18**, 30–3.

30 Kimura, Y. and Walco, G.A. (2007) Treatment of chronic pain in pediatric rheumatic disease. *Nature Clinical Practice. Rheumatology*, **3**, 210–18.

31 Palermo, T.M. and Chambers, C.T. (2005) Parent and family factors in pediatric chronic pain and disability: an integrative approach. *Pain*, **119**, 1–4.

32 Chambers, C.T. (2003) The role of family factors in pediatric pain, in *Pediatric Pain: Biological and Social Context* (eds P.J. McGrath and G.A. Finley), IASP Press, Seattle, pp. 26–37.

33 David, T.J. and Patel, L. (1995) Adult learning theory, problem based learning and paediatrics. *Archives of Disease in Childhood*, **73**, 357–63.

34 Ozuah, P., Curtis, J. and Stein, R.E. (2001) Impact of problem based learning on residents self-directed learning. *Archives of Pediatrics and Adolescent Medicine*, **155**, 669–72.

35 Miaskowski, C., Crews, J., Ready, L.B. *et al.* (1999) Anesthesia-based pain services improve the quality of postoperative pain management. *Pain*, **80**, 23–9.

36 Perquin, C.W., Hazebroek-Kampschreur, A.A., Hunfeld, J.A., Bohnen, A.M., van Suijlekom-Smit, L.W., Passchier, J. and van der Wouden, J.C. (2000) Pain in children and adolescents: a common experience. *Pain*, **87**, 51–8.

19
Chronic Pain in Elderly Persons

Thomas Hadjistavropoulos and Gregory P. Marchildon

- The Canadian population is aging and older people have high rates of pain.
- Pain is poorly assessed and managed in older people.
- Funding for better pain assessment and management in long-term care (LTC) facilities should be provided
- Specialized LTC pain assessment and management training should be expanded within health professional curricula.
- Pain management teams within LTC facilities should be established.
- Provinces should facilitate the education of nurse practitioners with special training in pain assessment and management for seniors.
- Public education encouraging older adults to seek treatment for persistent pain problems may help decrease the under-reporting of pain problems.

Changes in North America's demographic structure have been occurring over the last century and larger proportions of people are living past age 65. In the year 1900, only about 4% of the population were persons over 65 years of age. The number changed to over 10% in 1998 [1]. Currently, 14% of the Canadian population are persons over 65 years of age [2] and it is estimated that by the year 2021, the proportion of seniors will increase to 20% [3]. The changing demographic landscape adds urgency to the need to address concerns about pain management in older persons.

Pain is highly prevalent among older adults [4] and many health conditions that affect this population are associated with pain [5, 6]. The overall prevalence estimates for pain range from 25% to 65% for community dwelling older persons and up to 80% for seniors living in long-term care (LTC) facilities [7]. Despite its high prevalence, pain is undertreated in this population. In fact, a systematic survey of members of the American Pain Society [8] identified the undertreatment of pain among seniors and the inadequate pain management among persons with cognitive impairments as being among the most pressing

Chronic Pain: A Health Policy Perspective
Edited by S. Rashiq, D. Schopflocher, P. Taenzer, and E. Jonsson
Copyright © 2008 WILEY-VCH Verlag GmbH & Co. KGaA, Weinheim
ISBN: 978-3-527-32382-1

ethical concerns for pain clinicians. This concern is especially salient for patients with dementia [9, 10]. Moreover, older adults are less likely to be referred to pain clinics.

Many different reasons have been cited for the undermanagement of pain among seniors. In addition to challenges that are common among most age groups, the assessment of the older adult is complicated further by myths that having pain is "natural" for older persons, overestimation of the risk of addiction to opioids, sensory and cognitive impairments that affect some older persons, and an increased stoicism that makes many older adults less likely to report pain [9, 11–13]. Despite recent developments in pain assessment methodologies for seniors with and without cognitive impairments [14–16], nurses identify inadequate pain education, insufficient access to available assessment methodologies and sub-optimal levels of staffing as being barriers to adequate pain management in this population [9]. Concerns have also been raised about inadequate communication between long-term care facility (LTC) nurses and family physicians of patients [17].

The manner in which public healthcare has been historically organized in Canada contributes to the undertreatment of seniors suffering chronic pain. Public health insurance was first introduced for hospital care, first in Saskatchewan in 1947 followed by British Columbia (1949) and Alberta (1950), and then in the rest of Canada by the late 1950s through the spending power of the federal government [18]. The focus of government policy was on removing the financial barriers to access for hospital care, most of which focused on acute rather than chronic care. In the 1960s public health insurance was extended to physician care, in part to encourage early diagnosis of developing acute conditions [19]. While coverage for "medically necessary" and "medically required" services–as defined under the Canada Health Act–was placed on a first-dollar coverage basis through the intervention of both the federal and provincial governments, the care of chronic conditions outside of hospitals was left to the provinces and individuals [20].

Seniors with chronic conditions causing chronic pain require varying combinations of treatment at home, assisted-living accommodations, or LTC institutions offering continual care. The average age of seniors living in LTC facilities climbed from 75 to 85 between 1977 and 1995; and by the latter date, 46% of Canadians over the age of 85 were living in LTC institutions [21]. In addition, due to the demographic bulge caused by the baby boom in Canada, the percentage of Canadians who require LTC or home care services is expected to rise substantially in the coming years. As the average age of LTC residents increases, the intensity of care also increases, due largely to mental health problems. Based on a 1993 Canadian Study of Health and Aging, 58% of seniors in LTC suffered from dementia [22]. While the availability and quality of these services depends on numerous factors, the most important is the extent to which individual provinces have funded, subsidized and regulated these services. Since there is no equivalent of the Canada Health Act for chronic home, community, and long-term care, access and quality can vary considerably across provinces. Moreover, given the fact that home care, community care, and LTC services have been delegated by most provincial health ministries to regional health authorities (RHAs), there can be significant variations in practice and standards within provinces as well as among provinces.

In an effort to improve assessment and to create uniform standards, the Resident Assessment Instrument (RAI) (also referred to as Minimum Data Set 2.0) was developed [23, 24]. However, unlike the United States where RAI was first implemented in 1991 and is now required for virtually all nursing homes, only two provinces – Ontario and Saskatchewan – have implemented this uniform and comprehensive resident assessment system [25]. Although used more extensively in home care, Ontario introduced the first RAI instrument for chronic care/ nursing homes in 1996 [26]. Saskatchewan set up a pilot project in one health district (Prince Albert) before mandating RAI 2.0 in the rest of the province in 2001 [27]. Although this tool only involves a very cursory assessment of pain, with more research and refinement, it has the potential of contributing to better screening for pain problems in this population. It could also serve as a means for evaluating interventions in LTC and home care.

Prescription drug therapy is the most relied upon method to manage pain among seniors with chronic conditions. While physician care, and with it the right to prescribe, is a core part of Canadian medicare, prescription drugs themselves are outside medicare. This means that access to prescription drugs is determined by the benefit packages in employment-based private health insurance or the benefits offered in provincial drug plans. Due to their relative lack of attachment to the workforce, almost all seniors are reliant on public drug benefit plans, with coverage varying considerably from province to province, although there is a strong east–west gradient, with seniors resident in the Atlantic provinces facing far higher financial barriers to access than other Canadians [28].

Beyond access, there is an additional challenge faced by many seniors in LTC institutions. Most of the caregivers in such institutions are not properly educated or trained to recognize chronic conditions and pain symptoms among their residents. The vast majority of registered nurses continue to work in acute care environments, and for the most part LTC institutions are staffed by unregistered personal support workers and, to a lesser extent, by licensed/registered practical nurses [29]. The high incidence of complicating conditions such as dementia can sometimes mask the conditions or divert the attention of such caregivers. As a consequence, chronic condition pain suffered by seniors is often undermanaged in LTC institutions.

As pointed out by Fox *et al.* [30] in their review of the prevalence and treatment of pain in seniors receiving long-term care, professional LTC caregivers "receive only minimal training" in identifying and treating pain while physicians "are frequently unavailable" to provide oversight and training in LTC facilities. The situation is likely worst for home care given the number of informal caregivers with minimal professional education or training. One way to address this deficiency would be for provinces to facilitate the education and training of nurse practitioners (NP) with special training in pain assessment and management for seniors in LTC or home care. While current NPs are able to prescribe a narrow range of medications, a special class of NPs with more specialized training (and expanded scope of practice) could be given the right to prescribe a broader range of pain medications. This training would also include pain medication for end-of-life care [31]. This would address the issue identified by Ferrell [32] that while the demand for pain management among the frail elderly is growing, delays in pain treatment and pain medication adjustment are also growing because of the lack of availability of physicians.

At the same time, frail elderly persons in LTC suffer a disproportionately high incidence of adverse drug events [33]. This is due to a number of factors, including delays in getting physicians to alter medication regimens in response to changing needs, the weak linkage with pharmacists, the low level of education and training for professional caregivers in LTC settings, the lack of appropriate surveillance and reporting, and the absence of written policy within LTC homes concerning pain management. Here again, LTC nurse practitioners could play a major role. They would be on-site and therefore better able to identify in a timely way when changes need to be made to individual pain medication regiments. They could be made responsible for supervising other LTC professionals and care assistants to ensure proper surveillance and reporting, as well as play a leading role in on-site pain management training. The LTC team, led by the NP, could potentially include a consulting pharmacist who could provide, with minimal delay, advice on changing medications, combining medications, and how best to prevent potentially inappropriate medications [34]. This would be a reasonable way of leveling up the skills and knowledge in LTC given the limited resources and professional skills which are available in most LTC homes in Canada [32].

A nurse practitioner could play a similar role for high-risk seniors receiving home care. Through regular home visits, the NP could perform periodic pain assessments, mitigate the possibility of adverse drug reactions, and identify when changes in the medication regimen should be made. The NP could also work with the patient's informal caregivers so that they can help identify warning signs and ensure appropriate medication use.

In addition to the deployment of additional NPs, it is recommended that there be better integration on pain assessment and management in nursing and other health professional curricula. Several studies have identified gaps in education about pain assessment and management among LTC and home care staff (see e.g. [9, 35]). This can be corrected with better education. Funding for specialized pain assessment and management workshops within LTC facilities, as well as educational programing for informal caregivers providing home care, would also be important. However, the implementation of systematic pain assessment, which is a pre-requisite for good pain management, faces a number of systemic barriers (e.g. [9]). For example, due to limited personnel in many LTC facilities, nurses often indicate that they do not have sufficient time to engage in systematic and routine pain assessment in the nursing home. Increased funding for LTC facilities could help increase staff levels. For seniors getting their care at home, some provinces do not have sufficient resources to ensure systematic visits by trained home care personnel.

LTC facilities are accredited through the Canadian Council on Health Services Accreditation. In the future, accreditation should include an assessment of the

LTC facility's in-house education and training as it concerns managing pain, including the use of standardized pain assessment tools and written policy guidelines for pain management [36].

In summary, our key recommendations are the following:

1. Provinces should facilitate the education and training of nurse practitioners with special training in pain assessment and management for seniors in LTC. Such nurse practitioners should be on site to evaluate the need for medication changes (as well as for other interventions), work in concert with LTC nurses and communicate regularly with patients' physicians.

2. Pain management teams within LTC facilities should be established. Such teams should include a nurse practitioner, a pharmacist and LTC nurses.

3. Specialized LTC pain assessment and management training should be expanded within health professional curricula.

4. Funding from federal and provincial governments to encourage research on the evaluation and refinement of the RAI/MDS as a screening tool for pain in LTC and home care. Successful results would encourage wider use of this tool in Canada.

5. Funding for pain assessment and management workshops for LTC facility staff should be provided by regional health authorities.

6. Funding for increased staffing within LTC facilities should be provided by regional health authorities. Improved staffing levels should be accompanied by the incorporation of regular pain assessment protocols.

7. To address pain problems among seniors who do not reside in LTC facilities, we recommend that specific modules for pain assessment and management among older persons be introduced in medical education curricula. Moreover, given the aging of Canada's population and the high prevalence of pain problems, specialized geriatric pain assessment and management training should be specifically integrated in the experience of all health professionals.

8. Public education encouraging older adults to seek treatment for persistent pain problems may help decrease the underreporting of pain problems that we often see in this population.

References

1 Rowe, J.W. and Kahn, R.L. (1998) *Successful Aging*, Dell Publishing, New York.
2 Statistics Canada (2006) Age and Sex, 2006 Counts for Both Sexes, for Canada, Provinces and Territories – 100% Data. http://www12.statcan.ca/english/census06/data/highlights/agesex/pages/Page.cfm?Lang=E&Geo=PR&Code=01&Table=1&Data=Count&Sex=1&StartRec=1&Sort=2&Display=Page.
3 Tuokko, H. and Hadjistavropoulos, T. (1998) *An Assessment Guide to Geriatric Neuropsychology*, Lawrence Erlbaum Associates, Mahwah, New Jersey.
4 Charlton, J.E. (2005) *Core Curriculum for Professional Education in Pain*, IASP Press, Seattle.
5 Jones, G. and Macfarlane, G. (2005) Epidemiology of pain in older persons, in *Pain in Older Persons* (eds S. Gibson and D. Weiner), IASP Press, Seattle, pp. 3–22.
6 Proctor, W.R. and Hirdes, J.P. (2001) Pain and cognitive status among nursing home residents in Canada. *Pain Research and Management*, **6**, 119–25.
7 Gibson, S.J. (2003) Pain and aging: the pain experience over the adult life span, in *Proceedings of the 10th World Congress on Pain* (eds J.O. Dostrovsky, D.B. Carr and M. Koltzenburg), IASP Press, Seattle, pp. 767–90.
8 Ferrell, B.R., Novy, D., Sullivan, M.D., Banja, J., Dubois, M.Y., Gitlin, M.C., Hamaty, D., Lebovits, A. *et al.* (2001) Ethical dilemmas in pain management. *The Journal of Pain*, **2**, 171–80.
9 Martin, R., Williams, J., Hadjistavropoulos, T., Hadjistavropoulos, H.D. and MacLean, M. (2005) A qualitative investigation of seniors' and caregivers' views on pain assessment and management. *The Canadian Journal of Nursing Research*, **37**, 142–64.
10 Sengstaken, E.A. and King, S.A. (1993) The problems of pain and its detection among geriatric nursing home residents. *Journal of the American Geriatrics Society*, **41**, 541–4.
11 Machin, P. and Williams, A.C.d.C. (1998) Stiff upper lip: coping strategies of World War II veterans with phantom limb pain. *The Clinical Journal of Pain*, **14**, 290–4.
12 Klinger, L. and Spaulding, S.J. (1998) Chronic pain in the elderly: is silence really golden? *Physical and Occupational Therapy in Geriatrics*, **15**, 1–17.
13 Morley, S., Doyle, K. and Beese, A. (2000) Talking to others about pain: suffering in silence, in *Progress in Pain Research and Management* (eds M. Devor, M.C. Rowbotham and Z. Wiesenfeld-Hallin), IASP Press, Seattle, pp. 1123–9.
14 Hadjistavropoulos, T. (2005) Assessing pain in older persons with severe limitations in ability to communicate, in *Pain in Older Persons* (eds S. Gibson and D. Weiner), IASP Press, Seattle, pp. 135–51.
15 Zwakhalen, S.M., Hamers, J.P., Abu-Saad, H.H. and Berger, M.P. (2006) Pain in elderly people with severe dementia: a systematic review of behavioural pain assessment tools. *BMC Geriatrics*, **6**, 3.
16 Hadjistavropoulos, T., Herr, K., Turk, D.C., Fine, P.G., Dworkin, R.H., Helme, R., Jackson, K., Parmelee, P.A. *et al.* (2007) An interdisciplinary expert consensus statement on assessment of pain in older persons. *The Clinical Journal of Pain*, **23**, S1–43.
17 Kaasalainen, S., Coker, E., Dolovich, L., Papaioannou, A., Hadjistavropoulos, T., Emili, A. and Ploeg, J. (2007) Pain management decision-making among long-term care physicians and nurses. *Western Journal of Nursing Research*, **29**, 561–80.
18 Taylor, M.G. (1987) *Health Insurance and Canadian Public Policy: the Seven Decisions That Created the Canadian Healthcare System*, McGill-Queen's University Press, Montreal and Kingston.
19 Naylor, C.D. (1986) *Private Practice, Public Payment: Canadian Medicine and the Politics of Health Insurance*, McGill-Queen's University Press, Montreal and Kingston, pp. 1911–66.
20 Marchildon, G.P. (2006) *Health Systems in Transition: Canada*, University of Toronto Press, Toronto.

21 Havens, B. (1995) Who uses long-term care? in *Continuing the Care: the Issues and Challenges for Long-Term Care* (eds E. Sawyer and M. Stephenson), CHA Press, Ottawa, pp. 77–96.

22 Chappell, N.L. (1995) Aging in Canada, in *Continuing Care: the Issues and Challanges for Long-Term Care* (eds E. Sawyer and M. Stephenson), CHA Press, Ottawa, pp. 227–46.

23 Hirdes, J.P., Fries, B.E., Morris, J.N., Steel, K., Mor, V., Frijters, D., LaBine, S., Schalm, C. *et al.* (1999) Integrated health information systems based on the RAI/MDS series of instruments. *Healthcare Management Forum*, **12**, 30–40.

24 Morris, J.N., Fries, B.E., Steel, K., Ikegami, N., Bernabei, R., Carpenter, G.I., Gilgen, R., Hirdes, J.P. *et al.* (1997) Comprehensive clinical assessment in community setting: applicability of the MDS-HC. *Journal of the American Geriatrics Society*, **45**, 1017–24.

25 Ikegami, N., Hirdes, J.P. and Carpenter, I. (2002) Measuring the quality of long-term care in institutional and community settings, in *Measuring Up: Improving Health System Performance in OECD Countries* (ed. OECD), OECD, Paris, pp. 277–93.

26 Hirdes, J.P. (2006) Addressing the health needs of frail elderly people: Ontario's experience with an integrated health information system. *Age Ageing*, **35**, 329–31.

27 Hirdes, J.P., Sinclair, D.G., King, J., Tuttle, P. and McKinley, J. (2003) From anecdotes to evidence: complex continuing care at the dawn of the information age in Ontario, in *Implementing the Resident Assessment Instrument: Case Studies of Policymaking for Long-Term Care* (eds B.E. Fries and C.J. Fahey), Milbank Memorial Fund, New York, http://milbank.org/reports/ interRAI/030222 interRAI.html (accessed Mai 21, 2006).

28 Anix, A.H., Guh, D. and Wang, X. (2004) A dog's breakfast: prescription drug coverage varies widely across Canada. *Medical Care*, **39**, 315–24.

29 Ostry, A., (2006) *Change and Continuity in Canada's Health Care System*, CHA Press, Ottawa.

30 Fox, P.L., Raina, P. and Jadad, A.R. (1999) Prevalence and treatment of pain in older adults in nursing homes and other long-term care institutions: a systematic review. *Canadian Medical Association Journal*, **160**, 329–33.

31 Hall, P., Schroder, C. and Weaver, L. (2002) The last 48 hours of life in long-term care: a focused chart audit. *Journal of the American Geriatrics Society*, **50**, 501–6.

32 Ferrell, B.A. (2004) The management of pain in long-term care. *The Clinical Journal of Pain*, **20**, 240–3.

33 Gurwitz, J.H., Field, T.S., Judge, J., Rochon, P., Harrold, L.R., Cadoret, C., Lee, M., White, K. *et al.* (2005) The incidence of adverse drug events in two large academic long-term care facilities. *American Journal of Medicine*, **118**, 251–8.

34 Gill, S.S., Misiaszek, B.C. and Brymer, C. (2001) Improving prescribing in the elderly: a study in the long term care setting. *The Canadian Journal of Clinical Pharmacology*, **8**, 78–83.

35 Zwakhalen, S.M., Hamers, J.P., Peijnenburg, R.H. and Berger, M.P. (2007) Nursing staff knowledge and beliefs about pain in elderly nursing home residents with dementia. *Pain Research and Management*, **12**, 177–84.

36 Allcock, N., McGarry, J. and Elkan, R. (2002) Management of pain in older people within the nursing home: a preliminary study. *Health and Social Care in the Community*, **10**, 464–71.

20
Chronic Pain in Indigenous Populations

Louise Crane

- Cultural perspectives and biases of both aboriginal patients and their non-aboriginal providers create barriers to effective care.

- Although there is little relevant Canadian data to draw from, best estimates suggest that chronic pain is even more common in the aboriginal community and the Canadian population as a whole.

- Educate health practitioners to acknowledge and respect the cultural practices and choices of indigenous clients.

- More data is needed about indigenous communities and about the determinants of health and chronic pain in order to plan effective care for these communities.

- Programs which stress interdisciplinary collaboration between western and traditional healers are needed to provide culturally sensitive management of chronic pain in the indigenous communities nationwide.

- Policies, programs and services must genuinely reflect the diversity of the indigenous communities.

- Broad community input and establishment of partnerships needed for successful health planning and resource management pertaining to indigenous policies, programs and services.

- Support education programs concerning inappropriate use of prescription drugs that do not stigmatize clients, and promote healthier use of prescription medications.

- Strive to create seamless, integrated support for healthy indigenous populations within and across all levels of government.

Chronic Pain: A Health Policy Perspective
Edited by S. Rashiq, D. Schopflocher, P. Taenzer, and E. Jonsson
Copyright © 2008 WILEY-VCH Verlag GmbH & Co. KGaA, Weinheim
ISBN: 978-3-527-32382-1

*My hand is not the color of your hand, but if I pierce it I shall
feel pain. The blood that will follow from mine will be the same
color as yours. The Great Spirit made us both.*
                                    Luther Standing Bear, Oglala Sioux

I want to walk you through my story. It is the story of an Aboriginal person going through assessment, education, and self-managed care for chronic pain. I hope I can provide insight into the challenges of working with an Aboriginal patient. I feel challenged to include as much information as possible about the Aboriginal community and to focus on those policies that will benefit our community.

I am an Aboriginal woman (as defined by the 1982 Constitution of Canada). I am middle aged, and I am overweight. I have chronic pain and I need a cane to walk. Now I am considering using a walker. I have osteoarthritis in my knees and I have an undiagnosed pain in my leg that seems unrelated. I have visited a number of doctors over the years trying to find out what is wrong with my legs.

I was full of apprehension during my first visits. All of the experiences my parents and I had with the medical profession were colored by our "colonized" status. We were wary of not being heard, of being "taken care of" in a patronizing way, and of having the diversity of our band, or clan, or indeed ourselves as individuals, ignored.

Doctors asked for a personal history, and for a family history, and I shared what I knew. My grandparents had their share of aches and pains I'm sure, but I had no idea what they were. My grandparents had been taught in residential school not to complain. They also believed what the elders taught: that pain came from an imbalance in their lives and was simply part of nature. They never complained. They winced and they slowed down but they went on with life; doing what they could.

The first doctor looked at my legs, and then said; "What pills are you on? Here is the prescription you want." No diagnosis – just a presumption that I was there for the pills. Did he think I was an addict looking for a fix? I left too quickly to find out.

The next doctor I tried gave a little more attention to trying to provide a diagnosis. He ordered imaging and analyzed my gait. He also attempted to do a pain scale. (Aboriginal people have had trouble relating to the current scales. How do they relate to our lifestyles? How would the scales used be impacted by our nuances and facial habits?). After a lengthy work up, he prescribed some pills that he said would help. He left me with a final warning; "I stick strictly to Non-Insured Health Benefit guidelines. I will not replace these pills if you decide to sell them." I asked him if he was joking but his eyes told me he wasn't.

From that moment, I decided that if I was going to deal with the chronic pain that I suffered, I would need to do my own research. I went to my computer and quickly found that there were many sites on arthritis, group supports, and research. I felt overwhelmed but knew that I was not alone.

I discovered that the prevalence of arthritis is two times as high in aboriginal populations as in non-aboriginal populations. *The Practical Guide to the Provision of Chronic Pain Services for Adults in Primary Care* suggests that the burden of chronic pain in an average primary care organization is 10–15% [1]. This would mean that up to 20–30% of the Aboriginal adult population suffered chronic pain. The power of doing my own research and having all this new knowledge gave me a rush of excitement. I felt less alone in my pain.

At this time, I was very fortunate to be hired to interview Elders as part of a bone health study in the Aboriginal community (Meadows, L. The Social and Cultural Contexts of Bone Health and Secondary Fractures Prevention of Aboriginal Women (AHFMR Health Research Fund), University of Calgary, 2007). Here are some of the insights the elders provided me for understanding chronic pain and for understanding healing in the Aboriginal community.

> Doctors should look at us as a whole: our history as well as the physical, mental, emotional and spiritual. All the grannies will get together if somebody is sick. It does not matter which clan you are from they will pray for you.

> I wish that hospitals and clinics would also include spiritual ceremonialists. When I get bad news I want to ensure that my spirit protectors are there for me. I need to know that I can do something to make myself better and that I am protected from "bad medicine" that may have been placed on me.

> I knew a man once who lost an arm. The hospital threw it away. He was in constant pain where there was no arm. In a ceremony the man dreamed of his lost arm. In our culture we must know where we are at all times. We need to dispose of body parts with ceremony. After this man's dreams his healer suggested a ceremony to call on the spirit of the arm to return. This was done. The man put the spirit of his arm back in the proper place and the pain went away. Even though he had one arm he was spiritually whole.

> I was using the treadmill and some equipment at the clinic. It did help but was my healing slowed down when I know that the spirit of the other people's pain is still on the equipment? They need to smudge or cleanse the equipment daily. I am sure we will heal better.

> My doctor asked me to describe my pain. What did it look like? Did he understand how it affected my hunting? Does a number describe how I feel when I step over a large log and then have to sit on it for 10 minutes waiting for the pain to subside? Meanwhile the game is gone and I need to walk further.

> My doctor provides me the space and respects my choices around my use of traditional practices. I find it easier to be compliant to his action plan he participates in culturally safe teamwork.

These words from my elders inspired me to participate in a series of traditional sun dances and sweats. (A sweat is a sauna-like ceremony with songs, herbs and prayers that lasts from two to three hours. A sun dance is a two- to four-day ceremony of cleansing, healing and renewal. The chants and songs provide background to the inner circle participants fasting and using a bouncing step in time to the beat. These actions provide participants with a deep connection to the spiritual healers). Both activities involved the bending of my sore legs, so my faith was sorely tested. As well, the sun dances kept me on my feet for hours at a time. But I made it through the ceremonies. The healing prayers and the collective voice from my community had helped me and I was now mobile. My first thought was "What will my doctor think? In the city I could not walk the block around my house."

Some comments from the elders raised questions about government policies that hinder our healing.

> I found that working with a mental health worker and counselor helped me manage my pain better. Our Non-Insured Health will only cover a limited number of sessions. My pain is chronic and ongoing. How does someone in Ottawa decide what works or doesn't work for me?

> I was part of a drug trial and the medications worked. When the study stopped so did my new meds. The old ones were not effective and the non-insured policy will not cover drugs not approved by Health Canada. I will be dead before that process happens

> The only drugs that were on the non-insured list were highly addictive. There are new ones on the market that are not covered but far too expensive for me. Will they ever get on "the list"?

One of the elders I visited was house bound, confined to a wheel chair with severe Rheumatoid Arthritis. She found that home visits with a mental health worker and her ceremonialist really helped to relieve her pain and provided important social support. But current policy allows transportation of a client to the health center and does not cover home visits by workers. The roads on the reserve are major barriers for those with mobility issues. This elder finds the extra travel painful, humiliating and tiring; so instead she sits isolated in her home.

During my research for this project, I discovered a report about a possible relationship between arthritis and childhood trauma [2]. This reminded me of talks with the elders who commented on residential schools.

> *My arthritis must have come from standing out in cold.*
> *All lined up wearing thin school uniforms.*

I often wonder if the pain in my knees is from the hours of scrubbing the long halls at residential school.

The beds were awful. Did my back pain start there? I have
always had a weak back.

I never visit the doctor any more. I would rather put up with
the pain. The doctor at the residential school abused many of us
children.

Soon afterward I began to explore my own childhood trauma with a counselor. I spent two years slowly healing emotionally. Unfortunately, this treatment was not covered by insurance. At the same time, I saw healers in my community for spiritual healing. I also saw a specialist for physical healing and continued my research on chronic pain management.

I needed to work on my own to integrate my healing with my work, home, and my social life even though my healing activities covered the full spectrum of the medicine wheel. (The Medicine Wheel is used by Cree and Blackfoot Nations. The wheel is a metaphor for the circle of life that consists of a person's mental, spiritual, emotional and mental self. It is surrounded by the four stages of life; Infancy, Youth, Adult and Elder). I am now happier, sleeping better, being more active, and taking less medication. I feel blessed that I was able to create my own team to manage my own pain in a holistic way mixing traditional ways with western medicine to improve the quality of my life.

As an educated person, I have had many advantages in doing this personal research and in discovering a healing direction. Consider the plight of those in my community without the resources or the cognitive abilities to consider full self-managed care of their chronic pain. Consider the person with Fetal Alcohol Spectrum Disorder [3]. Would a physician have the time to understand the range of this problem in our communities? Would the clinic have the resources to follow a patient to ensure they are using the medication and treatments appropriately?

There are many other determinants related to our communities that would impact the ability to follow through on a course of treatment. These include poverty and poor housing. The World Health Organization also list social exclusions such as racism, stigmatization, hostility and unemployment [4] as social determinants of health. And these are challenges not only for Canadian aboriginals, but also for many other indigenous groups around the world.

At a recent summer institute (Center for American Indian Health, Johns Hopkins School of Public Health, Summer Indigenous Institute, July 2007) I talked with indigenous people from around the world and we were all in agreement that adopting the following policies could improve chronic pain care for Aboriginal peoples:

1. Educate health practitioners to acknowledge and respect the cultural practices and choices of indigenous clients.

2. Create true interdisciplinary collaborations between western and traditional healers.

3. Continue to collect and distribute data about indigenous communities and about the determinants of health and chronic pain in these communities.

4. Develop programs to address the management of chronic pain in the indigenous communities (and nationwide).

5. Ensure that policies, programs and services genuinely reflect the diversity of the indigenous communities.

6. Seek broad community input and establish partnerships when developing policies.

7. Include indigenous people in all health planning and resource management pertaining to indigenous policies, programs and services.

8. Support education programs concerning inappropriate use of prescription drugs that do not stigmatize clients, and promote healthier use of prescription medications.

9. Strive to create seamless, integrated support for healthy indigenous populations within and across all levels of government.

References

1 The British Pain Society and the Royal College of General Practitioners (2004) *A Practical Guide to the Provision of Chronic Pain Services for Adults in Primary Care*, The British Pain Society and the Royal College of General Practitioners, London.

2 Malleson, P.N., Oen, K., Cabral, D.A., Petty, R.E., Rosenberg, A.M. and Cheang, M. (2004) Predictors of pain in children with juvenile rheumatoid arthritis. *Arthritis Care and Research*, **51** (2), 222–7.

3 Godel, J. (2002) Fetal alcohol syndrome. *Paediatrics and Child Health*, **7** (3), 161–74.

4 Wilkinson, R. and Marmot, M. (eds) (2003) *Social Determinants of Health. The Solid Facts*, 2nd edn, World Health Organization.

Part Four
Health Systems and Chronic Pain Management

21
Mass Media Campaigns for the Prevention of Back Pain-Related Disability

Rachelle Buchbinder, Douglas P. Gross, Erik L. Werner, and Jill A. Hayden

- Due to the major societal burden of low back pain disability, population-based interventions targeting this conditon have the potential for huge societal improvements in health and well being.

- Mass media campaigns designed to alter societal views about back pain and promote behavior change have now been performed in several countries with mixed results.

- More intensive and expensive media campaigns may be more effective than low-budget campaigns.

- There is limited empirical understanding of the characteristics of effective (or ineffective) health campaigns based upon underlying general theories of health behavior change.

- Use of a theoretical framework to identify important issues may improve the planning and evaluation of mass media interventions for low back pain.

21.1
Introduction

Mass media campaigns are increasingly being used to help prevent and manage various health conditions. They involve the use of mass media and other commercial marketing techniques to promote behavior changes that will improve the health or well-being of a target audience or society as a whole [1–3]. Other terms that have been used for mass media campaigns include social marketing or media interventions/campaigns, public health interventions/campaigns and public education campaigns. Their use is based on the premise that targeting the population as a whole has the advantage of potentially modifying the knowledge or attitudes of a large proportion of the community simultaneously, thereby providing social support for behavioral change [4].

Chronic Pain: A Health Policy Perspective
Edited by S. Rashiq, D. Schopflocher, P. Taenzer, and E. Jonsson
Copyright © 2008 WILEY-VCH Verlag GmbH & Co. KGaA, Weinheim
ISBN: 978-3-527-32382-1

The World Health Organization's Ottawa Charter for Health Promotion has identified media advocacy as an established health promotion strategy [5]. Mass media campaigns have been used with some success to change population beliefs and behavior related to a variety of health conditions, including campaigns to reduce sun exposure, promote the use of cancer screening tests, reduce cigarette consumption, and increase physical activity [5]. Widespread primary prevention public health campaigns relating to sunlight exposure have been operating in Australia for twenty years [6]. The effect of these programs has been to create shifts in knowledge, attitudes and beliefs about health conditions, along with major shifts in related behavior. Research has shown they can be effective in changing healthcare utilization, promoting the use of effective interventions and health promotion strategies while discouraging ineffective ones [5].

In the area of pain-related disability, public health interventions designed to alter societal views about pain may be a highly effective method for reducing its overall burden. As low back pain is a leading cause of work disability and several mass media campaigns conducted to date have addressed this painful condition, back pain will be the focus of this chapter. Based on population-based surveys, the lifetime prevalence of back pain is reported to be 60 to 85%, with a point prevalence of 15% to 30% and a 1-year prevalence of 50% [7, 8]. Back pain has been reported as the second most common symptom leading to general practitioner consultation after upper respiratory complaints [9–11]. It is the most common cause of activity limitation in adults aged less than 45 years and the fourth most common in those aged 45 to 64 years [12]. Most episodes of acute low back pain improve, although longitudinal studies suggest that back pain is often persistent and typically a recurrent condition [13–17]. Nevertheless, the majority of people with acute back pain are able to resume normal function and return to work quickly whether the pain has fully resolved or not [13, 14, 16].

Despite the good prognosis of acute low back pain and increasing scientific evidence for simple conservative management [18, 19], back pain related disability continues to be a public health concern in developed countries worldwide. This is highlighted by the fact that back pain is one of the major issues being addressed in the Bone and Joint Decade of 2000–2010 [20].

Back pain also places a significant socioeconomic burden upon individuals. Workers' compensation estimates suggest that the costs for employers have been at least matched by similar amounts for the individual and the community [21]. Chronic back pain may also lead to isolation, depression, lengthy absences from work and unemployment in the long term [22]. Due to the major burden of back pain at the population level, broadly based interventions with even small or modest impacts have the potential for huge improvements in societal health and well being.

21.1.1
Primary Prevention

To date most primary prevention interventions for back pain that have been studied have attempted to reduce the *incidence* of occupational low back pain in the workplace [23, 24]. However, such primary prevention strategies for back pain have generally been unsuccessful. In one recent review the only intervention supported by adequate evidence was exercise therapy [25]. Recognizing that back pain is nearly ubiquitous and difficult to prevent, one of the main aims of public health campaigns for back pain is to prevent *disability* due to the condition. This may be considered secondary prevention of disability with the aim of reducing the risk of chronicity.

21.1.2
Why Public Health Campaigns?

Several recent studies performed in different countries have shown a mismatch between public beliefs about back pain and current evidence about the condition's prognosis and management [8, 26, 27]. Authors of recent evidence-based clinical practice guidelines advocate that back pain is a benign, self-limiting condition and recommend that early management include minimal medical intervention, reassurance and advice to stay active [28, 29]. In contrast, results of an on-the-street survey undertaken in the United Kingdom indicate widely held beliefs that back pain most often results from major pathology such as a "slipped disc" or "trapped nerve" [27]. This is consistent with public beliefs in Norway [30]. In Canada, the general public also holds overly pessimistic views about back pain, believing that back pain will eventually stop one from working, becomes progressively worse with age, and makes everything in life worse [8]. As beliefs and attitudes about back pain are associated with the development of chronicity [31, 32] strategies that align public views about back pain with current evidence are required. The need for re-education of the public was suggested nearly a decade ago by Deyo, a leading back pain researcher [33].

Recently several countries have initiated mass media campaigns designed to alter societal views about back pain. A population-based media campaign, which provided simple evidence-based positive messages about back pain, took place in Victoria, Australia between 1997 and 1999. This campaign resulted in dramatic improvements in both community and primary care physicians' beliefs [34, 35]. These belief changes were accompanied by a decline in number of workers' compensation claims for back pain and health care utilization over the duration of the campaign. The positive results of the Australian campaign encouraged other countries, including Scotland [36] Norway [37] and Canada [38] to adopt similar public health campaigns in their own settings.

However, while mass media interventions are an established strategy for delivering preventive health messages, there is limited empirical understanding of the characteristics of effective (or ineffective) health campaigns [39]. For example, little is known about which communication strategies or media formats are most effective or whether simultaneous targeted interventions enhance their effect. In some instances campaigns have been implemented without an accompanying evaluation strategy, limiting their usefulness in terms of guiding others contemplating such an approach.

The objectives of this overview are to describe and compare characteristics and outcomes of back pain media campaigns that have taken place internationally where this data is available, examine general theories of health behavior change from the mass media literature to determine whether it is possible to develop a theoretical framework to explain the observed outcomes, describe the outcome of discussion and expert consensus around lessons learned from these campaigns that may inform the planning and evaluation of future campaigns, and identify priorities for future research.

21.2
Methods

We used the Cochrane Effective Practice and Organization of Care Review Group (EPOC) data collection checklist [40] to summarize the key campaign elements and evaluation strategies of three back pain mass media campaigns that had formal scientific evaluations of effectiveness as part of their processes (Australia, Norway and Canada). The data extraction checklist provides a guide about relevant campaign and study characteristics including: the setting (e.g. country and reimbursement system), the population and healthcare provision for the general population and for injured workers, details about the campaign itself including the rationale for the campaign and any underlying basis for the campaign, the period of time over which it ran, who performed the campaign, who provided input into the content; endorsements if any; the major media of the campaign, the overall costs, the intensity and frequency of the campaign, the intended audience, the main messages, the messengers, any additional interventions that were performed, and the marketing and scientific evaluations that were performed.

General theories of health behavior change that have been considered in the context of the mass media literature were reviewed to determine whether it is possible to develop a theoretical framework to explain the observed outcomes. We circulated a draft of this paper to a group of international back pain experts selected to have international representation and multidisciplinary expertise, including epidemiologists, health professionals, economists and statisticians. We also presented a summary of this paper as a basis for discussion in a Workshop at the Low Back Pain Forum VIII: Primary Care Research on Low Back Pain held in Amsterdam in June 2006. Expert comments and those of workshop participants were synthesized and incorporated into the final manuscript. A briefer version of this chapter has been accepted for publication in the journal Spine [41].

21.3
Mass Media Campaigns for Back Pain

A summary of the key campaign elements and evaluation strategies of four back pain mass media campaigns are shown in Table 21.1 and are described below.

Table 21.1 Description of mass media campaigns for back pain.

	"Back pain, don't take it lying down"	"Working backs scotland"	"Back@It"	"Active back"
Setting	Australia	Scotland	Canada	Norway
Population	State of Victoria ~4.3 million	Entire country ~5.0 million	Province of Alberta (~3.3 million)	Vestfold and Aust-Agder counties (325 000)
Healthcare provision for the general population	Dual system of universal healthcare (Medicare) and private health insurance	Publicly funded health system	Dual system of universal healthcare and private health insurance	Medicare covers all inhabitants. A small fee comes in addition for each visit to health practitioner
Healthcare provision for injured workers	State-based work cover insurance paid for by employers, managed by VWA for the state government, provided by several insurance companies	Both private insurance and public pensions available	Province-based Workers' Compensation Board legally mandated to provide care for injured workers. Administration is paid for through employer premiums	Employers cover full salary the first 16 days of sickness, Medicare thereafter, for the employees
Period of campaign	Sept. 1997 to Dec. 1999	Oct. 2000 to Feb. 2003	May 6, 2005 to 2008 (ongoing)	April 2002 to June 2005
Who performed the campaign?	Victorian WorkCover Authority	United Kingdom Health and Safety Executive, National Health System Health Scotland	Multiple funding partners including: Alberta Government, WCB-Alberta, local industrial safety associations	The Hospital of Rehabilitation, Stavern and The Norwegian Back Pain Network, The Communication Unit

Table 21.1 Continued

	"Back pain, don't take it lying down"	"Working backs scotland"	"Back@It"	"Active back"
Rationale for campaign	Rising cost of back pain claims; recognition that educational interventions directed at general practice likely to be ineffective without concomitant education of the public and employers; and recognition of importance of attitudes and beliefs in development of disability from back pain	Rising cost of costs associated with back pain and reversal in the management strategy of back pain. Perceived need for public education about the condition	Rising costs associated with back pain. Perceived need for public education about the condition	Rising cost of disability and use of healthcare due to low back pain; great confusion and divergence of beliefs about management among the public and among different health professionals; multidisciplinary guidelines for acute back pain launched April 2002
Who had input into content?	Consulted widely with international and local experts, multidisciplinary committee composed of representatives from national or state professional organizations with an interest in back pain, medical defence organization, employer and employee groups	National partnership including all health professionals who treat back pain in primary care and occupational health, employers, unions, and patients' organizations	Organizing committee composed of representatives from the funding organizations. Consulted widely with local and international experts	Steering committee composed of the owners of the campaign, and reflecting all health professional groups
Endorsements	Widespread endorsement from relevant national or state professional medical bodies (incl. general practice, orthopaedic surgery, rheumatology, rehabilitation, physiotherapy, chiropractic, osteopathy, sports and occupational medicine)	NHS Health Scotland and UK Health and Safety Executive	Widespread endorsement from local health associations (physicians, surgeons, physiotherapy, and chiropractic)	The National Medical Association, The Norwegian Physiotherapist Association, The Norwegian Chiropractic Association, The Directorate for Health and Social Affairs
Primary medium	Television commercials aired during prime time	Radio ads	Radio ads and website	4 issues of a 16 page information paper to all households, local TV, radio and cinema commercials, specific web page for the campaign

Overall cost	$A10.1million over three years	Unknown	~$1 million CDN over three years	NOK 2 mill (USD 315 000) in direct cost, one full time project leader in addition on the expences of the owner of the campaign
Intensity and frequency if known	Intense campaign for 12 months, followed by less intense period for 12 months and then final intense campaign for three months. "Top-up" low intensity yearly ads were planned but never implemented	Continuous website. Radio ads during peak listening months only	Continuous website. Radio ads during peak listening months only.	Website continuously throughout the period, 4 one-month campaign periods during the period
Other media	Radio, billboard and print advertisements, posters, seminars, visits by well-known personalities to workplaces, publicity articles and publications	Website, practice guidelines distributed to health professionals treating patients with back pain, pamphlets and posters aimed at the general population	Website (www.wcb.ab.ca/back@it) Posters, pamphlets, bus and billboard advertisements and informational articles in the public and industry news publications. Some television public service announcements	Website (www.aktivrygg.no) Posters with the messages of the campaign at healthcare clinics
Additional interventions	*The Back Book*, was made widely available with translations in 16 languages, and all Victorian doctors sent evidence-based guidelines for the management of employees with compensatable low back pain	Focus on re-educating health professionals including orthopedic surgeons	Specific focus on employers and healthcare providers to distribute posters and pamphlets.	All primary care doctors, physiotherapists and chiropractors sent copy of Guidelines, and invited to specific courses – In addition, a specific intervention in six cooperating workplaces

Table 21.1 Continued

	"Back pain, don't take it lying down"	"Working backs scotland"	"Back@It"	"Active back"
Basis of campaign	Simple evidence-based messages derived from *The Back Book*	UK Clinical Guidelines and Occupational Health Guidelines	Simple evidence-based messages derived from *The Back Book*	Five specific statements based on the Norwegian Guidelines
Intended Audience	General population, healthcare providers (particularly general practitioners), employers	General population and healthcare providers	General population, general practitioners, employers	General population, healthcare providers in primary care, employers and employees
Main messages	Back pain is not a serious problem; positive attitudes are important and it is up to you; continue usual activities, don't rest for prolonged periods, continue exercising and remain at work if possible; X-rays are not useful; surgery may not be the answer; keep employees at work	1. Stay active. 2. Try simple pain relief. 3. If you need it, get advice	Back Pain: Don't Take it Lying Down The key to feeling better sooner is to stay active	Low back pain is not dangerous, X-ray is not useful, activity makes improvement, surgery is rarely necessary
Messengers	International back pain experts, sports personalities who had successfully managed back pain, actors, comedians, healthcare professionals, Minister for Health	Well-known Scottish sports personality	Local health care professionals and organizations, Olympic Gold Medalist	Animation figure (humorous)
Marketing Evaluation	Focus groups to measure community awareness, public opinion	Monthly awareness surveys	Awareness measured on an annual basis	Consulted at halfway to determine general awareness
Scientific evaluation	Independent evaluation funded by Victorian WorkCover Authority	Supported by NHS Health Scotland, Health and Safety Executive	Independent multidisciplinary research team funded by Alberta Heritage Foundation for Medical Research	Conducted by one of the founders of the campaign, but replaced on the steering committee halfway; funded by the project and its owner

21.3.1
The Victoria WorkCover Authority Back Campaign, Australia, 1997–1999

In 1997 the Victoria WorkCover Authority (VWA), the manager of the workers' compensation system in the Australian state of Victoria, began a state-wide public health campaign aimed at shifting the general population's attitudes and beliefs about back pain (henceforth called the VWA Back Campaign). The VWA Back Campaign was developed according to messages in The Back Book, an educational booklet for patients based on the biopsychosocial model [42]. Various unambiguous messages were included, focusing on staying active, exercising, not resting for prolonged periods, and staying at work. The campaign ran with varying intensity over three years. Television commercials were aired in prime-time slots and included dialogue by recognized national and international medical experts, sports and television personalities, and were endorsed by relevant professional bodies. The television campaign was supported by other mass media including extensive outdoor billboards and radio but minimal printed advertisements. As part of the VWA Back Campaign strategy, The Back Book [42], was made widely available with translations in 16 languages. Initially, copies were sent to all doctors, physiotherapists, chiropractors, osteopaths and massage therapists with the aim of providing it to patients presenting with back pain. Later, copies were provided to workers' compensation case managers with the aim of providing it to those with a new back pain claim. All Victorian doctors received evidence-based guidelines for the management of workers with compensatable low back pain.

Evaluation studies measured the effectiveness of the VWA Back Campaign on population beliefs about back pain and the knowledge and attitudes of general practitioners in telephone and mailed surveys, respectively [34, 35]. As the campaign was administered to the entire population over a period of time, a strict random controlled trial design was not feasible. Therefore, a quasi-experimental, non-randomized before–after study design with an adjacent state as control group was used [43]. These studies found that in the intervention state beliefs about back pain became more positive in the general population and among doctors. Analysis of the VWA claims database also found that over the duration of the campaign there was a clear decline in the number of claims for back pain, rates of days compensated, and costs of medical care [34, 35]. Significant sustained improvements in both population and general practitioner beliefs have also been observed 3 and 4.5 years after the cessation of the campaign [44–46], although there has been some decay.

The cost of the Australian campaign (US dollars $7.6 mill.: $4.4 mill. Yr 1, $1.7 mill. Yr 2, $1.5 mill. Yr 3), largely due to television advertising time (as well as the printing and distribution of The Back Book), combined with uncertainty regarding effectiveness in other jurisdictions and cultures has posed somewhat of a barrier to more widespread use. On the other hand, the cost-savings even in terms of reduction in compensation claims, duration of time off work and health-care utilization were estimated to be many times greater than the cost of the campaign. It had been intended to continue top-up low level advertisements at a fractional cost of approximately $188000 US dollars per annum to prevent decay of the effect prior to the change in government that prompted cessation of the campaign.

A potential limitation to evaluating the singular effects of the VWA media campaign was the concurrent wide availability of *The Back Book* and the mail-out of low back pain evidence-based guidelines to all Victorian doctors. Although designed as part of the overall strategy to complement and support the mass media campaign, it was not possible to separately quantify the effects of these interventions from the media campaign. However, negligible awareness of the guidelines among Victorian doctors suggests that their impact was likely minimal.

Another important limitation of the Australian campaign's evaluation of disability and healthcare utilization was that it was limited to information extracted from a workers' compensation database. While this data used non-back claims as a control group, an interstate control using workers' compensation data from other states was not possible. In addition, only healthcare data from the onset of the campaign was available and the downward trends observed could have predated the campaign. The authors considered this unlikely as there had been a steady rise in the number of back claims prior to the campaign followed by a clear decline relative to other claims over the duration of the campaign (representing a reduction of over 15% in absolute numbers). This decline in back claims was also not observed in either New South Wales or South Australia, two adjoining states with similar populations. However, further evaluation of such costly media campaigns is clearly needed.

21.3.2
The Working Backs Scotland Campaign, Scotland, 2000–2003

In Scotland, the Health Education Board for Scotland (HEBS) (www.healthscotland.com) and the Health and Safety Executive (HSE) (www.hse.gov.uk) launched a major public education campaign about back pain in October 2000. The primary messages were: (i) Stay active; (ii) try simple pain relief; (iii) if you need it, get advice. Specific recommendations regarding work were not presented as preliminary focus showed that advice about work was viewed with suspicion and perceived to be provided on behalf of government or employers. Twenty organizations representing health professionals, employers and unions were involved. The Scottish campaign used radio as opposed to television to reduce costs. The radio campaign broadcast 1777 15-second advertisements on all 15 commercial radio stations. Skilled packaging and support from a well-known Scottish sports personality were also used and garnered attention from the free press. Leaflets and posters were provided to employers and treating health professionals, and 35000 information packs were distributed to all health professionals treating back pain in Scotland. The organizers also developed and distributed further material for specific professional groups including occupational health, general practitioners and orthopedic surgeons.

Population surveys showed a 60% awareness rate of the radio advertisements and a 15% positive shift in public beliefs regarding staying active despite the pain [36]. This was accompanied by a change in advice provided by health professionals as determined by patient reports. However, evaluation of work-related low back pain disability outcomes indicates little impact was made on disability behaviors. The researchers report no effect on sickness absence or new awards of social security benefits for back pain over the course of the campaign. They hypothesized that the lack of effect on work-related outcomes was due to a lack of explicit recommendations about work.

21.3.3
The Norwegian Back Pain Network "Active Back" Campaign, Norway, 2002–2005

In 2002, to coincide with the launch of the multidisciplinary Norwegian guidelines for acute low back pain, the Norwegian Back Pain Network initiated a broad implementation project in two counties (Vestfold and Aust-Agder). Like the Australian campaign, the Norway project combined a media campaign with additional targeted strategies. The project consisted of four separate activities: a media campaign directed towards the general public, an information campaign directed towards physicians, physiotherapists and chiropractors in primary health care, an information campaign directed towards social security officers and a practical intervention in six co-operating workplaces.

The media campaign was concentrated in four one-month periods and included advertisements in cinemas, on local TV and radio, and in the newspapers. At the same time a 16-page information pamphlet was distributed to all households in the area. Throughout the campaign period, a website (www.aktivrygg.no) was updated with information about the campaign, the background to the messages of the campaign, and it was possible to read, listen to and watch the different commercials.

The media campaign focused on five statements which were the main messages of the project:

- Back pain is rarely caused by any dangerous illness.
- X-rays rarely show the reason for back pain.
- A back in motion improves faster.
- Work with your back! – One recovers faster if one returns to work as soon as possible, even if the back is still hurting.
- Only a few persons with back pain need surgery.

All the commercials included an easily recognizable, and humorous, animation figure that was used to underscore one of these statements.

In addition to the media campaign, the project also specifically targeted health professionals and social security officers, providing these groups with the same messages. They all received full and short form versions of the acute low back pain guidelines, and were invited to specific courses.

The fourth intervention took place in three workplaces in each of the two counties. One or several employees in each workplace were trained to support colleagues who developed acute back pain. The task of these "peer advisors" was to reduce fear avoidance and, in order to facilitate keeping their colleagues at work rather than being sick listed, they also had the ability to provide temporary light duties [47].

The Norwegian mass media campaign was studied by two independent evaluations. One focused on the process of the campaign, including the initial formation of the idea, the development of the components of the project and its execution, and the decisions and discussions that took place along the way. The other, a scientific evaluation, determined the results of the project in terms of shift in beliefs in all target groups, and actual change in behavior among those with back pain.

The scientific evaluation used quasi-experimental methodology with a concurrent control group formed from the unexposed population of a neighboring county (Telemark). To ascertain population beliefs, telephone interviews of a random sample of 500 people in both project counties and a control county, were performed before, during and after the campaign. In addition, mailed questionnaires were sent to all health professionals in the three counties. The 3500 employees in six workplaces in the project counties were also asked to complete a questionnaire. Finally, information about sick listing, surgeries, and imaging examinations for all three counties were collected before, during and after the campaign.

The results suggest that a statistically significant shift in beliefs towards more optimistic, self-coping attitudes in the public have occurred in the two intervention counties, but not in the control county. Considering that the maximum awareness of the general public of the campaign was 41%, the change in attitudes was impressive. Awareness of the campaign among health professionals was close to 100%. While the total number of imaging examinations remained stable in the intervention counties, there was an increase of 57% in the control county. There was a small increase in all three counties in surgical interventions. During the study period, there was a general reduction in sick listing in all of Norway, and no significant differences between the intervention counties and the control county was found. However, in the six cooperating workplaces the additional intervention with the "peer advisors" produced a reduction in overall sick listing by 27% and the low back pain related sickness absence was reduced by 49% [48].

This was a low budget project, with a total cost of approximately $350 000 US per year (total = $1.1 mill. US). The direct cost of the media campaign was $550 000 US. The use of local county broadcasts rather than national TV and radio for the media campaign was thought to be responsible for the low public awareness of the campaign. As there are only a small number of inhabitants in each of the Norwegian counties, local TV broadcasts occur for only a few hours every night, mostly with local news and therefore the ratings of local TV and radio are small. The website and the four distributions of the information pamphlets to all households were thought to explain most of the public awareness of the campaign.

21.3.4
The Alberta "Back@It" Campaign, Canada, 2005–2008 (ongoing)

The apparent success of the Australian and Scottish campaigns and the fact that back pain continues to be one of the most prevalent and costly medical conditions stimulated a related intervention in Alberta, Canada. The Alberta campaign builds on the Australian and Scottish experiences, maintaining the general themes of the Australian campaign while focusing on radio to reduce costs. The campaign initially aired between May 2005 and April 2006 but was extended two more years due to positive feedback and overwhelming support from the organizers and health professional groups. The campaign is sponsored by the Alberta Government (Alberta Human Resources and Employment, Workplace Health and Safety), the Workers' Compensation Board–Alberta (WCB-Alberta) and local safety associations (Alberta Hotel Safety Association, Manufacturers' Health and Safety Association, Alberta Construction Safety Association). It has also received widespread endorsement from local health associations (physicians, surgeons, physiotherapy, and chiropractic). Similar to the Norwegian campaign, the Alberta campaign is low budget with a total cost of approximately $310 000 US per year.

The Alberta campaign media advertisements were planned and created by the Communications Department of the WCB-Alberta in conjunction with a private marketing firm. The designers consulted widely with local and international experts in the field of back pain and with those responsible for the Australian and Scottish campaigns. Similar to the Australian campaign, it is based on *The Back Book* and uses the same slogan (*Back pain – Don't take it lying down*). Health professional associations, respected physicians, and an Olympic gold medalist who has successfully managed back pain through activity are being used to support the key messaging. The primary medium is radio and advertisements were aired in all major Alberta cities on key stations during prime listening months. An attempt was made to include stations of all music formats for broad penetration to all segments of the target population. In addition to the radio advertisements, posters and pamphlets were distributed widely to employers, city facilities, medical clinics, hospitals, and other health units. The radio and print advertisements can be accessed via the project's website (www.backactive.ca).

The Canadian campaign is currently being evaluated using quasi-experimental methodology. Similar to the Australian evaluation, changes in population back pain beliefs, healthcare utilization, and compensation claims are being evaluated with a concurrent control group formed from the unexposed population of a neighboring province (Saskatchewan). In terms of healthcare utilization, the evaluation incorporates indicators of health visits of individuals receiving care within the public delivery system (i.e. not indemnified by workers' compensation or private insurance systems). Beliefs are being measured before, during and after the campaign using telephone surveys of representative samples, whereas an interrupted time series design is being used to evaluate indicators of healthcare utilization and work-related disability.

The time series will be comprised of sequential measurement of outcome variables before and after the intervention, with the beginning of the campaign considered the "interruption" in the time series. Data will be extracted from government and workers' compensation administrative databases in monthly intervals for the period five years before the campaign until its conclusion. Primary outcomes will be indicators of healthcare use (visits to physicians and specialists, diagnostic imaging) and workers' compensation claim incidence and duration (number and duration of time loss back pain claims). Time series analysis including ARIMA techniques will be used to determine the impact of the intervention. ARIMA techniques take into account the auto-correlation inherent in outcomes, and will allow statistical evaluation of whether adding an intervention variable adds significantly to the prediction model of the post-intervention time series plots, indicative of an influence of the intervention.

21.4
Considering the Design and Evaluation of Back Pain Media Campaigns

Communication campaigns are a well accepted means for increasing knowledge, forming attitudes and achieving social and behavioral change [49]. In the past these were based on a formulaic rather than theory-driven approach. Yet theoretical approaches may provide a better understanding of the mechanisms that activate campaign effects. They may provide a theoretical basis for choosing the best message design to produce the intended outcome. For example, they may identify specific beliefs that need to be challenged to change or maintain a given behavior. Furthermore, theory may provide a better approach to evaluating the success of a campaign by ensuring that relevant and responsive outcomes are measured.

There are several general theories of health behavior change that have been advocated. However, a detailed discussion of these theories is beyond the scope of this paper. Several authors have incorporated these theories as an integrated model [50, 51]. From this we identified five considerations relevant to the planning and evaluation of mass media interventions for low back pain (Figure 21.1). We have used these as a framework for discussing the Australian, Scottish, Norwegian and Canadian back campaigns and to guide future planning.

1. What are current back pain attitudes and behaviors, or management practices that need to be changed and why? The rationale on which existing back campaigns have been based to date is that public attitudes and behaviors related to low back pain and disability are out of keeping with the currently available best evidence. To a varying degree they have also recognized the need to change primary care providers' attitudes and behavior in view of the lack of success with traditional educational means. Both the Alberta and Norway campaigns were informed by determining the current state of public perceptions about back pain in their settings in order to identify the most pertinent misconceptions. Some of the campaigns also considered other relevant subgroups such as employees, employers,

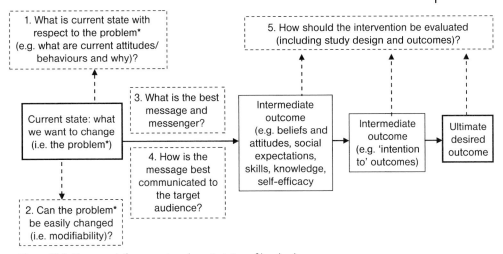

Figure 21.1 Framework for assessing characteristics of low back pain public health intervention campaigns based upon the theoretical framework of Hornick and Yanovitzky [50].

other health professionals, and workplaces. The context of the problem may also vary by setting – for example, one setting may wish to modify excessive disability from back pain whereas in another setting the major problem may be excessive health care utilization rather than disability. Identification of these issues would help to clarify the purpose of the campaign and the messages to be delivered.

Future campaign planning should consider what are the major problems that needs to be addressed? What are the main public misconceptions? Is there up-to-date evidence regarding current public understanding, attitudes, and management of back pain in various settings? Are there differences across countries, cultures or settings that should be considered? Are there relevant subgroups that should be targeted (for example, individuals, health practitioners or policy makers)?

2. How easy is it to change back pain attitudes and behaviors, and what are some of the barriers to changing patient management? The ease with which it is possible to change attitudes and behaviors for back pain may vary according to the setting and population, jurisdiction, typical practices, and type(s) of behavior or attitude that is being considered. The available outcomes of existing back campaigns to date have shown concordant results with respect to changing public attitudes and beliefs about back pain towards a more positive, self-coping approach [34, 36]. In Australia and Scotland there were similar improvements in doctors' attitudes and beliefs, and management intentions including reduced intentions to perform imaging for acute low back pain (actual behavior has not been measured). In Norway, while awareness of the campaign was close to 100% among health professionals, the intervention seemed to have impact on the use of imaging, but not on surgery rates [48]. However, the surgery rates were already low from the beginning in both intervention and control counties, suggesting that modification were not needed or, if desirable, may have been more difficult to achieve.

Future planning should consider the evidence on the modifiability of attitudes, perceptions and behaviors regarding low back pain and disability. Does this evidence apply to the population and setting of the proposed campaign? How easy is it to change attitudes and behaviors around back pain? Is the rationale for the intended impact of the back campaigns still relevant and "the best" approach for modifying behaviors? What are the barriers that need to be overcome? Are there other approaches to get change?

3. What are the best messages about back pain, and who should deliver the messages? The effectiveness of the campaign will also depend upon the specific messages and who delivers those messages. Messages in the Australian campaign, based on *The Back Book*, can be broadly categorized as encouraging exercise and limiting rest, keeping people at work, shifting responsibility of control to the individual with low back pain, explaining that investigations may not be helpful in finding a cause for the pain, and that spinal surgery may not be the answer to back pain [52]. The messengers included international back pain experts, local comic personalities, well-known Australian sportsmen (cricketer and Australian-rules footballer), and health professionals. From focus groups, the most memorable messengers included one of the international back experts who was seen as credible but was also remembered for his Scottish accent; and a successful Australian cricketer who is widely known to have overcome chronic back pain.

To try to address the issue of receiving different messages from various stakeholders, the Australian campaign used the same messages with different messengers to target specific groups such as the public, health professionals, employers and employees. For example, it included advertisements aimed at employers and showed employers re-integrating employees and stressed the importance of modified work. They also included the penalties involved for non-compliance (monetary fine). However, both the Scottish and the Alberta campaigns only provide generic advice about staying active and how to manage things from the patient's side with no messages aimed at employers. Preliminary focus groups undertaken in Scotland indicated that the public deemed advice about work with suspicion, and perceived it to be on behalf of government or employers. In Norway, although there were no specific advertisements directed at employers, the campaign specifically included workplace initiatives at six sites. "Peer advisors" were trained whose task was to prevent sicklisting of employees during an episode of back pain by assuring them that the pain was not harmful and that the individual would recover faster by staying at work and offering temporary adjustments at work. This resulted in change in beliefs which exceeded the change seen in the general population, and seemed also to produce the desired change in sickness behavior. In order to have an impact on work disability outcomes, future campaign messaging will likely need to include explicit advice about continuing work despite back pain (as part of the stay active message).

Future planning should consider that different messages and messengers may be perceived as more or less acceptable/persuasive. For example, would simple lay words and messages be more acceptable than "medical" messages? Could some messages be acted on more easily? Are some messengers more credible? How important are endorsements by professional and other organizations? Are different messages and delivery mechanisms needed for different groups? Do different groups respond differently to different messages and/or messengers? What is the effect of direct or primary exposure to the messages (e.g. individuals seeing the campaign firsthand) versus indirect or secondary exposure (e.g. doctors talking to patients about the campaign)?

4. What is the best way to deliver back pain messages to the target audience? The Australian and Norway campaigns used state-based and local county television for delivering the campaign messages to the target audience whereas both Alberta and Scotland primarily used radio. The level and saturation of the campaigns also varied. It is also important to know what other messages are being delivered in competition or in support of the campaign at the time. Prior to the onset of the campaign in Australia, 50% of the public were already aware of advertising about back pain, indicating that there were lots of other messages being delivered by others at the time.

Future planning should consider the advantages and disadvantages of different delivery strategies (for example cost and effectiveness). What considerations guide the intensity, frequency and duration of the intervention; the importance of the media used (e.g. television, radio, paper, leaflets, etc.); consideration of different cultures and settings? Are multi-faceted approaches better? Community characteristics should be considered in choice of media used; where do people get their daily information from–national or local media? The Norway campaign seemed to reach most people through a 16-page pamphlet delivered to each household in the area four times during the campaign.

5. How do we best study the effects of back pain mass media campaigns? Evaluating the effects of communication campaigns is complex and multifaceted. For example, decisions need to be made about the study design and measurement of intermediate or final outcomes. Also, the interpretation of the study results should include separating campaign effects from those of other sources of influence, and expectations for differential campaign effects across subpopulations [50]. To date, previous evaluations of back campaigns have measured their success by studying their impact on public and health professional beliefs and attitudes. Positive shifts in these intermediate outcomes have now been observed in several countries although there have been only variable improvements in final desired outcomes, such as improved health outcomes and reduced healthcare use. Reasons for these differences may relate to the characteristics of the campaigns themselves, or other contextual factors that interfere with attitude change resulting in a change in behavior, for example at the policy level. Furthermore, there may be a lag before attitudinal changes result in behavior change; the timing of measurement of outcome may therefore be crucial. The Australian campaign has demonstrated that positive shifts in public and doctors' beliefs can be sustained for at least several years after cessation of the campaign.

While some would argue that clear evidence is still needed that communication campaigns about back pain lead to economic and clinical benefits, we already have strong observational evidence that beliefs predict outcome[53]. It may therefore be more important to target our efforts on ensuring that campaigns can achieve sustained shifts in health professional and population beliefs, ideally for several years, with smaller parallel studies in individuals to see whether change in beliefs has led to actual behavior change.

Future planning should consider the advantages and disadvantages of different study designs and strategies to evaluate community interventions (control groups, follow-up periods and timing). What is the evidence of a link between intermediate (e.g. information reception, knowledge, intention, beliefs, etc.) and final desired outcomes (e.g. healthcare use and specific health outcomes)? What are the best outcomes to measure?

21.5
Conclusion

Due to the major societal burden of low back pain disability, population-based interventions targeting this conditon have the potential for large improvements in health and well-being. Mass media campaigns designed to alter societal views about back pain and promote behavior change have now been performed in several countries with mixed results. It appears that more intensive and expensive media campaigns may be more effective than low-budget campaigns and that if changes in work-related outcomes are desired then explicit recommendations regarding work may be needed. We have described and compared the characteristics and outcomes of four media campaigns designed to positively influence the public's attitudes and beliefs about back pain. In order to inform the design, planning and evaluation of future campaigns we have presented the results of expert commentary and discussion built around the lessons we have learned from our campaign involvement. We have also discussed broader models of health behavior change which should guide future research in this area.

Acknowledgments

We would like to acknowledge the insightful comments and suggestions of Drs. Kim Burton, Peter Croft, Glenn Pransky, Mark White, and the workshop participants at the Low Back Pain Forum VIII: Primary Care Research on Low Back Pain held in Amsterdam, June 2006.

References

1 Weinreich, N. (2007) *Hands on Social Marketing: a Step by Step Guide*, Sage Publications, Thousand Oaks, CA.

2 Andreasen, A. (2006) *Social Marketing in the 21st Century*, Sage Publications, Thousand Oaks, CA.

3 Kotler, P., Roberto, N. and Lee, N. (2002) *Social Marketing: Improving the Quality of Life*, 2nd edn, Sage Publications, Thousand Oaks, CA.

4 Redman, S., Spencer, E. and Sanson-Fisher, R. (1990) The role of mass media in changing health-related behaviour: a critical appraisal of two methods. *Health Promotion International*, **5**, 85–101.

5 Grilli, R., Ramsay, C. and Minozzi, S. (2001) Mass media interventions: effects on health services utilisation. *The Cochrane Database of Systematic Reviews*, **1**, CD000389.

6 Marks, R. (1999) Two decades of the public health approach to skin cancer control in Australia: why, how, and where are we now? *Australasian Journal of Dermatology*, **40**, 1–5.

7 Nachemson, A., Waddell, G. and Norlund, A. (2000) Epidemiology of neck and back pain, in *The Scientific Evidence of Causes, Diagnosis, and Treatment*, Lipppincott Williams & Wilkins, Philadelphia, pp. 165–88.

8 Gross, D., Ferrari, R., Russell, A.S., Battie, M.C., Schopflocher, D., Hu, R.W. *et al.* (2006) A population-based survey of back pain beliefs in Canada. *Spine*, **31**, 2142–5.

9 Bridges-Webb, C., Britt, H., Miles, D.A., Neary, S., Charles, J. and Traynor, V. (1992) Morbidity and treatment in general practice in Australia 1990–1991. *Medical Journal of Australia*, **157** (Suppl.), S1–56.

10 Cypress, B. (1983) Characteristics of physician visits for back pain symptoms: a national perspective. *American Journal Public Health*, **73**, 389–95.

11 Hart, L., Deyo, R. and Cherkin, D. (1995) Physician office visits for low back pain. Frequency, clinical evaluation, and treatment patterns from a US national survey. *Spine*, **20**, 11–19.

12 Andersson, G. (1997) The epidemiology of spinal disorders, in *The Adult Spine: Principles and Practice* (ed. F. JW.), Raven Press, New York, pp. 93–141.

13 von Korff, M. (1994) Studying the natural history of back pain. *Spine*, **19**, 2041–5.

14 van den Hoogen, H., Koes, B.W., van Eijk, J.T., Bouter, L.M. and Deville, W. (1998) On the course of low back pain in general practice: a one year follow up study. *Annals of the Rheumatic Diseases*, **57**, 13–19.

15 Croft, P., Macfarlane, G.J., Papageorgiou, A.C., Thomas, E. and Silman, A.J. (1998) Outcome of low back pain in general practice: a prospective study. *British Medical Journal*, **316**, 1356–9.

16 Schiottz-Christensen, B., Nielsen, G.L., Hansen, V.K., Schodt, T., Sorensen, H.T. and Olesen, F. (1999) Long-term prognosis of acute low back pain in patients seen in general practice: a 1-year prospective follow-up study. *Family Practice*, **16**, 223–32.

17 Cherkin, D., Deyo, R.A., Street, J.H. and Barlow, W. (1996) Predicting poor outcomes for back pain seen in primary care using patients' own criteria. *Spine*, **21**, 2900–7.

18 van Tulder, M., Koes, B. and Bouter, L. (1997) Conservative treatment of acute and chronic nonspecific low back pain. A systematic review of randomized controlled trials of the most common interventions. *Spine*, **22**, 2128–56.

19 Waddell, G., Feder, G. and Lewis, M. (1997) Systematic reviews of bed rest and advice to stay active for acute low back pain. *The British Journal of General Practice*, **47**, 647–52.

20 Brooks, P. and Hart, J. (2000) The bone and joint decade: 2000–2010. *Medical Journal of Australia*, **172**, 307–8.

21 Rae, J. and Easson, M. (1995) Work safety and health: an inquiry into occupational health and safety, Industry Commission Report, Commonwealth of Australia.

22 Hendler, N. (1984) Depression caused by chronic pain. *The Journal of Clinical Psychiatry*, **45**, 30–6.

23 Frank, J., Kerr, M.S., Brooker, A.S., DeMaio, S.E., Maetzel, A., Shannon, H.S. *et al.* (1996) Disability resulting from occupational low back pain. Part I: what do we know about primary prevention? A review of the scientific evidence on prevention before disability begins. *Spine*, **21**, 2908–17.

24 van Poppel, M., Koes, B.W., Smid, T. and Bouter, L.M. (1997) A systematic review of controlled clinical trials on the prevention of back pain in industry. *Occupational and Environmental Medicine*, **54**, 841–7.

25 Burton, A.K., Balague, F., Cardon, G., Eriksen, H.R., Henrotin, Y., Lahad, A. *et al.* (2006) Chapter 2. European guidelines for prevention in low back pain: November 2004. *European Spine Journal*, **15** (Suppl. 2), S136–68.

26 Ihlebæk, C. and Eriksen, H. (2003) Are the "myths" of low back pain alive in the general Norwegian population? *Scandinavian Journal of Public Health*, **31**, 395–8.

27 Klaber Moffett, J., Newbronner, E., Waddell, G., Croucher, K. and Spear, S. (2000) Public perceptions about low back pain and its management: a gap between expectations and reality? *Health Expect*, **3**, 161–8.

28 General, E.C.R.D. (2006) Low back pain: guidelines for its management. www.backpaineurope.org (accessed 2 November 2007).

29 van Tulder, M., Tuut, M., Pennick, V., Bombardier, C. and Assendelft, W.J. (2004) Quality of primary care guidelines for acute low back pain. *Spine*, **29**, E357–62.

30 Ihlebæk, C. and Eriksen, H. (2005) Myths and perseptions of back pain in the Norwegian population, before and after the introduction of guidelines for acute low back pain. *Scandinavian Journal of Public Health*, **33**, 401–6.

31 Symonds, T., Burton, A.K., Tillotson, K.M. and Main, C.J. (1995) Absence resulting from low back trouble can be reduced by psychosocial intervention at the work place. *Spine*, **20**, 2738–44.

32 Burton, A., Waddell, G., Tillotson, K.M. and Summerton, N. (1999) Information and advice to patients with back pain can

have a positive effect. A randomised controlled trial of a novel educational booklet in primary care. *Spine*, **24**, 1–8.

33 Deyo, R. (1996) Acute low back pain: a new paradigm for management. *British Medical Journal*, **313**, 1343–4.

34 Buchbinder, R., Jolley, D. and Wyatt, M. (2001) Population based intervention to change back pain beliefs and disability: three part evaluation. *British Medical Journal*, **322**, 1516–20.

35 Buchbinder, R., Jolley, D. and Wyatt, M. (2001) 2001 volvo award winner in clinical studies: effects of a media campaign on back pain beliefs and its potential influence on management of low back pain in general practice. *Spine*, **26**, 2535–42.

36 Waddell, G., O'Connor, M., Boorman, S. and Torsney, B. (2007) Working Backs Scotland: a public and professional health education campaign for back pain. *Spine*, **32**, 2139–43.

37 Network, T.N.B.P. (2002) Active Back. www.aktivrygg.no (accessed 16 May 2006).

38 Alberta Back Pain Campaign. (2006) Back Active. www.backactive.ca (accessed 2 November 2007).

39 Morton, T. and Duck, J. (2001) Communication and Health Beliefs. *Communication Research*, **28**, 602–26.

40 Effective Practice and Organisation of Care. (2002) The data collection checklist 2002. http://www.epoc.uottawa.ca/checklist2002.doc (accessed 2 November 2007).

41 Buchbinder, R., Gross, D.P., Werner, E.L. and Hayden, J.A. (2008) Understanding the characteristics of effective public health interventions for back pain and methodological challenges in evaluating their effects. *Spine*, **33** (1): 74–80.

42 Roland, M., Bigos, S., Roland, M., Waddell, G., Klaber-Moffat, J., Burton, A.K. and Main, C. (1996) *The Back Book*, The Stationary Office, United Kingdom.

43 Cook, T. and Campbell, D. (1979) *Quasi-Experimentation: Design and Analysis Issues for Field Settings*, Houghton Mifflin, Boston.

44 Buchbinder, R. and Jolley, D. (2004) Population-based intervention to change back pain beliefs: a three-year follow up study. *British Medical Journal*, **328**, 321.

45 Buchbinder, R. and Jolley, D. (2005) Effects of a media campaign on back beliefs is sustained three years after its cessation. *Spine*, **30**, 1323–30.

46 Buchbinder, R. and Jolley, D. (2007) Improvements in general practitioner beliefs and stated management of back pain persist four and a half years after the cessation of a public health media campaign. *Spine*, **32**, E156–62.

47 Werner, E., Laerum, E., Wormgoor, M.E., Lindh, E. and Indahl, A. (2007) Peer support in an occupational setting preventing LBP-related sick leave. *Occup Med* (Lond.), **57**, 590–5.

48 Werner, E.L., Ihlebæk, C., Laerum, E., Wormgoor, M.E., Indahl, A. (2008) Low back pain media campaign: no effect on sickness behaviour. *Patient Educ Couns* **71**(2), 198–203.

49 Capella, J. (2003) Editor's Introduction: theoretical approaches to communication campaigns. *Communication Theory*, **13** (2), 160–3.

50 Hornick, R. and Yanovitzky, I. (2003) Using theory to design evaluations of communication campaigns: the case of the National Youth Anti-Drug Media Campaign. *Communication Theory*, **13**, 204–24.

51 Fishbein, M. and Yzer, M. (2003) Using theory to design effective health behaviour interventions. *Communication Theory*, **13**, 164–83.

52 Buchbinder, R., Jolley, D. and Wyatt, M. (2003) Role of the media in disability management, in *Preventing and Managing Disabling Injury at Work* (eds T. Sullivan and J. Frank), Taylor and Francis Ltd, Toronto, pp. 101–41.

53 Woby, S., Watson, P.J., Roach, N.K. and Urmstron, M. (2004) Are changes in fear-avoidance beliefs, catastrophizing, and appraisals of control, predictive of changes in chronic low back pain and disability? *European Journal of Pain*, **29**, 1818–22.

22
Improving Pain Management Education

Elizabeth Peter and Judy Watt-Watson

- Pain education for health professionals has been repeatedly identified as important for effective pain management practices, but gaps in knowledge have been reported for almost two decades for a variety of health professional groups.

- An integrated, interprofessional, pre-licensure undergraduate pain curriculum can lead to changed knowledge and beliefs and positive student experiences; as can team-based post-licensure workshops and programs.

- Consistent application of pain policies, standards, and guidelines in practice has been problematic, perhaps because they are difficult to enforce.

- Entry level practitioners should possess a sound knowledge of the basis of pain and pain relief; a sound knowledge of the consequences of persistent problems; the ability to complete a comprehensive pain assessment; the capacity to manage pain within an interprofessional team.

- Accreditation standards are likely the most meaningful policy mechanism to further increase educational offerings related to pain post-licensure.

The management of unrelieved pain continues to be a challenge in the twenty-first century [1]. Evidence from pain research is not always effectively and consistently applied in practice. Evidence is also clear that unrelieved acute pain, including acute post-surgical pain, can contribute to chronic pain long after the time of healing.

Pain education for health professionals at all levels has been repeatedly identified as an important step to changing ineffective pain management practices [2]. However, gaps in pain knowledge and/or problematic beliefs have been reported for almost two decades for a variety of health professional groups including Medicine, Pharmacy, and Nursing [3–6], Occupational Therapy [7, 8], and Physical Therapy [9]. Despite these deficiencies, educational programs, especially for pre-licensure students, have included minimal or no pain content [10–12]. Conse-

Chronic Pain: A Health Policy Perspective
Edited by S. Rashiq, D. Schopflocher, P. Taenzer, and E. Jonsson
Copyright © 2008 WILEY-VCH Verlag GmbH & Co. KGaA, Weinheim
ISBN: 978-3-527-32382-1

quently, students lack important pain knowledge at graduation [5, 7, 8, 13]. Moreover, the extent to which inservice and continuing education opportunities exist for post-licensure professionals to enhance their pain management skills and knowledge is unknown.

Because effective pain management can be complex, it often requires approaches that exceed the expertise of one profession. Students in health professional programs, however, usually have few collaborative learning experiences [14]. These opportunities have been shown to help students to balance socialization into their own profession along with learning about interprofessional collaboration [15–17]. Russell *et al.* [17] found that students had little understanding of interprofessional collaboration and learned disciplinary stereotypes through tacit observation of staff behaviors. It is noteworthy that several generic interprofessional models have been found to improve knowledge and beliefs in pre-post design for health science students [18–20]; outcomes were better for undergraduates than graduates [16, 21]. However, published evidence of required interprofessional pain curricula models for pre-licensure students has been minimal [6] despite research supporting interprofessional collaboration as being important to the successful implementation of evidence into practice [22].

In this chapter, we discuss the importance of improving pain management education of pre- and post-licensure health professionals, including from an interprofessional perspective. First, the results of pain educational interventions that have been formally evaluated are briefly discussed. Second, policy initiatives, both nationally and internationally, that have the potential to foster improved pain education are explored. Third, additional policy strategies are proposed to further the development of improved pain management education. A particular emphasis is placed on ensuring the further development of pain competencies for health professionals by regulatory bodies, given their influence on curricula.

22.1
Pre-Licensure Educational Initiatives: Existing Data

Educational initiatives that encourage clinicians to examine both their unique uniprofessional and common interprofessional strengths in providing high quality pain management may facilitate interprofessional collaboration. Educational initiatives can be most successful when integrated early in the socialization and educational experience of diverse professionals and negative attitudes reinforced in the pre-licensure years are more challenging to change later [23]. However, pain management and interprofessional educational efforts have been mainly targeted towards health professionals after graduation. Future educational initiatives therefore need to extend their reach to include pre-licensure health professional students as well.

Watt-Watson and colleagues [6] were the first to provide data on the development, implementation and evaluation of an integrated, interprofessional, pre-licensure undergraduate pain curriculum. This initiative was led by the University of Toronto Center for the Study of Pain and organized by 16 members who represented the six Faculties/Departments of Dentistry, Medicine, Nursing, Pharmacy, Physical Therapy and Occupational Therapy. Based on the *International Association for the Study of Pain* curricula [24] and the *Position Statement on Pain Relief* from the *Canadian Pain Society* [25], the aim of the curriculum was to ensure a common basic understanding of pain assessment and management principles upon which to build profession-specific pain knowledge within an interprofessional context. The mandatory 20-hour curriculum was delivered to 540 students from the six Health Science Faculties/Departments during a one-week period. Teaching strategies included large multiprofessional presentations, small interprofessional group sessions to work on patient case-based pain assessment and management plans, Standardized Patients and 63 clinician-facilitators representing all of the participating professions. Overall, student evaluations of the curriculum were positive, and statistically significant changes were demonstrated in their pain knowledge and beliefs. This curriculum uses an iterative design based on extensive evaluation and will be implemented for the eighth time in 2008 with over 800 students.

The model provided by Watt-Watson and colleagues [6] may be useful for those wishing to develop an interactive educational initiative based on current pain research evidence that is applicable in pre-licensure or clinical settings. However, it also highlights the need for organized and cooperative interprofessional efforts to design and deliver pain education to both developing health professionals and those already working in the clinical setting. Regulatory bodies and academic institutions have an important role to play in making pain education a required element of pre-licensure education requirements. Health care organizations and policy makers can also support accountability for improved pain practices through institutional support for education and accreditation requirements. Ultimately, interactive, contextually-relevant strategies at all levels, which focus on building health professionals' knowledge and translating this knowledge into practice could contribute to improving both healthcare outcomes.

22.2
Post-Licensure Educational Initiatives: Existing Data

Pain education for health professionals in practice has been delivered with varying degrees of success in a variety of formats, such as workshops of different lengths and content and individual academic detailing with performance feedback [5, 26–29] (see also Chapter 23). For example, Simpson *et al.* [5] evaluated a pain management program developed at the South Georgia Medical Center. A number of educational and practice initiatives were adopted, including an organizational policy for pain management, standardized rating scales for pain, an annual day-long pain conference, quarterly hour-long pain management class, workshops, and educational materials. After two years significant changes in knowledge and attitudes were found in nurses, pharmacists, physiotherapists and physicians. This study, like the study by Watt-Watson *et al.* [6], demonstrates the importance of taking a team approach to education and of adopting both policy and educational initiatives to improve pain management. Zwarenstein, Reeves and Perrier [30] in their review of post-licensure collaborative interventions in clinical settings found very little research in this area.

22.3
Policy Initiatives Internationally

Unrelieved pain is an international challenge. The International Association for the Study of Pain in collaboration with the World Health Organization has established a yearly pain awareness day stating that "the relief of pain should be a human right" [24]. The United States Congress passed the *Decade of Pain Control and Research Bill*, beginning in January 2001, which declared "adequate pain care research, education and treatment as national public health priorities" [31]. The Canadian Senate also established a "National Pain Awareness Week" that occurs yearly in November [32]. As well, considerable effort has been directed internationally toward the evidence-based practice agenda [22] and its relevance to changing pain management practices. The Canadian Council on Health Services Accreditation (CCHSA) [33] in their Achieving Improved Measurement (AIM) standards incorporated an evidence-based pain management criterion that underlines the need for interprofessional collaboration and the organization's accountability for ongoing pain education of clinicians, patients and families about pain management options. The Canadian Pain Society Special Interest Group, Nursing Issues, has subsequently developed a manual to facilitate implementation of the criterion requirements [25].

While pain policies, standards, and guidelines have been available in various forms for over a decade their consistent application in practice has been problematic as evidenced by the prevalence of unrelieved pain [1]. Availability of a guideline does not insure it will be used. The challenge may be that although *guidelines* may be official policy or position statements, they are not law and have no enforceable authority [34]. Moreover, questions need to be asked about accountability for implementing guidelines and whether they can be enforced/mandated. Monitoring is also important and there may be gaps between what clinicians say they are doing and their actual practices [35–36].

22.4
Policy Strategies: Pre-Licensure Competencies

A number of elements shape policy choices, including institutions, ideas and interests. Institutions are the rules used to deliver policy choices, such as legislation, guidelines and regulations. Ideas are the beliefs, values or knowledge that shape a policy field, and interests represent the groups who are in a position to influence policy choices [37]. Understanding how these elements shape policy choices in the current policy environment of Canadian universities helps ensure that the possible policy strategies are meaningful and robust in changing the curricula of healthcare professionals.

The curricula of healthcare professionals are determined by a number of influences. Internal influences tend to consist of ideas and interests, such as the interests, knowledge and expertise of faculty or academic administrators, and student feedback. Formal governance structures, that is, the internal institutions, such as curriculum committees, also shape the eventual curricula. Externally, academic accrediting bodies and professional regulatory bodies strongly shape curricula through the regulations they impose. Schools must be accredited and students must acquire the necessary professional competencies to eventually become licensed by their respective colleges. Interest groups, such as patient groups, and representatives from clinical settings may also at times lobby professional schools for changes to curricula. Generally more distant authorities are provincial/territorial ministries of health and education whose various regulations must also be adhered to when delivering academic programs. To what extent they have an effect on curricula would be highly variable.

It is important to recognize that these curricula are increasingly under pressure to add or to change content as scientific knowledge grows and to accommodate the ever changing needs of the healthcare system. Resources, such as time, money and faculty expertise, are often in short supply and must be allocated judiciously. Influencing professional bodies to increase the number of pain management competencies may ultimately have the greatest impact on curricula. These competencies tend to be given high priority by academic administrators and curriculum committees.

A competency has been defined as the integrated knowledge, skills and judgment expected of a practitioner that must be measurable and observable [38]. Entry-to-practice competency requirements from national documents were surveyed including those for Dentistry, Medicine, Nursing, Pharmacy, Occupational Therapy and Physiotherapy [39–44]. Nursing and Dentistry [39, 41] each have two clearly identified pain statements; Physiotherapy [44] has one, and in the remainder pain is not mentioned. In contrast, the national competencies for Veterinary Medicine [45] include nine competencies related to analgesic and anesthetic management with one additional competency related to alleviating suffering.

Competencies related to pain management need to ensure that entry level practitioners possess the following: (i) a sound knowledge of current theories of the anatomical, pathophysiological, and psychological bases of pain and pain relief; (ii) a sound knowledge of the consequences of persistent problems related to activity limitation, physical impairment, and social participation, including an understanding that preventative pain management can reduce treatment complications and persistent pain; (iii) the ability to complete a comprehensive pain assessment, including the ability to identify common factors that influence pain expression and clinician response and the ability to use pain assessment tools; (iv) the capacity to manage pain within an interprofessional team, including monitoring effectiveness

[6]. Specific collaborative competencies suggested by Barr [46] are also relevant and include recognizing and respecting the roles, responsibilities, and competence of others in relation to one's own, and knowing when, where, and how to involve these other professionals. Building these into pre-licensure curricula as well as the early clinical experiences of recently qualified professionals is recommended to reinforce the importance of collaborative efforts to changing practices [47].

22.5
Policy Strategies: Post Licensure

Post-licensure education, in the form of in-service education provided by hospitals and other agencies, is also determined by a number of influences. While far less formally regulated than pre-licensure education, these policy elements need to be understood to create the best possible strategies to enhance health professionals' knowledge of pain management. Knowledge of assessment and management strategies needs to relate to the diversity of pain issues, including persistent problems. Like educators in universities and colleges, educators and administrators in clinical settings have their own values and knowledge which in part shape post-licensure educational opportunities. More important are patients who, through satisfaction surveys, often act as powerful interest groups affecting the decisions educators and administrators make regarding the content of in-service education. Public education to raise patients' expectations regarding the relief of pain could result in further pressure by patients to receive better care.

Accreditation standards, however, are likely the most meaningful mechanism to further increase educational offerings related to pain post-licensure. Since 2005 the Canadian Council on Health Services Accreditation (CCHSA) [33] has included pain assessment and management in its standards. Specific actions are suggested to meet the standard and relate to assessment, management, monitoring, organizational responsibility, and assessment measures. The criterion is evidence-based and includes the organization's accountability to train and update staff, patients and families on pain management options and strategies. This standard can be found in Acute Care Standard 7.0, under the sub-section Addressing Needs. Criterion 7.4 specifically addresses the team's processes for assessing and managing the client's pain. This criterion is relevant to all care sections of the standards where appropriate for the management of pain, from Cancer Care, Maternal/Child, Rehabilitation and Long-Term Care, to Acquired Brain Injury, Ambulatory Care and Critical Care. Processes addressed in this criterion are as follows:

- All clients receive a pain assessment on admission and routinely thereafter.
- The team assesses pain using standardized clinical measures.
- The team manages pain appropriately, and routinely monitors the effectiveness of pain management strategies.
- The team identifies and consults with pain management experts when a complex problem occurs.

- The team educates patients and families on pain management strategies.
- The team documents and shares the results of pain management strategies.
- The organization trains and updates staff on evidence-based strategies to prevent, minimize, or relieve pain.

These processes are comprehensive, emphasize a team approach and are explicit regarding the need for on-going organizational responsibilities related to staff education. However, improving pain management throughout an organization cannot be accomplished all at once. Key steps that have been identified as essential to this process have been articulated in the Building an Institutional Commitment to Pain Management Wisconsin Resource Manual [34]. These steps include: developing an interprofessional workgroup, together analyzing current pain management practices (e.g. chart audits), articulating and implementing a standard of practice, establishing accountability for pain management, deciding how to make pain a priority and visible within the organization, providing information about interventions (pharmacologic and non-pharmacologic) to healthcare practitioners, providing education for all healthcare professional's and patient-families, and continually evaluating and working to improve the quality of pain management.

In addition, while not directly related to chronic care, the CCHSA palliative care standards released in May 2006 are also a step in the right direction. Under Performance Measure #3: Degree and Management of Symptom Distress there is repeated reference to the need to assess and manage pain appropriately.

In conclusion, pain is the most common symptom experienced by people seeking help from health professionals of all disciplines. Many people continue to experience considerable unrelieved pain, acute and/or chronic, despite available treatment strategies that can help most people. However, mandatory formal pain education, particularly using an interprofessional approach, has not been a priority for institutions responsible for education and healthcare delivery. This chapter suggests that policy changes in education, both in pre-licensure pain curricula and post-licensure organizational educational initiatives and patient care monitoring, would facilitate better pain management for patients. These changes are most likely to be successful if further professional competencies related to pain are developed and implemented and if accreditation standards are met within healthcare provider agencies.

References

1 Brennan, F., Carr, D. and Cousins, M. (2007) Pain management: a fundamental human right. *Anesthesia and Analgesia*, **105** (1), 205–21.

2 Sessle, B. (2000) Incoming President's address: looking back, looking forward, in *Proceedings of the 9th World Congress on Pain. Progress in Pain Research and Management*, Vol. **16** (eds M. Devor, M.C. Rowbotham and Z. Wisenfield-Hallin), IASP Press, Seattle, pp. 9–18.

3 Lebovitz, A., Florence, I., Bathina, R., Hunko, V., Fox, M. and Bramble, C. (1997) Pain knowledge and attitudes of health care providers: practice characteristic differences. *The Clinical Journal of Pain*, **13**, 237–43.

4 Furstenberg, C. et al. (1998) Knowledge and attitudes of health-care providers toward cancer pain management: a comparison of physicians, nurses, and pharmacists in the State of New Hampshire. *Journal of Pain and Symptom Management*, **16**, 335–49.

5 Simpson, K., Kautzman, L. and Dodd, S. (2002) The effects of a pain management education program on the knowledge level and attitudes of clinical staff. *Pain Management Nursing*, **3**, 87–93.

6 Watt-Watson, J. et al. (2004) An integrated undergraduate curriculum, based on IASP curricula, for six health science faculties. *Pain*, **110**, 140–8.

7 Unruh, A. (1995) Teaching student occupational therapists about pain: a course evaluation. *Canadian Journal of Occupational Therapy*, **62**, 30–6.

8 Strong, J., Tooth, L. and Unruh, A. (1999) Knowledge of pain among newly graduated occupational therapists: relevance for curriculum development. *Canadian Journal of Occupational Therapy*, **66**, 221–8.

9 Scudds, R. and Solomon, P. (1995) Pain and its management: a new pain curriculum for occupational therapists and physical therapists. *Physiotherapy Canada*, **47**, 77–8.

10 Marcer, D. and Deighton, S. (1988) Intractable pain: a neglected area of medical education in the UK. *Journal of the Royal Society of Medicine*, **81**, 698–700.

11 Watt-Watson, J. and Watson, C.P.N. (1989) Inadequate teaching about pain. *Canadian Medical Association Journal*, **141**, 21–2.

12 Graffam, S. (1990) Pain content in the curriculum-A survey. *Nurse Educator*, **15**, 20–3.

13 Rochman, D.L. (1998) Student's knowledge of pain: a survey of four schools. *Occupational Therapy International*, **5**, 140–54.

14 Baldwin, D.W.C. (2001) Some historical notes on interdisciplinary and inter-professional education and practice in health care in the USA. *Journal of Interprofessional Care*, **10**, 173–87.

15 Roberston, P. and McCroskey, J. (1996) *Interprofessional Education and the Interprofessional Initiative at the University of Southern California: an Evaluation of the Interprofessional Initiative*, University of Southern California.

16 Zlotik, J.L. (1998) Myths and opportunities: an examination of the impact of discipline-specific accreditation on interprofessional education, in *Executive Summary Report, Preparing Human Service Workers for Interprofessional Practice: Accreditation Strategies for Effective Interprofessional Education*, Annie E. Casey Foundation, California.

17 Russell, L., Nyhof-Young, J., Abosh, B. and Robinson, S. (2006) An exploratory analysis of an interprofessional learning environment in two hospital teaching units. *Journal of Interprofessional Care*, **20** (1), 29–39.

18 Barr, H. (1996) End and means in interprofessional education: towards a typology, *Education for Health*, **9**, 341–52.

19 Cooper, H., Carlisle, C., Gibbs, T. and Watkins, C. (2001) Developing an evidence base for interdisciplinary learning: a systematic review. *Journal of Advanced Nursing*, **35**, 228–37.

20 Ponzer, S. and Hylin, U. (2004) Standardized patients-good help in teaching. *Lakartidningen*, **101** (42), 3240–2 and 3244.

21 Slack, M., Coyle, R. and Draugalis, J. (2001) An evaluation of instruments used to assess the impact of interdisciplinary training on health professions students. *Issues in Interdisciplinary Care*, **3**, 59–67.

22 Rycroft-Malone, J., Harvey, G., Seers, K., Kitson, A., McCormack, B. and Titchen, A. (2004) An exploration of the factors that influence the implementation of evidence into practice. *Journal of Clinical Nursing*, **13**, 913–24.

23 Barr, H., Freeth, D., Hammick, M., Koppel, I. and Reeves, S. (2000) *Evaluations of Interprofessional Education: a United Kingdom Review for Health and Social Care*, United Kingdom Centre for the Advancement of Interprofessional Education with the British Educational Research Association, London.

24 International Association for the Study of Pain (2005) Curricula. http://www.iasp-pain.org/curropen.html (accessed 16 October 2007).

25 Canadian Pain Society (1999) Position Statement on Pain Relief. http://www.canadianpainsociety.ca/policy.html (accessed 16 October 2007).

26 Duncan, K. and Pozehl, B. (2000) Effects of performance feedback on patient pain outcomes. *Clinical Nursing Research*, **9** (4), 379–97.

27 Ferrell, B.R., Barneman, T., Juarez, G. and Virani, R. (2002) Strategies for effective continuing education by oncology nurses. *Oncology Nursing Forum*, **29** (6), 907–9.

28 Solomon, P. and Geddes, E.L. (2001) A systematic process for content review in a problem-based learning curriculum. *Medical Teacher*, **23** (6), 556–60.

29 van Eijk, M.E., Avorn, J., Porsius, A.J. and de Boer, A. (2001) Reducing prescribing of highly anticholinergic antidepressants for elderly people: randomised trial of group versus individual academic detailing. *British Medical Journal*, **322** (7287), 654–7.

30 Zwarenstein, M., Reeves, S. and Perrier, L. (2005) Effectiveness of pre-licensure interprofessional education and post-licensure collaborative interventions. *Journal of Interprofessional Care Supplement*, **1**, 146–65.

31 National Dental Examining Board of Canada (2007) http://www.ndeb.ca/en/accredited/competencies.htm (accessed 16 October 2007).

32 Canadian Pain Coalition (2007) www.canadianpaincoalition.ca (accessed 16 October 2007).

33 Canadian Council on Health Services Accreditation (2007) http://www.cchsa.ca/Default.aspx (accessed 16 October 2007).

34 Wisconsin Pain Initiative (2007) http://www.medsch.wisc.edu/painpolicy (accessed 16 October 2007).

35 Dihle, A., Bjolseth, G. and Helseth, S. (2006) The gap between saying and doing in postoperative pain management. *Journal of Clinical Nursing*, **15**, 469–79.

36 Watt-Watson, J., Stevens, B., Streiner, D., Garfinkel, P. and Gallop, R. (2001) Relationship between pain knowledge and pain management outcomes for their postoperative cardiac patients. *Journal of Advanced Nursing*, **36** (4), 535–45.

37 Doern, G.B. and Phidd, R.W. (1992) *Canadian Public Policy: Ideas, Structure, Process*, 2nd edn, Nelson Canada, Scarborough, Ontario.

38 Canadian Nurse Association (2007) The Canadian Nurses Association Competencies, http://www.cona-nurse.org/standards/glossary.htm, http://online.nmtc.educ/vrc/curric/Glossary.html (accessed 16 October 2007).

39 The National Dental Examining Board of Canada (Gerrow, J.D., Chambers, D.W., Henderson, B.J., Boyd, M.A.) (1998) Competencies for A Beginning Dental Practitioner in Canada http://www.ndeb.ca/en/accredited/competencies.htm (accessed 16 October 2007).

40 The Royal College of Physicians and Surgeons of Canada (The Royal College of Physicians and Surgeons of Canada) (2005) http://rcpsc.medical.org/canmeds/about_e.php (accessed 16 October 2007).

41 Canadian Nurses Association (Canadian Nurses Association) (2007) http://www.can-aiic.ca/CNA/nursing/rnexam/competencies/default_e.aspx (accessed 16 October 2007).

42 National Association of Pharmacy Regulatory Authorities National Association of Pharmacy Regulatory Authorities (2003) http://www.napra.ca/pdfs/professional/competencies.pdf (accessed 16 October 2007).

43 Association of Canadian Occupational Therapy Regulatory Organizations (2000) *The Essential Competencies for Occupational Therapists in Canada*, Association of Canadian Occupational Therapy Regulatory Authorities, Toronto.

44 Canadian Physiotherapy Association Essential Competencies of Physiotherapists in Canada Canadian Physiotherapy Association (2004) http://www.physiotherapy.ca/PublicUploads/224032EssentialCompetency%5B1%5D.pdf (accessed 16 October 2007).

45 Professional Competencies of Canadian Veterinarians (Canadian Veterinary Medicine Association) (2007) A Basis for Curriculum Development, http://www.ovc.uoguelph.ca/services/college/dvm/

dvm2000/compet.pdf (accessed 16 October 2007).

46 Barr, H. (1998) Competent to collaborate: towards a competency-based model for interprofessional education. *Journal of Interprofessional Care*, **12**, 181–8.

47 Barr, H., Freeth, D., Hammick, D., Koppel, I. and Reeves, S. (2006) The evidence base and recommendations for interprofessional education in health and social care. *Journal of Interprofessional Care*, **20** (1), 75–8.

23
The Alberta Chronic Pain Ambassador Program

Paul Taenzer, Donald Schopflocher, Saifudin Rashiq, and Christa Harstall

- The Alberta Chronic Pain Ambassador Program addresses primary healthcare provider care education on evidence-based clinical best practices using a case-based multidisciplinary workshop format.

- Health Technology Assessment specialists worked with senior clinical health educators to develop and deliver the workshops.

- Independent evaluation indicated that clinicians report increased knowledge and changes in their clinical practices.

- This approach appears to be generalizible to other jurisdictions.

As discussed in the previous chapter, most health care professionals have little training in effective strategies for treating chronic pain. In surveys of medical school curricula, training in chronic pain management is either absent or minimal [1]. With the exception of physiotherapists, most allied health professionals also have little or no professional training in pain management. It is, therefore, not surprising that surveys of the effectiveness of pain management across the spectrum of acute, chronic nonmalignant, and cancer pain continue to indicate poor patient outcomes [2–4].

The Alberta Chronic Pain Ambassador Program [5–7] was conceived as a strategy for informing community clinicians about current research evidence on the management of chronic low back pain. It built on established relationships among researchers from the Health Technology Assessment Unit of the Alberta Heritage Foundation for Medical Research (AHFMR), senior clinicians from the University of Calgary and University of Alberta and senior government officials from Alberta Health and Wellness. As a group, we have been involved in health technology assessments (HTAs) and administrative research related to chronic pain [8–10], some of which has been presented in Chapters 3 and 4. The Program was initially funded through a capacity building grant by the Canadian Agency for Drugs and Technologies in Health (CADTH) and trialed in Alberta, Canada in 2004–2005.

Chronic Pain: A Health Policy Perspective
Edited by S. Rashiq, D. Schopflocher, P. Taenzer, and E. Jonsson
Copyright © 2008 WILEY-VCH Verlag GmbH & Co. KGaA, Weinheim
ISBN: 978-3-527-32382-1

23.1
Health Technology Assessment and Knowledge Translation

Health technology assessment uses the methods of systematic review [11] to provide credible, comprehensive, and contextualized research evidence to influence healthcare decisions. HTA distinguishes itself from health-related research by acting as a bridge between the worlds of research and decision making [12]. When HTA reports have been specifically requested by decision makers, dissemination is a relatively straightforward process. In contrast, when potential decision makers, such as health care administrators or clinicians, are unaware of HTA reports, dissemination is far more challenging. Simply providing evidence through the traditional strategy of publishing monographs or journal articles is not sufficient to influence clinical practice and improve patient care [13].

A fresh approach to disseminating HTA-derived evidence was pioneered by the Swedish Council on Technology Assessment in Health Care (SBU) (http://www.sbu.se/www/index.asp). SBU engaged senior clinicians representing each of the health councils to serve as HTA ambassadors to facilitate uptake of HTA reports by health councils in Sweden. The ambassadors met regularly with HTA researchers in Stockholm to learn about ongoing or recently completed HTA research projects. Upon returning to their health councils, the ambassadors arranged presentations at administrative meetings, clinical rounds, and local conferences to inform colleagues about current HTA research evidence that was relevant to local practice environments. The Alberta Chronic Pain Ambassador Program built upon the idea of using senior clinicians as HTA ambassadors.

23.2
Description of the Program

The Alberta Chronic Pain Ambassador Program involved an interactive case-based workshop delivered locally to clinical practitioners who were known to have a special interest in chronic pain management. These workshops combined several crucial dissemination strategies, including bringing the workshop to the participants, inviting multidisciplinary participation, and using highly credible facilitators. The interactive case-based, small-group learning approach ensured the active engagement of the participants, and education credits were available for the family physicians and pharmacists who participated.

The intent was to keep the workshops small in order to maximize interaction, with an average of 16 participants (range 8 to 28) per workshop. Once the participants took ownership of the "case", they became very enthusiastic in identifying treatment options and referring the case to participants in other specialties for more treatment suggestions. As participants identified interventions, the research evidence on that particular treatment was provided in the form of an Evidence in Brief summary (Figure 23.1). These single-sheet summaries consisted of three

Information presented in the far right column of the **Evidence Table** in many of the 'Evidence in Brief' summaries was graded according to the following criteria (adapted from 1).

Strong – consistent findings from at least two good quality randomised controlled trials (RCTs);

Moderate – consistent findings from one good quality RCT and/or at least two average quality RCTs and/or at least two poor quality trials (RCT or controlled clinical trial (CCT)) and/or one average and one poor quality trial (RCT or CCT);

Limited – findings from one average quality RCT or one poor quality trial (RCT or CCT);

Conflicting – inconsistent findings among multiple trials (RCT or CCT) of any quality.

AHFMR | EVIDENCE IN BRIEF | REPORT/SERIES XX

INTERVENTION A

CLINICAL QUESTION

Is Intervention A safe and effective in the management of chronic non-malignant low back pain **≥ 3 months' duration**?

THE EVIDENCE

Treatment	Condition	Comparator	Relevant Results/Authors' Conclusions
Intervention A[1]	Chronic non-malignant, non-specific low back pain	Control	**Limited** evidence that Intervention A is more effective than control.
Intervention A plus exercise[2]	Chronic non-malignant, non-specific low back pain	Intervention A	**Moderate** evidence that Intervention A plus exercise is more effective than control.

[1]Based on one POOR* quality randomised controlled trial published in 1987. [2]Based on two AVERAGE* quality randomised controlled trials published in 1987 and 2004

IMPLICATIONS FOR PRACTICE

What we don't know:

- Is Intervention A more effective than no treatment?
- Is Intervention A more effective in particular patient subgroups?

Research Evidence: What we know

In patients with chronic low back pain, evidence indicates that Intervention A:

- is not effective as a sole treatment;
- may be effective when used in conjunction with exercise.

Recommendation from Clinical Ambassadors

Intervention A is most appropriate and effective in carefully selected and monitored patients who are participating in an appropriate program of exercise.

The Clinical Ambassadors: Dr Pamela Barton, Dr Saifee Rashiq, Dr Paul Taenzer — September 2004

Reference: This Evidence Brief is based on results from a GOOD* quality systematic review (SR). Smith MJ, Jones C. Intervention A for chronic low back pain. In: *The Cochrane Library* Issue 3, 2004. Chichester, UK: John Wiley & Sons, Ltd.

*Quality ratings for RCTs and SR: Good ● Average ● Poor ●

The Ambassador Project is funded by a one time grant from the Canadian Coordinating Office for Health Technology Assessment

The RCTs were rated with respect to quality criteria as follows:

Good – at least 80% of criteria met;

Average – between 50% and 80% of criteria met;

Poor – ≤ 50% of criteria met.

Quality assessment of the systematic review (SR): Published systematic reviews were rated on how well their methods excluded bias and confounding by examining the inclusion/exclusion criteria and search strategy used; how the data extraction, quality assessment of the included studies, and data analysis/synthesis were conducted; whether the conclusions of the review matched the results; and if conflicts of interest and funding sources were reported. The reviews were rated with respect to six essential quality criteria as follows:

Good – six criteria met, or five criteria met and one criterion only partially met;

Average – one criterion not met, or one criterion not met and one criterion only partially met, or two criteria only partially met;

Poor – at least two criteria not met.

1. van Tulder et al. Updated method guidelines for systematic reviews in the Cochrane Collaboration Back Review Group. *Spine* 2003;28(12):1290-1299.

The Ambassador Project is funded by a one time grant from the Canadian Coordinating Office for Health Technology Assessment

Figure 23.1 Key to evidence gradings used in the *Evidence in Brief* summaries.

main components: the clinical question, a description and categorical assessment of the quality and strength of the best evidence available for the intervention, and implications for practice. The last component, implications for practice, outlined what is known and unknown about the intervention and provided pragmatic recommendations from the clinical ambassadors about its utility. The clinical ambassador facilitated a discussion on the benefits and side effects of each treatment. The workshop ended with the development of an action plan for improving the management of chronic pain patients in local practice. Feedback was provided to the participants one week after the workshop in the form of a summary of their preliminary action plan.

All of the materials presented at the workshop, as well as the protocol and methods for identifying the interventions and developing the Evidence in Brief summaries, were made freely available on the Internet (http://www.ihe.ca/hta/ambassador/index.html).

23.3
Credible and Reliable Research Evidence

The effectiveness of the workshops was partly due to having successfully anticipated the treatments most frequently suggested by the workshop participants. A panel of pain specialists and a primary care physician identified interventions for which the HTA researchers were requested to find the best available research evidence. The focus was on treatments provided or recommended by community physicians.

A comprehensive and objective review of the available evidence was undertaken by information specialists. Selection criteria were developed to retrieve the highest level of evidence, which according to the hierarchy of study design is a systematic review [14]. The definition developed by Cook *et al.* [15] was used to identify systematic reviews. Six quality criteria were used to rate the quality of the systematic reviews as good, average, or poor [16]. In addition, guidelines published by the Cochrane Back Review group were used to rate the strength of the evidence presented within the systematic reviews [17].

A total of 13 systematic reviews on chronic nonspecific low back pain met our inclusion criteria. An additional five systematic reviews were included on interventions that were not specifically targeted to nonspecific low back pain. From these 18 reviews, the clinical and HTA research ambassadors developed the *Evidence in Brief* summaries.

The process of creating these summaries was far from straightforward, and most of the interactions and debates between the HTA researchers and clinicians centered on their development. The main confusion for the clinical ambassadors was how all the rating scales fitted together, but applying the grading scales in a standardized manner reassured the clinical ambassadors, even when this meant that the assessment of the strength of the evidence presented in the Evidence in Brief summaries differed from the source systematic review. Using an open, transparent process and making all of the technical documents available on the web site also helped assuage difficulties with this process.

The Implications for Practice section of the Evidence in Brief summaries was the one most read by the workshop participants, since this is where the clinical ambassadors made treatment recommendations based on their interpretation of the research evidence and tacit clinical knowledge [18]. There were some intriguing exchanges between the research and clinical ambassadors about the contents of this section. On occasion, the available research evidence, which the HTA research ambassadors wanted to accurately reflect, fell short of answering specific clinical questions. Consequently, the clinical advice was sometimes at variance with the research evidence.

23.4
What Were the Outcomes?

The project was assessed by an independent evaluator who solicited participant opinions on various aspects of the workshop process via pre- and post-workshop questionnaires. Simple descriptive statistics were calculated for quantitative responses, while qualitative data from open-ended items were analyzed using traditional content analysis techniques [19].

Nearly all of the participants (99%) indicated that the workshops were a useful way of linking research to practice. In most areas relating to content and presentation, satisfaction scores were high. The Evidence in Brief summaries were particularly well received. The effect of the workshops on the knowledge of participants in five sample topic areas in chronic pain management was assessed. The participants recorded an increase in perceived knowledge in each area.

Of the respondents, 85% indicated that they had downloaded or planned to download information from the ambassador program web site. The most popular download was the technical document outlining the evidence-gathering process, followed by the Evidence in Brief summaries on muscle relaxants, exercise therapy, and long-acting opioids.

Nearly a third (30%) of the respondents reported that the workshop had changed the way they manage chronic pain. This percentage is very encouraging, given that the participants were local opinion leaders in chronic pain management and had established interest and expertise in the area.

23.5
Limitations

One glaring area for improvement within the ambassador program is the scope of its evaluation. The project began as a feasibility study. Viewed through that lens, the evaluation provided sufficient information to justify expanding the project. However, the evaluation provided no data on whether the participants actually changed their practices, or if their dissemination of the Evidence in Brief summaries led to any detectable change in the practice patterns of their colleagues. Before interventions like the Ambassador Program can themselves be counted as evidence-based, much more research will be required. It may be fair to suggest that greater priority will need to be focused on health technology assessment, knowledge translation, and health services research in general.

Another question relates to the generalizability of the ambassador approach. It could be argued that our team was ideal to conduct such a project since we had both the required breadth of skills and a long history of colleagueship. If the ambassador model is to be adopted elsewhere, other teams of researchers and senior clinicians will be needed. These teams will face the challenges of resolving scientific and clinical gaps to develop consensus on how best to communicate practice-relevant research evidence and adopt a successful interactive, flexible teaching style. A longer-term commitment will also be required in order to maintain ongoing educational and support services to community-based clinicians.

Yet another limitation of this project was that input from consumers and the public was not sought. This must be rectified in future iterations of the program to ensure that all stakeholder views are adequately addressed, particularly now that the doctor–patient relationship has shifted to a more cooperative partnership, and patient involvement is considered an essential part of any endeavor to improve the quality of health care [20, 21].

23.6
What Is in the Future?

The success of the initial Ambassador Program has led to additional funding for two years. The program is being expanded to include headache pain, which is also subject to widely varying treatment patterns in community practice. The project team is also working with a multidisciplinary group of local opinion leaders from across the province to develop clinical pathways for low back pain and headache that will be adapted to the treatment options available in local communities. These pathways will be converted into paper-based and electronic point-of-care tools. Dissemination strategies for these tools will be developed by local opinion leaders, with support from administrative leaders and project staff.

The project web site is a valued legacy of the program. However, the utility of the Evidence in Brief summaries will decline rapidly if they are not kept up to date. Thus, a 4-month cycle has been established for updating the research evidence and recommendations for the majority of the summaries.

Acknowledgments

The Canadian Agency for Drugs and Technologies in Health (formerly the Canadian Coordinating Office for Health Technology Assessment) provided a one-time, one-year grant that made this initiative possible. The Ambassador Program team includes the authors and the following individuals who participated in the planning and execution of the study: Donna Angus, Gail Barrington, Pamela Barton, Liza Chan, Seana Collins, David Cook, Liz Dennett, Sarah Hayward, Don Juzwishin, Jacques Magnan, Carmen Moga, Jennifer Rees, Ann Scott, Tara Schuller, Richard Thornley, Luxie Trachsel, Margaret Wanke, Bryan Ward, Rob Wedel, Joan Welch, Valerie Wiebe, Deb Wilson, and Kirby Wright.

References

1 Watt-Watson, J., Hunter, J., Pennefather, P. *et al.* (2004) An integrated undergraduate pain curriculum, based on IASP curricula, for six health science faculties. *Pain*, **110**, 140–8.

2 Moulin, D.E., Clark, A.J., Speechley, M. and Morley-Forster, P.K. (2002) Chronic pain in Canada: prevalence, treatment, impact and the role of opioid analgesia. *Pain Research and Management*, **7**, 179–84.

3 Brievik, H., Collette, G., Ventafridda, V., Cohen, R. and Gallacher, D. (2006) Survey of chronic pain in Europe: prevalence, impact on daily life and treatment. *European Journal of Pain*, **10**, 287–333.

4 McDermott, A.M., Toelle, T.R., Fowbotham, D.J., Schaefer, C.P. and Dukes, E.M. (2006) The burden of neuropathic pain: results from a cross-sectional survey. *European Journal of Pain*, **19**, 127–35.

5 Rashiq, S, Barton, P, Harstall, C, Schopflocher, D and Taenzer, P. (2006) The alberta ambassador program: delivering health technology assessment results to rural practitioners. BMC Medical Education, 6, 21. Available at: http://www.biomedcentral.com/content/pdf/1472-6920-6-21.pdf (accessed 12 October 2007).

6 Taenzer, P, Harstall, C, Rashiq, S, Barton, P and Schopflocher, D. (2006) Using an Ambassador Program to improve the management of chronic pain, in *Canadian Institutes of Health Research. Evidence in Action, Acting on Evidence. A Casebook of Health Services and Policy Research Knowledge Translation Stories*, Canadian Institutes of Health Research, Ottawa, pp. 113–16. http://www.cihr-irsc.gc.ca/e/documents/ihspr_ktcasebook_e.pdf (accessed 12 October 2007).

7 Scott, N.A., Moga, C., Barton, P., Rashiq, S., Schopflocher, D., Taenzer, P. and Harstall, C. (2007) Creating clinically relevant knowledge from systematic reviews: the challenges of knowledge translation. *Journal of Evaluation in Clinical Practice*, **13**, 681–8.

8 Ospina M and Harstall, C. (2002) *Prevalence of Chronic Pain: an Overview*. HTA 29. Alberta Heritage Foundation for Medical Research, Edmonton. http://www.ihe.ca/hta/publications.html (accessed 12 October 2007).

9 Ospina M and Harstall, C. (2003) *Multidisciplinary Pain Programs for Chronic Pain: Evidence from Systematic Reviews*. HTA 30. Alberta Heritage Foundation for Medical Research, Edmonton. http://www.ihe.ca/hta/publications.html (accessed 12 October 2007).

10 Schopflocher, D. (2003) *Chronic Pain in Alberta: a Portrait From the 1996 National Population Health Survey and the 2001 Canadian Community Health Survey*, Alberta Health and Wellness, Edmonton, Alberta. http://www.health.gov.ab.ca/public/dis_chronicpain.pdf (accessed 12 October 2007).

11 Egger, M., Smith, G.S. and Altman, D.G. (eds) (2001) *Systematic Reviews in Health Care*, BMJ Books, London.

12 Battista, R.N. and Hodge, M.J. (1999) The evolving paradigm of health technology assessment: reflections for the millennium. *Canadian Medical Association Journal*, **160**, 1464–7.

13 Grol, R. and Grimshaw, J. (2003) From best evidence to best practice: effective implementation of change in patient's care. *Lancet*, **362**, 1225–30.

14 Glasziou, P., Vandenbroucke, J. and Chalmers, I. (2004) Assessing the quality of research. *British Medical Journal*, **328**, 39–41.

15 Cook, D.J., Mulrow, C.W. and Haynes, R.B. (1997) Systematic reviews: synthesis of best evidence for clinical decisions. *Annals of Internal Medicine*, **126**, 376–80.

16 Institute of Health Economics. (2006) The Ambassador Program: Generating the Evidence, revised November 2006, Edmonton, Alberta. Available at: http://www.ihe.ca/hta/ambassador/index.html (accessed 12 October 2007).

17 van Tulder, M., Furlan, A., Bombardier, C. and Bouter, L. (2003) Editorial board of the cochrane collaboration back review group. Updated method guidelines for systematic reviews in the cochrane collaboration back review group. *Spine*, **28**, 1290–9.

18 Thornton, T. (2006) Tacit knowledge as the unifying factor in evidence based medicine and clinical judgement. *Philosophy, Ethics, and Humanities in Medicine*, **1**, 2.

19 Barrington Research Group, Inc. for the Calgary Health Region Chronic Pain Centre (2005) *Final Report. Evaluation of the Alberta Health Technology Assessment (HTA) Ambassador Program*, Calgary Health Region Chronic Pain Centre, Edmonton. Available at http://www.ihe.ca/hta/ambassador/doc/050503%20HTA%20FINAL%20Report%20March%2031%20%202005.pdf (accessed 12 October 2007).

20 Mechanic, D. (1998) Public trust and initiatives for new health care partnerships. *The Milbank Quarterly*, **76**, 281–302.

21 Say, R.E. and Thomson, R. (2003) The importance of patient preferences in treatment decisions – challenges for doctors. *British Medical Journal*, **327**, 542–5.

24
Health System Organization

Patricia L. Dobkin and Lucy J. Boothroyd

- Two healthcare systems, the national healthcare system in France and the Veterans Administration in the United States, have successfully reorganized pain services into integrated systems of care for this population.

- Common structural features include a hierarchy of services, primary care services for timely diagnosis and treatment, a gradation of specialized services including comprehensive multidisciplinary care, and education programs.

- Common process of care features include an interdisciplinary approach, use of care pathways and discharge protocols, an emphasis on facilitating continuity of care, and educational processes including professional education and patient self-care.

Given that chronic pain varies greatly in type, intensity, frequency, and prognosis, patients are found at all levels of the healthcare system and are treated by various health professionals. Several studies have shown that the clinical management of various chronic, non-cancer pain conditions remains unsatisfactory [1–6]. As the population ages, chronic pain will increase in prevalence, highlighting the urgency to ensure health systems are organized and delivered appropriately to optimize patient care and subsequent outcomes. One needs to be aware of certain challenges inherent in the management of chronic pain when considering the organization of healthcare services for chronic pain patients. These include:

1. **Timing.** Early intervention is required for some types of chronic pain to prevent the development of disability. A tendency to provide "reactive" rather than "preventive" care contributes to chronicity.

2. **Access.** Chronic pain patients generally require access to several healthcare services throughout their care process, and these often need to be provided concurrently. Access to care can be limited, waiting times are typically long and regional variations in availability of diagnostic and treatment services are common. Psychological and other allied health services are often unavailable.

Chronic Pain: A Health Policy Perspective
Edited by S. Rashiq, D. Schopflocher, P. Taenzer, and E. Jonsson
Copyright © 2008 WILEY-VCH Verlag GmbH & Co. KGaA, Weinheim
ISBN: 978-3-527-32382-1

3. **Continuity.** Services for those with chronic pain are often fragmented, without multidisciplinary integration. Gaps in communication and differences in care models can exist between medical and rehabilitative disciplines, leading to isolation of practitioners and healthcare establishments.

This chapter is based on a Health Technology Assessment (HTA) report written for the Quebec Health Services and Technology Assessment Agency at the request of the Ministry of Health and Social Services in Quebec, Canada [7]. The HTA report examined organizational issues in the management of chronic pain patients in order to inform policy-makers at various levels of the healthcare system. A literature search of published scientific and "gray" literature (i.e. HTA, web site and government documents) from January 1990 to May 2007 was used to identify organizational themes, and "real world" information was sought from two example jurisdictions: France and the Veterans Health Administration in the USA. In this chapter, we highlight organizational initiatives undertaken by these jurisdictions which have explicitly prioritized the improvement of health services for chronic pain patients. We particularly sought to identify common themes and those linked to patient outcomes. Our conclusions point to key issues for policy makers to keep in mind when organizing health services for chronic pain patients.

24.1
Health System Organization and Care Pathways

In the study of health system organization for a particular condition, structure refers to how services are organized, and includes such elements as types and sizes of facilities, human resources, infrastructure, and equipment. Process refers to how services are delivered, including coordination of care, communication, and care pathways. Healthcare systems are generally structured according to a hierarchy of levels at which care is delivered, ordered with increasing specialization in terms of human resources, facilities, and equipment as one moves up the hierarchy. A hierarchy of services does not imply, however, that patient care pathways are unidirectional – in fact, patients may need to move from one level of service to another and back over the course of time – nor does it imply that one level is more important than another. Ideally, referral protocols are put in place to coordinate the movement of patients through the care levels. We begin this chapter with a brief introduction of the types of healthcare practitioners and structures implicated in management of chronic pain in general[1].

Healthcare services for chronic pain patients can be found in unidisciplinary or multidisciplinary settings where care providers may follow different "models

---

1) It should be noted that this chapter does not address complementary and alternative medicine (e.g. chiropracty, osteopathy, acupuncture, biofeedback, yoga, hypno-therapy, massage, herbal medicine) as these are not covered by public insurance in the province for which this work was originally commissioned.

of care", such as biomedical, psychosocial, or rehabilitative approaches (or a combination of these). The majority of chronic pain patients are managed in the "primary care" milieu–that is, the level of directly accessible, first contact medical services–by general practitioners (GPs). GPs may refer the patient to an allied health professional (e.g. physiotherapist, psychologist) or a medical specialist. Allied health professionals involved in chronic pain treatment are found in primary care, in institutional settings[2] (in rehabilitation centers and hospitals) and in specialized pain clinics. In Canada, hospitals usually offer allied health services, but these are generally reserved for inpatients and patients must seek private services if they require more care once discharged. Specialists that examine and/or treat chronic pain patients include anesthesiologists, pain specialists, neurologists, rheumatologists, neurosurgeons, orthopedic surgeons, general surgeons, and physiatrists.

Rehabilitation centers may offer specific programs for chronic pain patients. Such centers are often comprised of a team of health professionals (including occupational physicians, physiotherapists, and psychologists) and receive referrals from GPs or specialists. Patients referred to a multidisciplinary pain clinic (MPC) typically are the most challenging to treat as they have not obtained adequate pain relief elsewhere. MPCs employ an integrated team approach which includes care providers from a variety of disciplines, each with specialized training and experience in different aspects of pain management.

24.2
Two Jurisdictions That Have Prioritized Management of Chronic Pain

Jurisdictions (i.e. countries, states/provinces or specific health systems) differ with regard to how they administer and finance services for patients with chronic pain. In the following section we provide a description of the commitment made to the management of chronic pain in the two example jurisdictions, selected because (i) they have made a clear commitment to chronic pain health services and their quality control; (ii) they provide services to all members of a specific region or group; (iii) they have published or placed documents on the Internet in English or French which allowed us to examine their initiatives; and (iv) we found evidence of attempts to study patient outcomes or monitor implementation.

For more than a decade, pain, and chronic pain in particular, have been a national priority for France [8]. Through the establishment of Phase I (1998–2002) of a "National fight against pain", the Minister of Health committed to improving

2) Non-teaching hospitals, rehabilitation centers and specialty services can be considered "secondary care", requiring referral from the primary care level, while "tertiary care" can be used to denote highly specialized centers such as university teaching hospitals that include research and training activities and facilities.

the overall management and treatment of cronic pain, both within healthcare establishments and in outpatient settings [9]. The Ministry provided financing to set up multidisciplinary structures for the evaluation and treatment of persistent chronic pain throughout France, albeit mostly at the tertiary care level. During the second phase (2002–2005), the national program focused on three specific domains: pediatric pain, migraine headaches, and post-operative pain [10]. A hierarchical structure of services was put into place. Individual hospital pain committees were formed, responsible for ensuring continuing education of health professionals, supporting quality improvement programs, and coordinating pain services, among other activities [11]. A methodological guide for clinicians and policy makers on organizing management of pain in health establishments was published in 2002 [12]. France appears to rely heavily on the Internet to disseminate policy and procedures, including materials for patients and professionals (such as clinical guidelines). A third program phase (2006–2010) has commenced, dedicating 26.74 million euros, and prioritizing vulnerable populations (e.g. children, the elderly, terminally ill patients), the education of health professionals, improving the use of analgesics and non-pharmacological interventions, and the restructuring of pain care pathways [8].

The Veteran Health Administration (VHA) is responsible for healthcare for military veterans across the USA, which number about 25 million persons, and provides services in more than 163 facilities [13]. Beginning in 1998, the VHA, in collaboration with the Institute for Healthcare Improvement [14], undertook the development of the VHA National Pain Management Strategy [15]. The VHA's overall objective has been to develop a comprehensive, standardized, multicultural, integrated, monitored and system-wide approach to pain management for veterans with acute and chronic pain. In order to organize the structure for strategy implementation, a coordinating committee was formed, consisting of a chair, a national program coordinator, and a multidisciplinary group of experts from VHA facilities with expertise in pain management, education, research, and information technology [13, 16]. There are also several working groups, chaired by committee members, and responsible for an aspect central to the strategy's objectives: clinical guideline development, pharmacy guidelines, outcome measurement, research, and education.

Structures and processes have been put into place within the VHA to facilitate patients (i.e. patients' reception of care at various levels of the system, according to the type of pain and duration of the problem [17]. Clinics have been established, multidisciplinary pain teams have been formed, and increased access to specialty pain care has been offered [13, 16]. Furthering clinical competency through the dissemination of training materials and increasing awareness of pain management are key goals, targeting patients and families as well as staff. Interdisciplinary management of pain is promoted [18]. Importantly, the VHA has invested in an information technology platform to make computerized pain assessment templates available and to facilitate the systematic monitoring of patient outcomes and assessment of the quality of services [16]. Research is advocated as an integral part of the VHA's vision for pain management [18, 19]. In

2005, continued commitment to the National Pain Management Strategy was emphasized [13].

24.3
Structure Elements in the Example Jurisdictions

In France, pain "units" serve to evaluate and treat pain using a multidisciplinary structure, with access to hospital facilities, while multidisciplinary pain "centers" are found within university teaching hospitals or hospitals affiliated with a university and have additional training and medical research objectives [20]. The third structure present in France is the medical specialist "consultation service", a multidisciplinary entity that provides assessments and recommendations for patients with persistent pain [21]. The number of pain care structures doubled in the period 2000–2005 [8].[3]

In the VHA system, pain management services are provided to chronic pain patients via several models of multidisciplinary care delivery [17], although single discipline clinics also exist [22]. These structures, found mainly in outpatient settings, vary in available services, required resources, and types of patients treated. They display a gradation in specialization and in activities other than patient care. Single service or modality-oriented outpatient clinics are usually run by care providers from a single discipline, are appropriate for milder chronic pain, and aim for pain reduction through a specific type of treatment [17]. Outpatient pain clinics focus on the diagnosis and management of specific pain problems (e.g. headache, arthritis, low back pain) or more general pain conditions, and may include patients with moderately severe chronic pain. Staff from one or more disciplines can be involved. Comprehensive multidisciplinary pain clinics can include outpatient or inpatient components and are most appropriate for patients needing global and intensive treatment of pain and related dysfunction.

Multidisciplinary pain centers provide the most intensive patient treatment, or serve those that require closer monitoring during treatment [17]. Typically associated with a medical school or teaching hospital, these centers are appropriate for individuals with moderate to severe chronic pain, for those with less severe syndromes but very complex and refractory pain problems, or those with co-morbid conditions. These types of centers must engage in pain-related research and staff education. On the VHA web site we found information about the Chronic Pain Rehabilitation Program (CPRP) at James A. Haley Veterans Hospital, Tampa, Florida [23] which consists of inpatient treatment and associated outpatient clinics, meets the pain center criteria and is accredited by CARF (Commission on Accreditation of Rehabilitation Facilities).[4] There are also pain consultants in the VHA that assist outpatient clinics [17].

---

3) A list of structures by region can be found at http://www.sante.gouv.fr/htm/dossiers/ prog_douleur/ (accessed on 21 August 2007).

4) It is possible that there are other pain centers in the VHA that were not identified by our literature search.

24.4
Training of Healthcare Providers

As examined in Chapter 22 of this book, professional education is a basic building block for effective and efficient delivery of pain management. In France, pain management training is mandatory in medical schools [9] and physician training regarding pain treatment has been reinforced, particularly for specialists, since the national pain program was put into place [24]. Education for health professionals was harmonized by creating an inter-university pain management diploma [25]. Other professional education for pain management has been enhanced (e.g. for nurses, physiotherapists, psychologists [24]), and hospital-based continuing education has been promoted [11]. A practical handbook on pain management for health professionals was produced in 2004 and clinical guidelines, teaching documents, and a pain resource center have been placed on the Internet [8].

The VHA Office of Academic Affiliations, in collaboration with the National Pain Management Strategy Coordinating Committee, has set up advanced clinical training in pain management [16]. Specific residency training in pain management, which incorporates clinical research, is offered within anesthesiology, psychiatry, physiatry and neurology specialties. VHA personnel are encouraged to take part in continuing education opportunities such as national pain conferences, national satellite broadcasts, system-wide teleconferences, Web-based training, and an electronic mail group [13, 14, 18]. The VHA serves as a training site for nurses, psychologists, physical medicine and rehabilitation therapists, and medical students.

24.5
Process Elements in the Example Jurisdictions

24.5.1
Referral

Within healthcare systems in general, a chronic pain patient who wishes to access a specialist or be admitted to a pain clinic must be referred by a GP. Several management issues arise with the referral process, whether for diagnostic testing or treatment. One relates to the primary healthcare provider knowing when and where to refer a patient. A second relates to timely patient access to the next level of care, which can be problematic. For some types of chronic pain (e.g. back pain, complex regional pain syndrome), waiting too long for appropriate diagnosis and treatment contributes to the development of long-term disability. Ideally, a referring GP will explicitly commit to continuing to treat the patient once he/she is discharged from specialty care.

A key element of the VHA pain management strategy is to develop a national management system which, among other actions, will facilitate a national referral system to ensure access to appropriate services in every service network [15]. Multidisciplinary outpatient clinics receive referrals from primary care, general clinics, and other specialty pain clinics [22]. Guidelines for referrals are being developed at facility and regional levels [13].

24.5.2
Inter-Discipline and Inter-Level Communication

A consistent problem in many healthcare systems is lack of communication, both between healthcare professionals (e.g. between GP and physiotherapist) and between levels of care (e.g. between GP and medical specialist). This problem can negatively impact patient outcomes if diagnostic test results are not received in a timely manner, or if no one is tracking the various professionals a patient has seen, and the treatments tried, when making referrals or treatment decisions. In France, communication between the various levels of the healthcare system is considered indispensable for the management of chronic pain [26]. In the VHA, information is relayed back to the referring physician for patients treated in comprehensive multidisciplinary outpatient clinics [22].

Wiecha and Pollard [27] have suggested that the Internet is a logical platform for supporting interdisciplinary clinical teamwork. They state that the effectiveness and efficiency of how healthcare teams function rest on two factors: patient data and coordination of team members' activities, both of which can be handled using information technology. In one of the few investigations we found addressing this issue, Dobscha et al. [28] queried 21 primary care physicians with regard to preferences for communication with collaborative support teams in five urban and rural VHA primary care clinics. Most preferred e-mail (95%) or telephone calls (68%) as a means of sharing patient information.

24.5.3
Coordination of Care

Care coordination plays an important role in "stepped care", an approach to disease management in which patients progressively receive more complex, specialized and, often, costly interventions according to need [29]. We found little specific information pertaining to this aspect of management of chronic pain in the documents reviewed from France. In the VHA, an Office of Care Coordination was established in 2003 to support system-wide implementation of case management and to ensure "the right care at the right place at the right time" [30]. The home is recognized as the preferred place of care when possible, and services incorporate the use of computerized patient records, telehealth technologies and an emphasis on patient self-management [31]. The 2005–2009 care coordination strategic plan focuses on elderly veterans and those with mental health problems, as well as care coordination training and program evaluation [32]. The Care Coordination Home Telehealth program in one service region in particular has been a leader in this effort and includes chronic pain as a target condition [33]. In this region, service coordinators liaise with healthcare teams and patients in the home using telehealth tools.

24.5.4
Discharge and Continuity of Care Plans

"Discharge" is likely the wrong term to use for what happens to chronic pain patients after a period of care because very few are "cured". Rather, these patients may reach the limit of what a particular service offers them, and are usually expected to return to the GP for continued care. Clearly, standards for continuity of care need to include a discharge planning process [34].

24.5.5
Use of Innovative Technology

In the VHA system the use of videoconferencing to deliver clinic follow-up services was recently tested in a small group of 36 consecutive, stable chronic pain patients over a 29-month period [35]. Both patients and staff (a pain medicine physician, a behavioral medicine psychologist and a clinical nurse specialist) completed questionnaires pertaining to their satisfaction with this means of providing pain management. The majority of both groups found it to be a very good or excellent means of receiving or providing follow-up care. The viability of this approach was further supported by the fact that patients found the technology easy to manage and almost all were able to communicate adequately. The VHA Care Coordination Home Telehealth project in Florida uses an in-home messaging device for chronic pain patients to communicate with care coordinators [36].

24.6
Evaluation

24.6.1
Program Implementation

Program implementation refers to the establishment of both structure and process elements, and can be studied using data from the patient, provider or system level. Implementation is important to assess because failure to adequately administer interventions is one of the most frequent reasons that these do not work in real-world settings; implementation should also be documented to facilitate interpretation of program results [37]. In France, initial efforts were directed at tracking the implementation of the national pain program by a national committee, with representatives from various disciplines and interest groups [25]. For example, a steady rise in the use of analgesics since 1996 has been inferred by market data [38]. Hospital pain committees are responsible for coordinating activities to improve the quality of pain management in all healthcare establishments [11]. These activities are linked to quality assurance and accreditation of facilities.

A survey of the structures in place was conducted across France during phase II, including 207 structures, 20 interviews with physicians, and 28 interviews with chronic pain patients [39]. With a response rate of 72%, it was found that about half of the patients received multidisciplinary care and integration of chronic pain structures within institutions was relatively good, in terms of having access to hospital beds. The success of a program often depended on an institution-based "champion" of the need for chronic pain services. Multidisciplinary team meetings were considered indispensable for the functioning of pain structures.

In the VHA, the National Pain Management Strategy Coordinating Committee is mandated to establish target goals, mechanisms for accountability, and a timeline for implementation of the strategy [15]. The 2007 Military and Veterans Pain Care Act reinforces this committee's mandate through its recognition of the importance of acute and chronic pain and its specification that a pain care initiative in all healthcare facilities of the VHA should be implemented by 2008 for inpatients and 2009 for outpatients. [40] In this Act, pain care standards are clearly specified and include the use of recognized chronic pain guidelines and referral to specialist and multidisciplinary programs when appropriate. The Act also states that although pain programs have been implemented in some regions, the comprehensiveness of services provided throughout the VHA system needs to be further improved.

24.6.2
Process Evaluation

Process evaluation is an integral aspect of quality control of healthcare delivery, incorporating a broad range of study objectives and targets, such as: Do patients have timely access to treatment? Are health professionals using guidelines? Are services integrated between levels of care and between health professionals? [41]. VHA facilities are mandated to incorporate patient questionnaires in their services [15]. Pain assessment and pain care plans are to be documented, and patient education activities are to be included in treatment plans. The VHA Office of Quality and Performance is responsible for setting performance indicators consistent with the national pain strategy and assuring quality, access and patient satisfaction at the level of facilities and regional integrated "service networks" [42]. An External Peer Review Program monitors compliance with the Office's standards through quarterly review of samples of records and reporting of results to VHA leadership.

Based on a 9-month collaborative project between the VHA and the Institute for Healthcare Improvement, pain assessment increased once the "Pain as a fifth Vital Sign Toolkit" was implemented [14]. This toolkit was promoted across the system by teams representing the service networks [14] and through provider education [43]. The distribution of patient educational materials and documentation of care plans also increased. More recently, however, Mularski et al. retrospectively reviewed medical records in a single VHA medical center to assess the impact of the use of this toolkit on the quality of pain management by 15 primary care providers [43]. Six hundred records were examined in a general medicine outpatient clinic: 300 before and 300 after toolkit implementation. Quality of pain management appeared to be unchanged according to seven process measures (e.g. Did the physician document the pain examination?; Did the physician order new medications or other therapies for pain treatment?). Moreover, there were substantial deficits in the documentation of pain evaluation and in the treatment of patients who reported substantial pain levels (≥ 4 on the 0 to 10 scale). The authors concluded that assessing pain in and of itself was not sufficient to change provider behaviors.

24.7
Monitoring Patient Outcomes

While program evaluation and process implementation research focus on system changes, outcome studies address how well patients are doing once treatment is completed. This type of research asks the crucial question, "When treatments are provided as planned, do patients improve?" In 2000, Auquier and Arthuis noted that efficacy and effectiveness studies of pain centers were lacking in France. Recently, a study by Huas *et al.* [44], in a general practice setting, showed that the use of pain assessment scales was not associated with greater pain relief for 728 outpatients with chronic musculoskeletal pain in a randomized controlled multi-center trial evaluating the effectiveness of using pain guidelines. This conclusion concurs with Mularski *et al.* [43] in that measuring pain is necessary but insufficient to impact patient outcomes.

The guide for implementation of the French national pain program promotes the use and regular analysis of discharge questionnaires in health establishments that include patient satisfaction surveys [12]. In the small survey of 28 patients from five structures (representing centers, units, and consultations) during phase II, patients indicated that they were satisfied with how they were treated and with the information received, and that the doctors were responsive to their needs [39]. It is unknown, however, if the sample is representative of chronic pain patients in general given the limited number interviewed.

In the VHA, patient outcomes are explicitly linked to continual performance monitoring and improvement, and thus their measurement is central to an overall vision of accountability and a monitored approach to care [15]. Electronic monitoring of pain assessment and effectiveness of pain management interventions is being implemented in the VHA. The Pain Outcomes Toolkit details specifically how and when to measure patient status [42]. The Pain Outcomes Questionnaire can be requested on-line to collect data pertaining to a patient's progress, and includes measurement of pain intensity, pain interference, negative affect, vitality, pain-related fear, vocational functioning, patient satisfaction, and medical resource utilization. The psychometric properties of this instrument have been studied and demonstrate its validity and reliability [45].

Patient outcome data from the Tampa Chronic Pain Rehabilitation Program (CPRP) are available on-line for the years 2004–2005. [46] Generally, these outcomes look favorable in terms of reductions in pain and disability. A related slide presentation available on the Internet also shows decreases in healthcare visits and associated costs [47]. This appears to be the most relevant example of patient outcome monitoring we found in our literature search for these jurisdictions. CPRP patient outcomes are summarized quarterly for communication to all staff [23]. This is, however, a model program and it is unknown if other centers have a similar monitoring process or comparable results.

24.8
Is There Evidence for Chronic Pain Patient Improvements?

We did not find published studies explicitly studying the impact of changes made in the healthcare system in France on the mental and physical health of chronic pain patients, with the exception of a recent investigation by Allaria-Lapierre *et al.* [48]. This study specifically aimed to address the national program objective of increasing access to specialized services, in the context of outpatient management of chronic pain. A cohort of 172 chronic pain patients being treated in GPs' offices in five regions were examined and asked to complete questionnaires; 109 (63%) who had not previously received specialized treatment for chronic pain were followed over three to six months. Sixty-five patients initiated specialized treatment (administered by the GP for over 50%), whereas 20 had no specific pain therapy over the six months. The treated patients showed a significant decrease ($p = 0.027$) in their pain intensity score, from an average of 6.54 at three months to 5.97 at six months (on a 0–10 visual analog scale; scores for the untreated group were 5.55 and 5.05, respectively). Pain improved over the six months by at least 10% for 66% of treated and 29.8% of untreated patients ($p = 0.002$). While encouraging, these results may not be clinically significant. Regional differences regarding access to specialized pain treatment were found.

Consistent with the emphasis on quality control in the VHA system, research pertaining to patient outcomes is being published. For example, Tan *et al.* [49] conducted an observational study in a VHA medical center, in which about half of a chronic pain cohort was reassessed six to nine months following an outpatient, multidisciplinary "integrated pain management program". For the 122 patients, there were significant decreases in pain severity, pain interference, and depression after treatment, but the effect sizes were small in the latter two outcomes and medium for pain severity. There were no significant changes in disability and patients' satisfaction with treatment was variable, with 38% giving low satisfaction ratings. While these research projects are encouraging insofar as patient outcomes related to services are being considered, we did not find studies directly examining the impact of VHA system changes.

Despite the limited literature identified for France and the VHA addressing this issue, a broader search in other jurisdictions in the context of our original HTA report showed supportive evidence for a number of the structure and process elements developed and promoted by these jurisdictions. For example, increased access to specialized care and multidisciplinary teams [50, 51]; improvement of communication between care providers and coordination of care [51, 55]; use of guidelines [56, 57]; and training of healthcare providers [58] have all been investigated elsewhere with positive results.

24.9
Conclusions

Due to its magnitude as a health problem in the general population, its associated burden, and the viability of modes of intervention (despite their complexity in terms of need for coordination and integration), chronic pain should be considered a priority within healthcare systems. It is clear that appropriate resources are required to support the structures and processes involved in providing evidence-based management of chronic pain and monitoring outcomes. The example jurisdictions we examined have seen fit to invest in the organization and financing of chronic pain management. Our conclusions are grouped below according to theme.

24.9.1
Structure

A hierarchy of services is an appropriate structure to assist patients to be treated by the right health professional, at the right time. This structure facilitates stepped care.

Primary care structures need to be put in place to provide timely diagnosis and treatment.

Among specialized structures it is advisable to have a gradation of services, the most comprehensive type involving multidisciplinary patient care.

As highlighted in Chapter 22 of this book, education for physicians and allied health professionals at all levels of the healthcare system optimizes treatment of chronic pain patients.

24.9.2
Process

An interdisciplinary approach – in which health professionals from different disciplines work together to provide care, as needed for the individual case – is ideal for management of chronic pain at all levels of the healthcare system. This includes links between primary care physicians and physical medicine/rehabilitation practitioners, as well as collaboration between multiple care providers in specialized clinics.

Integration and coordination of services is important so that different types of health professionals, from various disciplines and levels of care, can be involved in seamless delivery of care. Care pathways and discharge protocols facilitate continuity of care.

Chronic pain patients need to be viewed as part of the solution in that they require education about pain, including self-management strategies.

24.9.3
Outcomes

As for all other types of patients, it is important that outcomes experienced by chronic pain patients are systematically assessed. Outcomes that should be considered in the case of chronic pain may be different to those traditionally used for other patient groups where "cure" is more likely.

Information technology and the Internet are useful tools for tracking outcomes.

When chronic pain structures and processes of care are altered, there is a need to collect relevant baseline and follow-up data in order to monitor program implementation and examine the impact of these elements on outcomes for patients and care providers.

Acknowledgments

The authors thank Pierre Vincent for literature searching and Denis Santerre for bibliographic management at AETMIS.

References

1 Collett, B.J. (2004) The current state of pain management. *Hospital Medicine*, **65**, 70–1.

2 Dr Foster, Long-term Medical Conditions Alliance (LMCA) and UK Patients Association (2004) *Adult Chronic Pain Management Services in Primary Care*, Dr Foster, London, England. Available at http://www.drfoster.co.uk/library/reports/painManagement.pdf (accessed 29 August 2007).

3 Dewar, A., White, M., Posade, S.T. and Dillon, W. (2003) Using nominal group technique to assess chronic pain, patients' perceived challenges and needs in a community health region. *Health Expect*, **6**, 44–52.

4 Dr Foster and The Pain Society (2003) *Adult Chronic Pain Management Services in the UK*. Dr Foster, London, England. Available at http://www.britishpainsociety.org/dr_foster.pdf (accessed 29 August 2007).

5 Jensen, T.S., Wilson, P.R. and Rice, A.S.C. (2003) Chronic pain, in *Clinical Pain Management* (eds A.S.C. Rice, C.A. Warfield, D. Justins and C. Eccleston), Arnold, London, England.

6 Clinical Standards Advisory Group (CSAG) (2000) *Services for Patients with Pain*, HMSO, Department of Health, London, England. Available at http://www.healthinparliament.org.uk/Pain/CSAG%20part%201.pdf (accessed 29 August 2007).

7 Agence d'Évaluation des Technologies et des Modes d'Intervention en Santé (AETMIS) (2006) Management of chronic (non-cancer) pain: organization of health services. Report prepared by Patricia L. Dobkin and Lucy J. Boothroyd. AETMIS, Montreal. Available at http://www.aetmis.gouv.qc.ca/site/en_publications_2006.phtml (accessed 29 August 2007).

8 Ministère de la Santé et des Solidarités (2006) *Plan d'Amélioration de la Prise en Charge de la Douleur 2006–2010.* Ministère de la Santé et des Solidarités, Paris, France. Available at http://www.sante.gouv.fr/htm/dossiers/prog_douleur/doc_pdf/plan_douleur06_2010.pdf (accessed 25 May 2007) (in French).

9 Ministère de l'Emploi et de la Solidarité (1998) *Circulaire DGS/DH No 98/586.* Ministère de l'Emploi et de la Solidarité, Paris, France. Available at http://www.sante.gouv.fr/htm/pointsur/douleur/6-reglementa/611-circ586.htm (accessed 21 August 2007) (in French).

10 Ministère de l'Emploi et de la Solidarité (2002) *Circulaire No DHOS/E2/2002/266.* Ministère de l'Emploi et de la Solidarité, Paris, France. Available at http://www.sante.gouv.fr/htm/dossiers/prog_douleur/ (accessed 21 August 2007) (in French).

11 Center National de Ressources Contre la Douleur (CNRD) (1998) *Recommandations Relatives à l'Organisation de la Lutte contre la Douleur (CLUD) dans les Établissements Publics et Privés Participant au Service Public Hospitalier,* CNRD, Paris, France. Available at http://www.cnrd.fr/article.php3?id_article=357 (accessed 20 August 2007) (in French).

12 Ministère de la Santé, de la Famille et des Personnes Handicapées (MSFPH) (2002) *Guide pour la Mise en Place d'un Programme de Lutte contre la Douleur dans les Établissements de Santé,* MSFPH, Paris, France. Available at http://www.sante.gouv.fr/htm/publication/dhos/douleur/guid_douleur.pdf (accessed 21 August 2007) (in French).

13 Kerns, R.D., Booss, J., Bryan, M., Clark, M.E., Drake, A.C., Gallagher, R.M., Green-Rashad, B., Markham, R. *et al.* (2006) Veterans Health Administration national pain management strategy: update and future directions. APS Bulletin, 16. Available at http://www.ampainsoc.org/pub/bulletin/win06/inno1.htm (accessed 29 August 2007).

14 Cleeland, C.S., Reyes-Gibby, C.C., Schall, M., Nolan, K., Paice, J., Rosenberg, J.M., Tollett, J.H. and Kerns, R.D. (2003) Rapid improvement in pain management: the Veterans Health Administration and the Institute for Healthcare Improvement collaborative. *The Clinical Journal of Pain,* **19**, 298–305.

15 Veterans Health Administration (VHA) (2003) *Pain Management. VHA Directive 2003-021,* Department of Veterans Affairs, Washington, DC. Available at http://www1.va.gov/pain_management/docs/VHAPainDirective_03.pdf (accessed 20 August 2007).

16 Craine, M. and Kerns, R.D. (2003) Pain management improvement strategies in the Veterans Health Administration. APS Bulletin, 13. Available at http://www.ampainsoc.org/pub/bulletin/sep03/article1.htm (accessed 21 August 2007).

17 Clark, M.E. (2004) *Chronic Pain Primer,* James A. Haley Veterans Hospital, Department of Veteran Affairs, Tampa, FL. Available at: http://www.vachronicpain.org/Downloads/ChronicPainPrimerrev.PDF (accessed 21 August 2007).

18 Kerns, R.D. (2003) Clinical research as a foundation for Veterans Health Administration pain management strategy. *Journal of Rehabilitation Research and Development,* **40**, 9–11.

19 Elnitsky, C., Bryan, M. and Kerns, R.D. (2007) Guest editorial: Veterans Health Administration's pain research portfolio and publications. *Journal of Rehabilitation Research and Development,* **44**, 11–18.

20 Ministère de la Santé et de la Protection Sociale (2004) *La Douleur en Questions,* Ministère de la Santé et de la Protection Sociale, Paris, France. Available at http://www.sante.gouv.fr/htm/pointsur/douleur/3-pratique/ladouleurenquestions.pdf (accessed 21 August 2007) (in French).

21 Auquier, L. and Arthuis, M. (2000) Les avancées dans le domaine des douleurs et de leur traitement chez l'adulte et chez l'enfant. Bulletin de l'Academie Nationale de Medecine, 184, 1907–41. Available at http://www.academie-medecine.fr/upload/base/rapports_65_fichier_lie.rtf (accessed 29 August 2007) (in French).

22 Clark, M.E. (2004) *Pain Treatment Program Models*, James A. Haley Veterans Hospital, Department of Veteran Affairs, Tampa FL. Available at http://www.vachronicpain.org/Downloads/Developing%20and%20Maintaining%20Pain%20Programs.PPT (accessed 21 August 2007).

23 Chronic Pain Rehabilitation Program (CPRP) (2005) *Chronic Pain Rehabilitation Program Manual*, 5th edn, James A. Haley Veterans Hospital, Department of Veteran Affairs, Tampa, FL. Available at http://www.vachronicpain.org/Downloads/CPRPMan2005%20for%20web.pdf (accessed 20 August 2007).

24 Cullet, D. and Tortay, I. (2002) *Bilan de l'Offre Universitaire en Matière de Prise en Charge de la Douleur (Diplômes Universitaires)*, Direction de l'Hospitalisation et de l'Organisation des Soins (DHOS), Paris, France. Available at http://www.sante.gouv.fr/htm/dossiers/prog_douleur/doc_pdf/bilan_u.pdf (accessed 29 August 2007).

25 Ministère de la Santé et des Solidarités (2004) *Programme National de Lutte contre la Douleur 2002-2005: État d'Avancement au Ler Février 2004*, Ministère de la Santé et des Solidarités, Paris, France. Available at http://www.sante.gouv.fr/htm/dossiers/prog_douleur/doc_pdf/suiviplan0104.pdf (accessed 21 August 2007) (in French).

26 Agence Régionale de l'Hospitalisation Rhône-Alpes (ARHRA) (1998) *Groupe de Travail Préparation au SROS II. Prise en Charge de la Douleur*, ARHRA, France. Available at http://www.satelnet.fr/arhra/douleur.htm (accessed 5 February 2005).

27 Wiecha, J. and Pollard, T. (2004) The interdisciplinary eHealth team: chronic care for the future. *Journal of Medical Internet Research*, **6**, e22.

28 Dobscha, S.K., Leibowitz, R.Q., Flores, J.A., Doak, M. and Gerrity, M.S. (2007) Primary care provider preferences for working with a collaborative support team. *Implementation Science*, **2**, 16.

29 Von Korff, M., Glasgow, R.E. and Sharpe, M. (2002) Organising care for chronic illness. *British Medical Journal*, **325**, 92–4.

30 The Office of Care Coordination in the VA (2007) Department of Veterans Affairs web site, available at http://www.va.gov/occ/OfficeCareCoord.asp (accessed 28 August 2007).

31 Veterans Health Administration (VHA) (2007) Care Coordination in the VA and History and Rationale for Care Coordination in the VA, Department of Veterans Affairs web site. Available at http://www.va.gov/occ/CCinVA.asp and http://www.va.gov/occ/history.asp (accessed 28 August 2007).

32 Office of Care Coordination Strategic Direction 2005–2009 (Veterans Health Administration) (2007) Department of Veterans Affairs web site. Available at http://www.va.gov/occ/strategy.asp (accessed 28 August 2007).

33 Principles of Care Coordination (Veterans Affairs Sunshine Healthcare Network) (2007) Department of Veterans Affairs web site. Available at http://www.visn8.med.va.gov/v8/Clinical/cccs/clinical/carecoordination.asp (accessed 28 August 2007).

34 Phillips, D.M. (2000) JCAHO pain management standards are unveiled. Joint Commission on Accreditation of Healthcare Organizations. *The Journal of the American Medical Association*, **284**, 428–9.

35 Elliott, J., Chapman, J. and Clark, D.J. (2007) Videoconferencing for a veteran's pain management follow-up clinic. *Pain Management Nursing*, **8**, 35–46.

36 Lake City–Tech Care Cancer/Pain Program (2007) Department of Veterans Affairs web site. Available at http://www.visn8.med.va.gov/v8/clinical/cccs/service/Field/LakeCity-TechCarePain.asp (accessed 28 August 2007).

37 Glasgow, R.E., Davidson, K.W., Dobkin, P.L., Ockene, J. and Spring, B. (2006) Practical behavioral trials to advance evidence-based behavioral medicine. *Annals of Behavioral Medicine*, **31**, 5–13.

38 Lothon-Demerliac, C., Laurent-Beq, A. and Marec, P. (2001) Évaluation du Plan Triennal de Lutte contre le Douleur, Société Française de Santé Publique, Vandoeuvre-lès-Nancy, France. Available at

http://www.sante.gouv.fr/htm/actu/douleur/sfsp.pdf (accessed 29 August 2007) (in French).

39 Duburcq, A. and Donio, V. (2004) *Étude sur la Situation des Structures de Prise en Charge de la Douleur chronique Rebelle en 2004 : Étude pour la Direction de l'Hospitalisation et de l'Organisation des Soins (DHOS)*, Cemka-Éval, DHOS, Paris, France. Available at http://www.sante.gouv.fr/htm/dossiers/prog_douleur/doc_pdf/etude_cemka.pdf (accessed 29 August 2007) (in French).

40 Military and Veterans Pain Care Act of 2007. American Pain Foundation web site. Available at http://www.painfoundation.org/page.asp?file=Action/MilVetBillDraft01-07.htm (accessed 13 June 2007).

41 Joint Commission on Accreditation of Healthcare Organizations (JCAHO) and National Pharmaceutical Council (NPC) (2003) Improving the Quality of Pain Management through Measurement and Action. JCAHO-NPC, Oakbrook Terrace, IL. Available at http://www.npcnow.org/resources/PDFs/PainMonograph2.pdf (accessed on 29 August 2007).

42 National VA Pain Outcomes Working Group and Pain Management Coordinating Committee (2003) *VHA Pain Outcomes Toolkit*. Department of Veterans Affairs, Washington, DC. Available at http://www.vachronicpain.org/Downloads/TOOL%20KIT%20OUTCOMES%20FINAL2.PDF (accessed 21 August 2007).

43 Mularski, R.A., White-Chu, F., Overbay, D., Miller, L., Asch, S.M. and Ganzini, L. (2006) Measuring pain as the 5th vital sign does not improve quality of pain management. *Journal of General Internal Medicine*, **21**, 607–12.

44 Huas, D., Pouchain, D., Gay, B., Avouac, B., Bouvenot, G. and the French College of Teachers in General Practice (2006) Assessing chronic pain in general practice: Are guidelines relevant? A cluster randomized controlled trial. *The European Journal of General Practice*, **12**, 52–7.

45 Clark, M.E., Gironda, R.J. and Young, R.W. (2003) Development and validation of the pain outcomes questionnaire-VA.

Journal of Rehabilitation Research and Development, **40**, 381–95.

46 Pain Resources Chronic Pain Rehabilitation Program (CPRP) (2007) Chronic Pain Rehabilitation Program (CPRP) web site, Pain Resources. Available at http://www.vachronicpain.org/Pages/pain_resources.htm (accessed 21 August 2007).

47 Clark, M.E. (2000) *Chronic Pain Syndromes vs Chronic Pain*, James A. Haley Veterans Hospital, Department of Veterans Affairs, Tampa, FL. Available at http://www.vachronicpain.org/Downloads/ChronicPainSyndromesrev.ppt (accessed 21 August 2007).

48 Allaria-Lapierre, V., Blanc, V., Jacquème, B., Horte, C. and Chanut, C. (2007) Prise en charge ambulatoire de la douleur chronique. *Pratiques et Organisation des Soins*, **38**, 21–9.

49 Tan, G., Jensen, M.P., Thornby, J.I. and Anderson, K.O. (2006) Are patient ratings of chronic pain services related to treatment outcome? *Journal of Rehabilitation Research and Development*, **43**, 451–60.

50 Johansen, B., Mainz, J., Sabroe, S., Manniche, C. and Leboeuf-Yde, C. (2004) Quality improvement in an outpatient department for subacute low back pain patients: prospective surveillance by outcome and performance measures in a health technology assessment perspective. *Spine*, **29**, 925–31.

51 Loisel, P., Abenhaim, L., Durand, P., Esdaile, J.M., Suissa, S., Gosselin, L., Simard, R., Turcotte, J. *et al.* (1997) A population-based, randomized clinical trial on back pain management. *Spine*, **22**, 2911–18.

52 Chelminski, P.R., Ives, T.J., Felix, K.M., Prakken, S.T., Miller, T.M., Perhac, J.S., Malone, R.M., Bryant, M.E. *et al.* (2005) A primary care, multi-disciplinary disease management program for opioid-treated patients with chronic non-cancer pain and a high burden of psychiatric comorbidity. *BMC Health Services Research*, **5**, 3.

53 Loisel, P., Durand, M.J., Diallo, B., Vachon, B., Charpentier, N. and Labelle, J. (2003) From evidence to community practice in work rehabilitation: the Quebec experience. *The Clinical Journal of Pain*, **19**, 105–13.

54 Loisel, P., Lemaire, J., Poitras, S., Durand, M.J., Champagne, F., Stock, S., Diallo, B. and Tremblay, C. (2002) Cost-benefit and cost-effectiveness analysis of a disability prevention model for back pain management: a six year follow up study. *Occupational and Environmental Medicine*, **59**, 807–15.

55 Ahles, T.A., Seville, J., Wasson, J., Johnson, D., Callahan, E. and Stukel, T.A. (2001) Panel-based pain management in primary care. A pilot study. *Journal of Pain and Symptom Management*, **22**, 584–90.

56 Masters, S. (2004) Back pain: the Australian experience. *Australian Family Physician*, **33**, 389.

57 McGuirk, B., King, W., Govind, J., Lowry, J. and Bogduk, N. (2001) Safety, efficacy, and cost effectiveness of evidence-based guidelines for the management of acute low back pain in primary care. *Spine*, **26**, 2615–22.

58 Tornkvist, L., Gardulf, A. and Strender, L.E. (2003) Effects of "pain-advisers": district nurses' opinions regarding their own knowledge, management and documentation of patients in chronic pain. *Scandinavian Journal of Caring Sciences*, **17**, 332–8.

25
The Quebec Experience

Pierre Bouchard

- In response to demands from users, clinicians and user advocacy groups, the Québec Ministry of Health and Social Services recently completed a planning process for the development of a provincial strategy for managing chronic pain. The key success factors were:

- The use of tested and recognized processes that have guided the development of similar continuums of services.

- Joint ministerial leadership fostering the emergence of a productive synergy among clinicians, researchers, managers and users of the anticipated services.

- The presence of credible practitioners on the advisory committee who actively strove to move the issue forward and championed the cause.

- Use of evidence-based data in the scientific literature and the involvement of an agency recognized for its assessment of healthcare technology and treatment methods to increase the credibility of the project.

- Access to the government's administrative apparatus, and capacity of the engaged patient organization to advocate at the political level.

In recent years, the Québec Ministry of Health and Social Services[1] has involved itself in developing a vision for a province-wide organization of services for all individuals suffering from chronic pain. This organization of services, which involves a very large number of clinicians, researchers and service organizers, as well as stakeholders from university and community resources, has recently begun to be implemented.

With a view to sharing what is now known, this chapter describes the environment encompassing this process, the main elements out of which the process emerged, and the path taken to carry it out.

1) Translator's note: The official title is the *Ministère de la Santé et des Services sociaux du Québec*. The English version is provided only for the convenience of the reader.

Chronic Pain: A Health Policy Perspective
Edited by S. Rashiq, D. Schopflocher, P. Taenzer, and E. Jonsson
Copyright © 2008 WILEY-VCH Verlag GmbH & Co. KGaA, Weinheim
ISBN: 978-3-527-32382-1

To conclude, some observations and future options for action are offered to the reader in order to facilitate the reproduction and adaptation of the lessons drawn from this experience to other health and social services.

25.1
Context

When one considers the expectations that citizens in any jurisdiction similar to ours have of their government, healthcare is at the top of the list. Accordingly, every agency and ministry active in a healthcare sector generally finds itself confronted with a multitude of public demands. It will come as no surprise to anyone to observe that these demands in most cases exceed the existing response capacity, whether from a financial, professional or other point of view.

In an attempt to prevent immediately adding new items to an already strained budget, (demands are usually far greater than the available budgets) these organizations generally establish a more or less formal process for screening and selecting the issues so as to maintain a balance between supply and demand at this level. In this kind of context, the largest share of funds allocated to these activity sectors is mainly earmarked for maintaining the services already in place. Any possible residual financial leeway is then allocated to fulfilling the commitments announced by the political party forming the government or even to assuaging strongly felt pressures that emerge during the government's term of office. Obvious examples of this would be issues that garner intense media coverage, mobilize public opinion and jeopardize the life or the long-term health of the affected citizens. It is for this reason that waiting times for accessing emergency hospital services and waiting lists for certain surgical operations, among other things, rank highly among the government's priorities.

Other problem areas are also given priority attention by political decision makers because they involve important social issues, such as the issues surrounding Québec's aging population that affect a broad client base.

In this kind of context, every proponent of a new health-related project must be prepared to deal with some very stiff competition, first to attract the attention of the relevant decision makers, and second to gain access to the always limited resources required for carrying out such projects.

In comparison, health conditions such as chronic pain, which at first glance do not seem to have a visible impact on the life or health of individuals, and which do not receive sufficient media coverage to provoke a society-wide reaction, are faced with greater difficulty from the outset as regards being recognized as a health problem that needs a solution.

Additionally, in this particular case, pain is still generally considered to be a widespread evil that will inevitably afflict every individual at some point in his or her life.

For the public decision makers concerned, who in all likelihood hold this view as well, the need to tackle the phenomenon of chronic pain is not immediately evident. Additionally, because the advances in effective treatments for chronic pain are quite recent, the knowledge now available in this regard is not widespread. For many people, a decision to invest in an apparently unmapped activity sector may be like opening Pandora's Box and setting loose all sorts of evils and expectations that may well never again be contained.

The decision is therefore a difficult one to make, and solid foundations must be laid to provide backing for the decision makers along the way and to ensure that concrete results are actually achieved.

25.2
Elements That Helped to Germinate the Québec Strategy

Several factors converged and provided an argument for recognizing the importance of addressing the issue of chronic pain in Quebec.

First, citizens wrote letters to the Ministry, relating their problems and requesting that action be taken to help them find relief for their pain. These letters sowed the seeds for an early awareness that chronic pain is an issue. This awareness was subsequently augmented by requests from hospitals and clinicians to the Ministry for funding to bring in the new expensive pain treatment technologies (e.g. intrathecal pumps and neurostimulators).

These separate events thus instilled an initial recognition within the Ministry's administrative services of the broader problem surrounding the treatment of chronic pain and fostered the informal opening of a file on this issue. This led to the appointment of a professional who would be responsible for the issue, who would thereafter consolidate the pertinent information originating from inside or outside the Ministry, and who would respond to the various questions being raised. Additionally, an acknowledgement of the scientific literature produced in recent years on chronic pain gradually helped to show the relevance of rethinking the still unquestioned views on chronic pain treatment.

Accordingly, when the Ministry was approached by a professional association of clinicians and researchers devoted to improving the services supplied to chronic pain sufferers, the situation was already conducive for them to receive a positive reply to their request that the Ministry consider improving this activity sector.

With all these factors being taken into consideration, coupled with a conviction that Québec's health system could improve its performance in this area, the Ministry then launched a formal process to do so. It was then decided to form a ministerial advisory committee comprised of, among others, clinical experts, researchers in the area of chronic pain, as well as user representatives, and mandate it to study the issue and provide Ministry authorities with a convincing argument, if any, for the appropriateness of embarking on improvements to this problem area and to specify the scope of such action.

In parallel, the Ministry asked Québec's Health Services and Technology Assessment Agency[2] (AETMIS) to carry out a meta-analysis of the treatment and management of chronic pain and of the best clinical practices developed elsewhere in the world.

---

2) Officially the *Agence d'évaluation des technologies et des modes d'intervention en santé.*

25.3
The Advocated Development Strategy

When selecting a strategy for completing its task, the ministerial committee drew on the experience acquired through the development of continuums of services for other client groups within Québec's network of health and social services. It therefore described its strategy as *"the made-to-measure application, based on the objective needs of the target patients and the established best practices, of a project management framework incorporating a continuum of services. This continuum is to cover all the steps of care required for the effective, efficient treatment of this particular health problem."*

To steer the process, the members of the ministerial advisory committee agreed to follow these guidelines:

- Organization of services based, above all else, on patient needs.
- Thoroughness, objectivity, and transparency.
- An approach that allows at various steps a decision whether or not to move on to the next step.
- Use of evidence-based data, best practices, and expert consensus to validate the advisory committee's positions.
- Basing the committee's positions on evidence-based data, best practices, and expert consensus

The main steps taken to effect this strategy were:

1. Define an optimal vision of the organization of services.
2. Validate this vision with the various stakeholders to ensure its relevance, practicality and accuracy.
3. Propose to decision makers the advocated organizational parameters and the relevant recommendations for action.
4. Prepare an operational plan to progressively transpose the vision and selected recommendations into the network of services and ensure that monitoring mechanisms are put in place.
5. Put quality-assurance and continual-improvement mechanisms in place for the services supplied.

To facilitate a fuller understanding of the process undertaken and its possible adaptation to other contexts, each of the above steps is briefly described below.

25.3.1
Define An Optimal Vision of the Organization of Services

Just as marketing and business plans are defined in the commercial sector, it appeared critical for our undertaking to properly define an overall vision of the desired organization of services. This vision took the form of an operational document entitled, *Programme national d'évaluation, de traitement et de gestion de la* *douleur chronique: paramètres d'organisation, 1er mars 2006,*[3] which is summarized on the following pages. It should be noted that this document takes into account the work done in 2006 by AETMIS to produce a clinical document entitled *Management of Chronic (Non-Cancer) Pain: Organization of Health Services.*[4] The fact that the authors of this second document were on the advisory committee as "researchers in residence" made it possible to contribute data from the scientific and "gray" literature in response to the specific concerns and needs raised during the advisory committee meetings. This interconnection brought added value to each of the two documents.

Given the importance of the document prepared by the advisory committee as a vehicle for the vision it advocates, a short description is provided below of the key elements of each chapter, whose (translated) title is shown in italics.

The document begins with a *description of the known issues,* using as points of view the three essential characteristics required of any effective and efficient organization of health and social services, namely accessibility, continuity and quality. Over this backdrop were superimposed the various steps of care that should form the continuum of services needed to adequately treat chronic pain.

Next, selected difficulties were reviewed concerning access to primary level services, specialized services and unidisciplinary or interdisciplinary treatment, be they care-oriented, pharmacological, or adaptation/rehabilitation-oriented. Also taken into account were the difficulties in continuity of services, patient follow-up, and service quality. The notion of quality was analyzed from the point of view of organization of services, the skills required of clinicians, and the continual improvement of services.

For each identified difficulty, some additional factors related to the *context for implementing a continuum of services* were considered because they could influence, facilitate or constrain the anticipated organization of services. Such factors included, among others, the courses of action established by the Ministry for addressing this particular issue, and environmental elements linked to the nature of the services supplied to individuals suffering from chronic pain Such factors included, among others, certain policies adopted by the Ministry regarding the organization of health and social services, including availability and access to the full spectrum of evidence-based pain treatment, the development and promotion of self-management approaches and access to relevant social services.

The breadth of service sectors involved and the diversity of the aspects of the continuum of services to be promoted required taking into account the various elements to be considered when setting the *program goals.* For this reason the goals were defined in terms of services supplied, education and training, research, assessment of technology and treatment methods, and services organization model.

---

3) National program for the evaluation, treatment, and
   management of chronic pain: organizational parameters,
   1 March 2006.

4) http://www.aetmis.gouv.qc.ca/site/en_publications_2006.phtml.

Given the large number and diversity of the needs presented by the many *targeted client groups* and the amount of effort needed to deploy a full range of services for everyone, it appeared necessary to segment the patient population. This decision was made to make it possible to set priority targets in terms of client groups and later to delimit the development of the initial supply of services. In so doing, due consideration was given to the fact that some client groups (pediatric patients, cancer-pain patients) already receive services which likely could be improved but which constitute a better service base than is available to other client groups. It was also decided to initially target the client groups offering a reasonably high likelihood of success in response to the effort made (adults whose functional capacities would permit a return to work, adults for whom the goal is to improve the quality of life and autonomy). This was done to create a precedent supporting the development of additional services for a more varied, more highly populated client group.

Once these basic elements were established, the advisory committee shifted its focus to defining the *overall organization of a continuum of services* deemed necessary given the need to integrate the clinical, education and training, and research sectors. To do so, we took the route described below.

First, we defined the guidelines for organizing the planned services. In so doing, we found it necessary to categorize these principles in terms of users, organization of services and clinical factors, as this would allow better defining of the expectations to be taken into account in program planning.

Next, we focused on a definition of a continuum of services architecture by integrating the several large wholes, namely promotion and prevention, assessment, diagnosis, treatment, follow-up and, lastly, the continual improvement of services.

Also, we specified the respective responsibilities of every service provider in the network of services, to ensure that their contributions are complementary and interdependent. Beginning with the principle of organizing services into a hierarchy, we clearly defined the mandate, client groups to be served and means to be used for each recommended level of services. For this we used the classification comprised of highly specialized services (with a provincial mandate), specialized services at the regional level, and primary services at the local level.

However, given the important prospective role of the highly specialized services in developing the other points along the continuum, special attention was given to describing *expert centers* and clarifying their expected roles and responsibilities. Briefly stated, a center of expertise is comprised of a university hospital center with a faculty of medicine offering training, education, research, and evaluation programs for all the disciplines involved in supplying services for the targeted client group. This center would also supply specialized rehabilitation services on site or through a formal service contract with a rehabilitation center supplying services on a supra-regional basis. In the latter case, the expert center would be a consortium whose responsibilities are jointly and severally held by its component institutions. Its mandates cover the provision of clinical services, education and training, research, development of the continuum of services, and coordination of services.

Once the services architecture was defined, continuity and integration mechanisms were proposed to foster the flow of the services supplied, both at a clinical and organizational level. The various measures to be implemented for this purpose relate primarily to standardizing the tools and procedures used at the different points along the continuum of services, and to putting in place shared strategies for treating the targeted client groups.

The next section of the document proposes measures for the *continual improvement of the services supplied all along the continuum*. These measures include provisions affecting support for the implementation and continual improvement of the program, quality assurance procedures, education and continuing education, and research and assessment of healthcare technology and treatment methods.

The last section covers the *recommended methodology for program implementation* by listing the main steps to be taken. The details for these steps are presented in an appendix along with the advisory committee's recommendations to the Québec Ministry of Health and Social Services.

25.3.2
Validate this Vision with the Various Stakeholders to Ensure Its Relevance, Practicality and Accuracy

The vision portrayed in the document prepared by the advisory committee was submitted to more than ten organizations with a province-wide mandate likely to be affected directly or indirectly by the proposed organization of services. They were invited to share their comments or concerns on the form, content and proposed method of implementation with a view to improving its impact. These organizations included associations of institutions, clinicians and users, and the various formal structures within the network of health and social services, the health-related faculties of universities and the health research networks.

This validation exercise showed there was a very broad consensus on the relevance of taking action to address chronic pain. In addition, the feedback proved to be supportive of the proposed form of organization (a hierarchical approach to service delivery) and the importance of discussing the training and support to be given to the primary care workers in order to prevent the transition from acute to chronic pain.

Analysis of the comments and suggestions received by the advisory committee led to improvements to the draft document in the form of clarifications and adjustments. The favorable comments received at this step encouraged the decision makers involved to continue the process as planned.

25.3.3
Inform Decision Makers of the Advocated Organizational Parameters and the Relevant Recommendations for Action

Decision makers were informed by means of periodic presentations of the results of the committee's work to the various hierarchical levels within the Ministry.

These presentations served to obtain backing in principle from the authorities for going ahead with the first steps for deploying the program proposed by the advisory committee. Ministry approval came with a request to prepare a strategic plan identifying all the required steps, the entities responsible for carrying them out, and an implementation calendar.

25.3.4
Prepare an Operational Plan to Progressively, Systematically and Carefully Implement. The Vision and Selected Recommendations into the Network of Services and Ensure that Monitoring Mechanisms Are Put in Place

The operational planning exercise was carried out based on the following development strategy:

1. Gradually deploy each link in the chain of services, beginning with the creation of the centers highest in the hierarchy.
2. Establish formal ties from one link to the next.
3. Deploy the means designed to enhance program effectiveness and efficiency and continually improve it.

This planning also required the contribution of the following stakeholders in accordance with their respective responsibilities.

First, the Minister is authorized by the *Act respecting health services and social services* to "determine the supra-regional mission, and mandate of an institution with regard to certain highly specialized services it offers" or "limit to certain institutions the function of offering certain services or dispensing certain medicines he determines." It is therefore up to the Minister to formally designate which expert centers are to supply highly specialized chronic pain services, based on the recommendations of an expert panel that has evaluated the services offered by these centers.

There are four integrated university health networks with a faculty of medicine at their hub that together cover all of Quebec. It is up to them to support the organization of clinical services, training and research in the expert centers, and to support the clinical organization of services in the regions within their respective territories.

The regional health and social services agencies are responsible for organizing services within their territory. The plan is, initially, to give the agencies whose region hosts the university-level faculties of medicine the responsibility of providing the leadership needed for developing the expert centers with input from the four university networks. Later, all the regional agencies will be asked to develop a supply of specialized services at the regional level, and primary services at the local level, within their territory.

In order to provide the global oversight needed to put this plan into practice, a ministerial monitoring unit will be created to support the deployment, follow-up and continual improvement of this program. This unit will include representatives from the planning, organizing, clinic, research, and user organizations.

The institutions shortlisted to become expert centers will be evaluated through an external audit conducted by an advisory group comprised of experts, user representatives and representatives of the aforementioned practitioners. After completing its audit, the advisory group will make its recommendations to the various relevant authorities, including the Minister.

In keeping with the established plan, the centers of expertise are now being designated. The subsequent steps will be directed by the designated expert centers, with a focus on consolidation and a better use of the resources already available in their environment, and thereafter on the possible development of the various resources.

25.3.5
Gradually Put Quality-Assurance and Continual-Improvement Mechanisms in Place for the Services Offered

The initial mechanisms put in place will be for monitoring the mandated activities of the expert centers and the recommendations made to them by the experts after evaluating their offer of services.

Their services will also be re-evaluated thereafter every three years, based on clearly defined deliverables, with adjustments for the new knowledge developed in the various activity sectors to be covered. The use of recognized accrediting agencies such as the Canadian Council of Accreditation of Health Services could then be recommended.

Continual improvement will occur through the progressive introduction of training, educational and research activities.

25.4
Observations and Avenues for Future Action

Although we are still early in the operational phase, which will shortly give birth to the expert centers for chronic pain, some mobilization is already observable among the universities and institutions shortlisted to become expert centers. Additionally, Québec's Health Research Fund[5] (FRSQ) has flagged pain research, and more particularly chronic pain research, as one of its priorities in its latest strategic development plan. This has led to the creation of a Quebec network for pain research (RQRD[6]) comprised of researchers active in the four Québec universities with a faculty of medicine. There are plans in the current process to develop a large clinical research component centered on the emerging service network.

The new knowledge that has emerged thus far from the development project is already being used to identify promising avenues that likely merit further development in terms of best clinical and administrative practices.

---

5) Officially the *Fonds de la recherche en santé du Québec*.
6) *Réseau québécois de recherche sur la douleur*.

With that in mind, some of the elements that helped to achieve the results are presented here. It should be noted that the following list is not based on a comprehensive selection process but rather on an informal consideration of the matter and observations gathered from similar experiences.

- An approach employed that made use of tested and recognized processes that have guided the development of similar continuums of services.
- Joint ministerial leadership fostering the emergence of a productive synergy among clinicians, researchers, managers and users of the anticipated services. The effectiveness of the leadership provided by the person assigned to the project by the Ministry was enhanced by this person's overall knowledge of the environment, mastery of the management processes involved in complex projects, and familiarity with the internal functioning of government organizations.
- Presence of credible practitioners on the advisory committee who actively strove to move the issue forward and championed the cause.
- Evidence-based data in the scientific literature and the involvement of an agency recognized for its assessment of healthcare technology and treatment methods increased the credibility of the project.
- Access to the government's administrative apparatus, and capacity of the engaged patient organization to advocate at the political level.
- Quality of management processes applied to the implementation of such programs.

In addition to the foregoing, it seems appropriate to submit, for consideration by the reader, the issues identified by Québec's Auditor General as part of a recent value-for-money audit designed to highlight healthy health management practices. Some of these issues overlap with the foregoing list. This audit targeted the continuum of traumatology services,[7] which served as a model in the development of the chronic care continuum. The criteria for success identified by the audit that seem applicable are:

- Leadership
- Clearly allocated responsibilities
- Mechanisms for collaboration, coordination and monitoring
- Practitioners involved in the project
- Accessibility and continuity of services
- Deployment of communication channels and bidirectional liaisons among practitioners
- Availability of clinical information and the fact it flows quickly and easily
- Sharing of protocols based on recognized practices or evidence-based data
- The presence of a continual quality improvement system.

7) Vérificateur général du Québec, *Rapport à l assemblée générale pour l année 2004-2005, Tome 2* http://communiques.gouv.qc. ca/gouvqc/communiques/GPQF/Decembre2005/13/c6465. html#contenu.

25.5
Conclusion

Although still unfinished, Quebec's program for the evaluation, treatment, and management of chronic pain will unquestionably be called upon to become an effective and efficient network of services that by its very nature is able to continually improve itself.

However, it is likely that its full potential will not be reached quickly because full deployment of the program will require a transformation and consolidation of the existing resources within Québec's network of health and social services. Of course, this transformation cannot be achieved without cost. Some additional investments are foreseeable for the participating institutions. But these additional investments will likely come in many forms, and will not be exclusively direct financial "add ons". For example, they can take the form of allowing, by the Ministry or regional agencies, the adjustment and increase of human resource planning to program needs, the development of immobilization investment plans for the institutions involved, or even the provision of additional equipment from the Ministry, regional agencies and philanthropic foundations.

The deep seated conviction that this approach is realistic stems from the existence of evidence-based precedents in related fields. For example, the continuum of traumatology services and its subprograms were implemented using this approach and they have in ten years served to reduce the rate of mortality of severe trauma victims from 52 to 8.6%. This development model was also successfully applied to the continuum of services for people with cancer and their families, and it is being applied to the continuum of services for people who have had a stroke or are at risk of having one.

In the light of the successful application of the same process in different contexts, we can only conclude that we are observing the emergence of an organizational model of services that has already proven itself and that continues to evolve. Indeed, last year the proponents of this model were awarded the *Prix d'excellence* in the public service category by Québec's Institute of Public Administration[8] for the creation of a consortium of traumatology services.

There is therefore no question that the approach advocated herein, as well as some of the steps, strategies and principles being used, can be successfully adapted for use in other contexts and for other client groups, not only in Quebec but also in other Canadian provinces and elsewhere around the world.

---

8) Officially the *Institut d administration publique du Québec.*

Conclusion – The Way Forward

Paul Taenzer, Donald Schopflocher, Saifudin Rashiq, and Egon Jonsson

> *You have to do the right thing . . . You may never know what*
> *results come from your action. But if you do nothing, there will*
> *be no result.*   M. Gandhi

In the opening chapter of this volume we were introduced to the voices of people in pain; voices that are all around us but rarely heard. In addition to expressions of suffering inherent in the pain experience we also heard a distinct message that our interactions as providers can add a further burden when people in pain feel unheard. Let us now return to the patients, and hear the voices of those who have had successful healthcare encounters for their chronic pain.

The following are extracts of feedback from patients who participated in the multidisciplinary chronic pain program in Calgary, Alberta, Canada. This program is a service offered by the Calgary Health Region, the public healthcare provider for this geographic area. Patients cared for through this program report that they have been in chronic pain for an average of eight years and that their average pain intensity is in the moderate to severe range. Upon completing the program, these patients participated in a focus group and responded to queries about their experiences.

> . . . they just address the situation here and now, whereas in my past experience, you really do begin to doubt yourself when you keep being referred to specialists who . . . look at you as though . . . it really is in your imagination, or ask "are you under any stress" . . .

> I think what struck me most the first day I was here during the first group I was in was I finally found some people who understand what it's like to have lost your life, to have your life come down around you and . . . to not know what to do about it . . . So that alone, just having a community where I could be . . . has been incredibly helpful to me . . . just putting the pieces back together . . .

Chronic Pain: A Health Policy Perspective
Edited by S. Rashiq, D. Schopflocher, P. Taenzer, and E. Jonsson
Copyright © 2008 WILEY-VCH Verlag GmbH & Co. KGaA, Weinheim
ISBN: 978-3-527-32382-1

. . . it was . . . a whole different scenery in my pain atmosphere. Like suddenly people are there and they acknowledge . . . that you're in pain. And that meant so much to me. Like they haven't seen me or treated me before, but right away, like I was someone, I was special . . .

my expectation before I came here is there's a cure, and why haven't I gotten it yet. But when [you come here] they give you that first realization that there may not be a cure, but they're going to help you so that you can live your life again . . .

at the beginning you're asked what are your goals, what do you . . . expect to get out of this program. And so there's . . . something that you're pushing toward, and I think . . . they're not saying . . . we can do this, this and this and that's the end of it, I think they're saying, Well what do you want to get out of it . . . we'll help to take you as far as we can get with that . . .

I think if you talk to lots of other patients here you're going to hear the same story; that you were at a pretty bad low when you come in the door and they kind of lift you back up and say you're going to be okay. And you get going again. And sometimes you don't get better completely here, but you're . . . able to keep moving on . . .

I think I went from being kind of . . . hopeless . . . almost like a victim of this disease to being back in the driver's seat here. They put me in the driver's seat . . .

for me it was really . . . a godsend to come here and be treated by these people . . . It's made such a difference in my whole life . . .

These experiences offer us a rich palette from which to form impressions of the benefits of quality chronic pain management. Several themes flavor these accounts. The first is implicit. Even for patients who have been suffering for many years, compassionate evidence-based care can and does lead to remarkable recovery of function and human spirit. A second theme is the importance of being heard, being believed and being encouraged by healthcare providers and a resulting resurgence of self-esteem. Another theme is the importance of successfully negotiating the shift from a frustrated quest for "the cure" to a sense of empowerment and re-engaging in life.

While these examples are drawn from one specific chronic pain program, it is clear from the evidence documented throughout this book that these benefits need not be limited to the lucky few who currently have access to state-of-the-art chronic pain care. Indeed, if we are to take up the challenge of addressing the ethical and moral obligation to relieve pain and suffering, we have the healthcare technology and expertise to do so now. The problem of chronic pain will not go away of its own accord. Chronic pain is widespread, and will very likely increase in prevalence as the population ages. The costs to society are staggering, and will certainly not decrease if action is not taken.

Patricia Dobkin and Lucy Boothroyd explored examples of healthcare systems that recognized this challenge a decade ago and have made significant strides in reorganizing their healthcare education and delivery systems to optimize chronic pain care. Pierre Bouchard has ably described a Canadian example where a broadly defined effort brought together all of the relevant stakeholders in a skillfully managed process to capitalize on their contextual advantages and needs to craft a provincial chronic pain strategy.

Here, then, is where to start in offering specific policy recommendations for the policy makers in our jurisdiction, the Government of Alberta Ministry of Health and Wellness, that stem from the evidence presented in this book. While the specific nature of other jurisdictions and the special considerations specific to their contexts will mean that the precise recommendations will change, we hope that these recommendations have much to offer others.

Improving Chronic Pain Management in Alberta – Policy Considerations

A clear health policy is needed to guide and implement specific measures. Such a policy must emphasize the importance of appropriate support for and treatment of people in chronic pain, and demand professional accountability for adequate and evidence-based services in the field. There are several ways to start:

- Continue to build awareness by convening a Consensus Conference to further explore the evidence surrounding interventions to treat and manage chronic pain. Such a Consensus Conference could be modeled on previous successful Consensus Conferences in Alberta [1, 2].

- Follow the lead of the government of France in publicly initiating a "Provincial Fight against Pain".

- Recognize chronic pain as a public health problem, and initiate a Chronic Pain Strategy modeled on other Alberta Health and Wellness Strategies such as the Alberta Diabetes Strategy [3].

Subsequently, or as an immediate alternative, we suggest that the Minister of Health of Alberta establish a provincial Chronic Pain Steering Commitee (CPSC) and commission it to

- Develop a comprehensive proposal for improved services for people in chronic pain in the province.

- Regularly report on the progress of that work, and of the quality, accessibility, and effectiveness of chronic pain services across the province.

The CPSC could be composed of representatives of Alberta Health and Wellness, the Regional Health Authorities, the professional training institutions, the professional colleges, and representatives of the general public. Ad hoc membership could be offered to other organizations where important issues of service quality have been identified.

Alberta has recently completed a strategic plan for Health Technology Assessment in which an Alberta network was identified as a main element. Subsequently, stakeholders for a Health Evidence Network of Alberta (HENA) have met to advance implementation of this plan. We believe that the CPSC should also work in close collaboration with HENA.

Based on the findings from the reviews presented here, the CPSC might attend to, but not be limited to, the following:

Monitoring

- Establishment of health surveillance activities. This might be facilitated through the development of a "chronic pain registry" similar to the Alberta Cancer Registry where patients are followed over the course of their illness and administrative data are available for research and disease surveillance.

- Establishment of, and monitoring of, provincial quality standards.

Prevention

- Prevention of chronic pain resulting from post-surgical pain by establishment of Acute Pain Services (APS) in each acute care hospital.

- Development and coordination of public health campaigns to improve pain management outcomes across the province such as the Alberta "Back@It" Campaign described in Chapter 21.

Treatment Services

- Establishment of chronic pain teams in each health region with ongoing funding contingent on meeting defined wait time standards and negotiated clinical outcome targets.

- Development of a forum for the propagation of best practices among the regional pain teams.

- Establishment of multidisciplinary CP care for severely affected patients and promotion of adequate access to these services.

- Increase in the priority of pain management in primary care through
  - Provision of timely access to specialist consultation via telephone/ telehealth or traditional "in person" consultation
  - Establishment of community-based pain self-management services including pain self-management and active rehabilitation program.

- Placement of special emphasis on the needs of vulnerable populations who require services adapted to their particular circumstances including infants and children, the frail elderly, those with significant co-morbid mental health disorders including addictions, and those in aboriginal communities.

- Establishment of provincial programs for low volume high cost treatments such as neuromodulation

Education

- Development of effective undergraduate and post-licensure clinical training programs in pain management.

- Encouragement of professional colleges to include competency in pain management as a requirement for ongoing licensure.

- Development and implementation of multidisciplinary evidence informed clinical practice guidelines for the primary care workforce, and delivery of programs such as the Alberta Chronic Pain HTA Ambassador Project for low back pain management described in Chapter 23.

Research

- Development of strategies for obtaining targeted funds for pain research and for health services research, including resources to continuously assess outcomes and quality improvements.

- Identification of clinically important research gaps through quality improvement activities, and confirmation of these gaps by secondary research activities such as Health Technology Assessment reports.

- Encouragement of pragmatic research trials conducted as needed.

- Development of strategies for research transfer undertaken with educational and service delivery stakeholders.

Lest one mistakenly acquire the notion that systemic solutions must begin on a provincial or national scale, it seems appropriate to briefly describe the more local solution arrived at in Calgary, Alberta, Canada, a healthcare environment of 1.2 million citizens with a publicly funded integrated healthcare system. Like the Quebec model, the leaders of the Calgary process engaged all of the stakeholders from across the education and service delivery sectors as well as patient representatives to identify gaps in service, and opportunities for further collaboration and integration of care. An advisory committee was established that reported at the senior executive level of the Health Region that included decision makers representing the stakeholder groups. The advisory committee considered the entire spectrum of care gaps and agreed on priorities for service development, most of which were funded for implementation. An innovative feature of this approach is that it brought together stakeholders from across the spectrum of pain education and care, from pediatrics to seniors' health, from rural outpatient and primary care to urban secondary and tertiary ambulatory care, to acute and long term hospital care, from acute, chronic and palliative pain care. The advisory committee provides a continuing forum for stakeholders to re-assess priorities and seek innovative solutions for further integration to foster seamless transitions for patients requiring multiple services.

To end, we want once again to emphasize that the accumulating research evidence signals a great potential for creating synergistic solutions to the challenges of chronic pain. It is now time to reach out to our fellow stakeholders and to act.

References

1 Institute of Health Economics (2006) Consensus Statement on Self-monitoring in Diabetes, Institute of Health Economics CONSENSUS STATEMENTS, Vol. 1. Available at http://www.ihe.ca/documents/ihe/consensus_statement_complete_nov17.pdf (accessed January 31, 2008)

2 Institute of Health Economics (2007) Consensus Statement on Healthy Mothers-Healthy Babies: How to Prevent Low Birth Weight, Institute of Health Economics CONSENSUS STATEMENTS, Vol. 2. Available at http://www.ihe.ca/documents/healthy%20moms%20statement%20revised.pdf (accessed January 31, 2008)

3 Alberta Health and Wellness (2003) Alberta Diabetes Strategy, Alberta Health and Wellness, Edmonton Alberta. ISBN Number: 0-7785-2673-9, PDF ISBN Number: 0-7785-2674-7 Available at http://www.health.alberta.ca/public/dis_diabetesstrategy.pdf (accessed January 31, 2008).

Index

Chronic Pain: A Health Policy Perspective
Edited by S. Rashiq, D. Schopflocher, P. Taenzer, and E. Jonsson
Copyright © 2008 WILEY-VCH Verlag GmbH & Co. KGaA, Weinheim
ISBN: 978-3-527-32382-1